BETWEEN LOGIC AND INTUITION

This collection of new essays offers a "state-of-the-art" conspectus of major trends in the philosophy of logic and philosophy of mathematics. A distinguished group of philosophers addresses issues at the center of contemporary debate: semantic and set-theoretic paradoxes, the set/class distinction, foundations of set theory, mathematical intuition and many others. The volume includes Hilary Putnam's 1995 Alfred Tarski lectures published here for the first time.

The essays are presented to honor the work of Charles Parsons.

CONTRIBUTORS:

Hilary Putnam
Arnold Koslow
Vann McGee
James Higginbotham
Gila Sher
Isaac Levi
Carl J. Posy
Michael Friedman
Michael D. Resnik
Richard Tieszen
George Boolos
W. W. Tait
Mark Steiner
Penelope Maddy
Solomon Feferman
Geoffrey Hellman

Gila Sher is Associate Professor of Philosophy at the University of California, San Diego.

Richard Tieszen is Professor of Philosophy at San Jose State University.

Between Logic and Intuition

Essays in Honor of Charles Parsons

Edited by

GILA SHER
University of California, San Diego

RICHARD TIESZEN
San Jose State University

CAMBRIDGE UNIVERSITY PRESS
Cambridge, New York, Melbourne, Madrid, Cape Town, Singapore, São Paulo

Cambridge University Press
The Edinburgh Building, Cambridge CB2 8RU, UK

Published in the United States of America by Cambridge University Press, New York

www.cambridge.org
Information on this title: www.cambridge.org/9780521650762

First published 2000
This digitally printed version 2007

A catalogue record for this publication is available from the British Library

ISBN 978-0-521-65076-2 hardback
ISBN 978-0-521-03825-6 paperback

Contents

Preface

We offer this collection in honor of the academic career and work of our teacher, colleague, and friend, Charles Parsons. Parsons is widely known for his original essays on mathematical intuition, the semantic paradoxes, the set-theoretical hierarchy, and Kant. His wide-ranging contributions also include works on the foundations of number theory, modal and intensional logic, mathematical structuralism, substitutional quantification, predicative and constructive mathematics, the theory of objects, philosophical and mathematical methodology, Godel, and Quine. His encyclopedia articles and book reviews have found many appreciative readers over the years. Parsons' work exhibits a systematic blend of rigor, attention to detail, and historical scholarship. His writings have exerted a considerable influence on the development of the philosophy of logic and the philosophy of mathematics in our time. He has placed a set of issues and concerns at the center of the field, and has presented a series of deep, illuminating, and sometimes provocative views on these issues. Most typically, his work is distinguished by an uncommon sense integrity, circumspection, and drive toward precision and truth.

As a teacher, Parsons has been a model of open-mindedness, breadth, and erudition. He would give a graduate seminar on set theory one year, Kant the next, and Husserl the year after that. His reading groups on the German edition of Kant's first Critique have become legendary. As his former students at Columbia University, we fondly remember his meticulously formulated lectures, his careful and thorough comments on our work, his dedication to our dissertations, and the feel he developed for our philosophical interests and identity. We still value the sense he inspired in us of the difficulty of philosophical questions, the seriousness, care, and modesty called for in dealing with them, and the importance of going beyond questions to answers, preferably positive, constructive answers.

In planning this volume we wished to honor Parsons with a collection of essays that, in addition to reflecting his influence, would stand on its own as representing major trends in turn-of-the-century philosophy of logic and mathematics. The volume opens with the Alfred Tarski Lectures delivered by Hilary Putnam in 1995 at Berkeley and published here for the first time. Putnam's lectures are devoted to the semantic and set theoretical paradoxes, and they are followed by a series of papers on the crossroads of logic, ontology, epistemology, and the philosophy of language. The middle section presents a collection of essays on mathematical intuition and related issues, viewed from historical as well as contemporary perspectives. It includes work on the notion of intuition in Kant as well as reflections on Parsons' account of intuition. The last section is devoted to foundational issues in set theory and number theory, their roots in Cantor and Frege, their epistemology, and the set-class distinction.

Two distinguished scholars who regarded Parsons with great affection and esteem did not live to see this volume. Hao Wang succumbed to an illness shortly after accepting our invitation, and George Boolos, who was an enthusiastic supporter of the project in its first stages, died after a short illness prior to its completion. Boolos' paper, "Must We Believe in Set Theory?" is one of the jewels of this collection.

It is with great pleasure that we present Charles with this volume. It is a small token of our appreciation of the many ways in which he enriched us through his research, his pursuit of academic excellence, and his devotion to philosophy and logic.

Gila Sher
Richard Tieszen

I. LOGIC

Paradox Revisited I: Truth

HILARY PUTNAM

In this pair of essays, I revisit the logical paradoxes. In the present essay I discuss the most famous of the so-called semantical paradoxes, the paradox of the Liar, the sentence that says of itself that it is not true, and in the essay that follows (Paradox Revisited II) I shall consider whether we should really accept a view once expressed by Gödel, the view that the paradoxes of set theory are ones that we can see through, can definitely and satisfactorily resolve, even if (as he conceded) the same cannot be said for the semantical paradoxes.

The Liar Paradox

The best presentation I know of the Liar Paradox is Charles Parsons', and in the end the view I shall defend is, I believe, an elaboration of his. In "The Liar Paradox,"[1] a paper I have thought about for almost twenty years, the paradox is stated in different ways. One of these ways is in terms of three alternatives: either a sentence expresses a true proposition, or it expresses a false proposition, or it does not express a proposition at all. A second way mentioned in that paper is the one I followed in my presentation of the Liar paradox in *Realism with a Human Face*,[2] in which talk of propositions is avoided, and I mostly employ that way here in order to facilitate comparison with Tarski's work.

It is an empirical fact that the one and only sentence numbered (I) on page 11 of my *Realism with a Human Face* is the following:

(I) The sentence (I) is false.

Is the sentence numbered (I) (on page 11 of my *Realism with a Human Face*) true? Tarski famously used "Snow is white" as his example of a typical sentence, and his "Convention T" requires that a satisfactory treatment of truth must enable us to show that

"Snow is white" is true if and only if snow is white.

If we suppose that sentence (I) has a truth value at all, it follows by Convention T that

(i) "The sentence (I) is false" is true if and only if the sentence (I) is false.

But, as just mentioned, sentence (I) = "The sentence (I) is false," and hence

(ii) Sentence (I) is true if and only if sentence (I) is false

which is a contradiction!

So far, we do not have an actual inconsistency. We assumed that sentence (I) has a truth value, and that assumption has now been refuted. We cannot consistently assert either that (I) is true or that (I) is false. But now we come to the "strong Liar." The form I considered in *Realism with a Human Face* (p. 12) is:

(II) The sentence (II) is either false or lacks a truth-value.

Sentence (II) is paradoxical because, if we try to avoid the previous argument by denying that (II) has a truth value, that is, by asserting

(II) lacks a truth value,

then it obviously follows that

(II) is either false or lacks a truth value,

and sentence (II) is one that we discover ourselves to have just asserted! So, we must agree that (II) is true, which means that we have contradicted ourselves.

Tarski showed us how to avoid such paradoxes by relativizing the predicate "is true" to whichever language we are speaking of, and by introducing a hierarchy of languages. If I say of a sentence in a language *L* that it is true or false, my assertion belongs to a language of a higher level – a meta-language. No language is allowed to contain its own truth predicate. The closest I can come to such sentences as (I) or (II) is to form a sentence (III) with a relativized truth predicate:

(III) The sentence (III) is not true-in-L,

but this sentence does not belong to *L* itself, only to meta-*L*. Since it does not belong to *L*, it is true that it is not true-in-L. And since this is exactly what it says in meta-*L*, it *is* true in meta-*L*. Sentence (III) is not even well formed in the "object language" *L*, and is true in the meta-language, meta-*L*, and this dissolves the paradox.

In *Realism with a Human Face*, I asked "if Tarski [had] succeeded, or if he ha[d] only pushed the antinomy out of the formal language and into the informal language which he himself employs when he explains the significance of his formal work." If each language has its own truth predicate, and the notion "true-in-L," where *L* is a language, is itself expressible in meta-*L*, but not in *L*, all of the semantical paradoxes can be avoided, then I agreed. "But in what language is Tarski himself supposed to be saying all this?" I asked. (p. 13)

"Tarski's theory introduces a "hierarchy of languages," I continued.

> There is the "object language" .... there is the meta-language, the meta-meta-language, and so on. For every finite number n, there is a meta-language of level n. Using the so-called transfinite numbers, one can even extend the hierarchy into the transfinite – there are meta-languages of higher and higher *infinite* orders. The paradoxical aspect of Tarski's theory, indeed of any hierarchical theory, is that one has to stand outside the whole hierarchy even to formulate the statement that the hierarchy exists. But what is this "outside place" – "informal language" – supposed to be? It cannot be "ordinary language," because ordinary language, according to Tarski, is semantically closed and hence inconsistent. But neither can it be a regimented language, for no regimented language can make semantic generalizations about itself or about languages on a higher level than itself. (pp. 13–14)

I also considered Parsons' way out; as I explained it then, this way involves the claim that the informal discourse in which we say such things as "every language has a meta-language, and the truth predicate for the language belongs to the meta-language and not to the language itself" is not part of any language but a kind of speech that is *sui generis* (call it, "systematic ambiguity"). And I found difficulty in seeing what this comes to. After all, one can formally escape the paradox by insisting that all languages properly so-called are to be written with ink other than red, I pointed out, and red ink reserved for discourse that generalizes about "languages properly so-called." Since generalizations about "all languages" would not include the Red Ink Language in which they are written (the Red Ink Language is *sui generis*), we cannot derive the Liar paradox. But is this not just a formalistic trick? How, I asked, does Parsons' "systematic ambiguity" differ from Red Ink Language? In this essay, I hope to answer my own objection, and thereby to deepen our understanding of what systematic ambiguity is and why it is necessary. In the essay that follows, I will, among other things, argue that something like systematic ambiguity is inevitable in set theory as well.

"Black Hole" Sentences?

When I wrote the title paper of *Realism with a Human Face*, I regarded Tarski's hierarchical solution as just a technical solution, a way of constructing restricted languages in which no paradox arises. It seemed to me that, as a *general* solution, the hierarchical solution can only be "shown but not said"; it is literally inexpressible. But I proposed no solution of my own. In an earlier paper, a memorial lecture for James Thompson, I did propose a solution, but I was dissatisfied and did not publish that lecture. In that unpublished lecture, I set up a language that is not hierarchical, and in which the truth predicate can be applied to any and all of the sentences of the language. Semantical paradoxes were avoided by assuming the Convention T only for a subset of the sentences of the language, the "Tarskian" sentences. I did not define the set of Tarskian sentences once and for all; instead there were axioms enabling us to prove that certain sentences (sentences that are sure to be paradox free and that are likely to be needed) are all Tarskian. The idea was that, just as we add stronger axioms of set existence to set theory when we discover that we need them, we could add axioms specifying that additional sentences are Tarskian as these become necessary. With respect to the obviously paradoxical sentences such as the Liar, the position I recommended was a sort of logical quietism; that is, "Don't say anything (semantical) about them at all!" (This idea was, perhaps, an anticipation of Haim Gaifman's idea that there are "black hole" sentences, that is, sentences that are paradoxical and, moreover, such that the application of any semantical predicate to one of them simply generates a further paradox.[3]) But this seemed to me desperately unsatisfactory, for if we are content not to say anything at all about the paradoxical sentences, why do we not just stick to Tarski's solution? The problem with that solution, after all, arises only if we try to state it as a *general* solution. One could just as well be a quietist about the principle *underlying* the Tarskian route to avoidance of the paradoxes in particular cases as about the Black Hole sentences. Both forms of "quietism" are *so* unsatisfactory that I want to make another attempt to see if we can find something more satisfactory to say about the paradoxes. But I must warn you in advance that what I will end up with will not be a "solution" to the paradoxes, in the sense of a point of view that simply makes all appearance of paradox go away. Indeed, I still agree with the main moral of "Realism with a Human Face," which is that such a solution does not seem to be possible.

Reconsideration of Parsons' Solution

As I indicated, I now accept Parsons' solution. As I explained a moment ago, my objection to that solution was that it rests on the notion of systematic ambiguity, but it wasn't clear to me why systematic ambiguity wasn't just another language, a language that was simply *stipulated* to be outside the hierarchy, and thus available to serve as a kind of Archimedean point. Another problem (one that I did not mention in "Realism with a Human Face") was that I had difficulty in understanding the following: when Parsons applies his solution to natural language, he asserts that

> However vaguely defined the schemes of interpretation of the ordinary (and also not so ordinary) use of language may be, they arrange themselves naturally into a hierarchy, though clearly not a linearly ordered one. A scheme of interpretation that is "more comprehensive" than another or involves "reflection" on another will involve either a

larger universe of discourse, or assignments of extensions or intensions to a broader body of discourse, or commitments as to the translations of more possible utterances. A less comprehensive interpretation can be appealed to in a discourse using the more comprehensive interpretation as a metadiscourse.

To many the hierarchical approach to the semantical paradoxes has seemed implausible in application to natural languages because there seemed to be no division of a natural language into a hierarchy of "languages" such that the higher ones contain the "semantics" of the lower ones. Indeed there is no such neat division of any language as a whole. What the objection fails to appreciate is just how far the variation in the truth-conditions of sentences of a natural language with the occasion of utterance can go, and in particular how this can arise for expressions that are crucially relevant to the semantic paradoxes: perhaps not "true," but at all events quantifiers, "say," "mean," and other expressions that involve indirect speech. (p. 250)

The fact is that I found it difficult to understand what Parsons meant by his claim that talk of "meaning" and "saying" and "expressing" in natural language (if not the use of the predicate "true" itself) presupposes interpretations that can be arranged into hierarchies. But what I want to do now is to apply Parsons' idea to a context in which it will be clear what the "interpretations" are and how they form hierarchies.

Parsons considered the following sentence (p. 227):

(2) The sentence written in the upper right-hand corner of the blackboard in Room 913-D South Laboratory, the Rockefeller University, at 3:15 P.M. on December 16, 1971, does not express a true proposition.

(Note for later use that I shall sometimes abbreviate the sentence-description in (2) as A.)

We are given that sentence (2) was, in fact, written in the upper right-hand corner of the blackboard in the room mentioned on December 16, 1971, and was the only sentence so written. The guiding idea behind Parsons' solution to the Liar is contained on page 230 of his paper, in which he says, in effect (I have slightly simplified the exposition):

(2) says of itself that it does not express a true proposition. Since it does not express *any* proposition, in particular it does not express a true one. Hence it seems to say something true. Must we then say that sentence (2) expresses a true proposition? In either case, we shall be landed in a contradiction. A simple observation that would avoid this is as follows: the quantifiers in one object language could be interpreted as ranging over a certain universe of discourse U. Then a sentence such as

$$(\exists x)(x \text{ is a proposition. } A \text{ expresses } x)$$

is true just in case U contains a proposition expressed by A, that is, by (2). But what reason do we have to conclude from the fact that we have made sense of

(2), and even determined its truth value, that it expresses a proposition that lies *in the universe U*?

It is this rhetorical question that leads Parsons to speak of a hierarchy of interpretations of paradoxical sentences such as (2). To generate a hierarchy of interpretations that can serve as a kind of formal model for what Parsons is suggesting here, I shall begin by using Saul Kripke's idea to generate an initial interpretation.

A Hierarchy Beginning with a Kripkean Interpretation

Saul Kripke is, of course, *the* contemporary logician who put the idea that there is an alternative to Tarski's method, that is, a way to construct a consistently interpreted formalized language so that the truth predicate for the sentences of the language belongs to the language itself, on the map. Since his famous paper,[4] research into that alternative has never ceased. And it should not be surprising that I drag Kripke in here, because sentences such as (II) cannot generate a paradox if we follow Tarski's method; the very fact that they are not in the language that they speak about aborts the paradox. If we wish to formalize the paradoxical reasoning at all, we need a language in which the truth predicate belongs to the language *L* and not only to a meta-language, and this is the kind of language Kripke showed us how to interpret.

Kripke himself admits that his solution does not wholly avoid hierarchy, for a reason that I shall mention soon, and, of course, the whole moral of Parsons' paper was that, even if the language itself (or natural language itself) is not stratified into object language, meta-language, and so on – that is, even if its *syntax* is not hierarchical – still the best way to think about what is going on with the Liar paradox is to think of a hierarchy: not a hierarchy of formalisms, but a hierarchy of *interpretations* of the syntactically unstratified formalism.

What Kripke achieved was to find a natural way (actually, a whole class of natural ways) to do the following: to assign to the predicate "true" (using a device from recursion theory called "monotone inductive definition") not simply an extension, but a *triple* of sets of sentences, say ⟨Trues, Undecideds, Falses⟩. The first set in the triple – let us call it the Trues – consists of sentences in the object language (henceforth simply *L*) that are assigned the truth-value "true"; the third set, the Falses, consists of sentences in *L* that are assigned the truth-value "false"; and the middle set, the Undecideds, contains the remaining sentences, the ones whose truth value is undefined. All this is done in such a way that, of course, the Liar sentence itself turns out to be one of the Undecideds.

I mentioned monotone inductive definition. In fact, the pair ⟨Trues, Falses⟩ is itself the limit of a monotone increasing sequence of (pairs of) sets ⟨Trues_α, Falses_α⟩, indexed by ordinals; and this whole sequence is a precise mathematical object, as is its limit. Thus, each of the sets Trues, Undecideds, and Falses is itself definable (explicitly and precisely) in a strong enough language, although not in L itself. This is why Kripke speaks of "the ghost of Tarski's hierarchy" as still being present in his construction.

We are going to formally model Parsons' remarks about the Liar paradox in the following way: To simplify matters, we shall not speak of sentences as expressing propositions, as Parsons did in the paragraphs I quoted. Instead, we shall simply think of sentences as true, false, or lacking in truth value. To say of a sentence S that it is not true, that is, to write "$-T(S)$," will be to say something that is (intuitively, as opposed to what happens in Kripke's scheme) *true* if S is either *definitely* false or *definitely* lacking in truth value. In short, "$-T(S)$" says "S is either false or lacking in truth value," or, in Parsons' formulation, "S does not express a true proposition."

We shall assume that, as is usual in formal work, sentences are identified with their Gödel numbers, and that the language L is rich enough to do elementary number theory. Then, by a familar Diagonal lemma, given any open sentence $P(x)$ of the language, we can effectively construct an arithmetical term σ, such that the numerical value of the term σ is the Gödel number of the very sentence $P(\sigma)$, that is, we have a uniform technique for constructing self-referring sentences. Since the language contains the predicate "T" for truth, that is, for truth in an interpretation (although the interpretation will be allowed to vary in the course of our discussion), and hence contains its negation "$-T$," we can effectively find a numerical term τ such that the numerical value of τ is the Gödel number of the very sentence

$$-T(\tau).$$

Speaking loosely, the sentence $-T(\tau)$ says of itself that it is not true or, in Parsons' language, that it does not express a true proposition, and this is precisely the Liar.

We will model Parsons' discussion as follows: we will suppose that when a student – call her Alice – first thinks about the Liar sentence, that is, about the sentence $-T(\tau)$, her first reaction is to say that this is a "meaningless" sentence, that is, not true or false. We shall also follow Parsons by supposing that Alice (implicitly) has an interpretation in mind. Parsons, reasonably enough, supposes that the schemes of interpretation that actual speakers have in mind are only "vaguely defined" (p. 250), but since we are idealizing, we will assume that Alice, implausibly, of course, has in mind precisely one scheme of interpretation, and that it is given by a Kripkean construction.[5] Thus when Alice says of the Liar sentence that it is neither true nor false, she means that it is one of Kripke's Undecideds. But, when we ask her, following Parsons' scenario, whether it follows from the fact that the Liar sentence is neither true nor false that *in particular it is not true*, and we bring her to say "yes," then what has happened (according to Parsons' analysis) – and this seems reasonable – is that she has subtly shifted her understanding (her "interpretation") of the predicate "true" ("T"). The sense in which the Liar, "$-T(\tau)$," is not true is that it does not belong to what we might call the "positive extension" of "T" in Alice's initial (Kripkean) interpretation, that is, to the set of Trues. She has now shifted to a *bivalent* interpretation of "T" under which "$T(\sigma)$" is true (where σ is any numerical term of language L) just in case the statement that the numerical value of σ lies in the set of Trues is a true sentence of meta-L, the language in which the Kripkean interpretation of L is explicitly defined. Parsons' own discussion ends at this point; he is content to point out that Alice need not be contradicting herself when she says that the Liar sentence is not true, because the interpretation presupposed by this second remark is not the interpretation of L presupposed by her initial statement that the Liar sentence is neither true nor false. But what interests me is something else – something pointed out some years ago by Professor Ulrich Blau of Munich University, who has not yet published the long work on the paradoxes on which he has been working for many years. What interests me is that the situation is now *unstable*.

Note first that the second interpretation – let us refer to the initial interpretation as Interpretation$_0$ and the second as Interpretation$_1$, henceforth – has a paradoxical feature. For, on the second interpretation, $T(\tau)$ is true just in case the numerical value of the term τ lies in the set of Trues (generated by Interpretation$_0$), and it does not. Hence, $T(\tau)$ is not true, and hence $-T(\tau)$ is true (since Interpretation$_1$ is *bivalent*). But Convention T requires that, if the numerical value of any term σ is (the Gödel number of) a sentence S, then

$$\mathrm{T}(\sigma) \Leftrightarrow \mathrm{S}$$

is true, and τ *is* (the Gödel number of) the sentence $-T(\tau)$. Hence

$$T(\tau) \Leftrightarrow -T(\tau)$$

should be true! Of course, this failure of Convention T is not surprising since, under Interpretation$_1$ 'T' does not refer to truth *under Interpretation*$_1$ *itself* (so Convention T does not really apply) but to truth under the initial interpretation.

The instability, of course, arises because reflection on this new interpretation will generate still another interpretation, and, by iteration, an infinite series of interpretations. To spell this out: under the next interpretation, Interpretation$_{n+1}$, $T(\tau)$ is true just in case the numerical value of the term τ lies in the set of sentences (identified, as we stipulated, with their Gödel numbers) that are true under Interpretation$_n$. Since Interpretation$_1$ is simply the bivalent interpretation of L generated by letting "T" stand for the set of Truths of Interpretation$_0$, and that set is definable in Meta-L, the set of sentences that are true under Interpretation$_1$ is itself definable in Meta-Meta-L (or Meta^2-L). As we have just seen, under Interpretation$_1$, the Liar sentence is true; hence $T(\tau)$ is true under Interpretation$_2$, and hence the Liar sentence is *false* under Interpretation$_2$. In short, the truth value of the Liar sentence flips when we go from Interpretation$_n$ to Interpretation$_{n+1}$, $n > 0$. (Interpretation$_n$ is, of course, definable in Meta^{n+1}-L.)

The series of interpretations can be extended into the transfinite. We shall define a sentence S to be true at a limit ordinal λ if it has *become stably true* at some ordinal $<\lambda$, that is, if there is an ordinal $\kappa < \lambda$, such S is true under Interpretation$_\gamma$ for every γ such that $\kappa < \gamma < \lambda$. Sentences that have become stably true at a stage before λ are true at λ. Similarly, sentences that have *become stably false* at some stage less than λ are false at λ, and sentences that have not become stably true or stably false (e.g., the Liar sentence) are undecided at λ. (Limit interpretations are not bivalent.)

What I want to come to now is the point hinted at in the closing sentences (before the Postscript) of Parsons' paper (p. 251):

> In a simple case, such as that of the word 'I', we can describe a function that gives it a reference, depending on some feature of the context of utterance (the speaker). We could treat the "scheme of interpretation" in this way as argument to a function, but that, of course, is to treat it as an object, for example a set. But a discourse quantifying over *all* schemes of interpretation, if not interpreted so that it did not really capture *all*, like talk of sets interpreted over a set, would have to have its quantifiers taken more absolutely, in which case it would not be covered by any scheme of interpretation in the sense in question. We could produce a "superliar" paradox: a sentence that says of itself that it is not true under any scheme of interpretation. We would either have to prohibit semantic reflection on this discourse or extend the notion of a scheme of interpretation to cover it. The most that can be claimed for the self-applicability of our discussion is that if it is given a precise sense by one scheme of interpretation, then there is *another* scheme of interpretation of our discourse which applies the discourse to itself under the *first* interpretation. But of course this remark applies to the concept "scheme of interpretation" itself. Of it one must say what Herzberger says about truth: in it "there is something schematic . . . which requires filling in."

The sequence of schemes of interpretation of the semantical paradoxes that I just described is a well-defined set-theoretic construction. So far, we have simply associated a scheme of interpretation with each ordinal. (Of course, if we continues it through all the ordinals, then, by cardinality considerations, at some point we will only get interpretations that are extensionally identical to ones already constructed.) But – this is the point that Parsons, citing Herzberger,[6] hints at in the paragraph I just quoted, and the point that Ulrich Blau emphasizes[7] – there is still another source of paradox here. To see this source of paradox, we need to imagine a different scenario than the one Parsons imagined earlier in his paper (our scenario with Alice). There (p. 227 *passim*) Parsons imagines someone who looks at the Liar sentence, decides that it is not true or false (that it is meaningless or, in Parsons' terminology, that it does not express a proposition), and then concludes from that very fact that it is *true* that it doesn't express a true proposition; and he is concerned to argue that that judgment may be totally in order provided we recognize that the scheme of interpretation has changed in the course of the reflection itself. But it seems to me unlikely that this could be the terminus of Alice's reflections. If she is sophisticated, Alice naturally will be led to investigate just the hierarchy of interpretations we constructed, the hierarchy that would result if her act of reflection were iterated through the transfinite.

At this point, a new temptation may arise for Alice, the temptation to land herself in what Parsons refers to as the "superliar paradox." This need not be a temptation to suppose that one can *stand outside* the hierarchy (although one can do that, since the whole inductive definition is carried out, so far at least, within set theory). It is the temptation to suppose that, even *standing within the hierarchy* (and "gazing up," as it were), one can define an *ultimate* sense of "stably true," namely, stably true with respect to the whole hierarchy, and see now that in an *ultimate* sense the Liar sentence is not true (does not express a true proposition), namely that it does not *ever* become stably true. But this, of course, will simply generate a new hierarchy.

Can we go still further? It seems to me that we can. To do so in an interesting way (there are some obvious but uninteresting ways of going further), we will need to use a phrase such as "all the hierarchies one might ever arrive at by continuing reflection," and that means we shall no longer be dealing with precisely defined set-theoretic constructions. This is important, because it may indicate what the answer to my question as to how systematic ambiguity is supposed to differ from just another language might be. When we imagine continuing reflection without limit, creating new hierarchies, and then summing them up – going to the "ultimate interpretation" with respect to a hierarchy, and then taking that ultimate interpretation as the zero stage of a new hierarchy, and so on – we are no longer in the realm of the mathematically well-defined, and hence we cannot assume bivalence (or classical logic). Nevertheless, it does seem that there are things that can be seen to be true in the sense of *provable from the very description of the procedure*. (Compare, in the Tarskian hierarchy, which also can be imagined as extended without limit in similar fashion, the way in which we can see the truth of "For every language *L*, there is a meta-language *ML* that contains a truth predicate for *L*.") For example, we see from the very description of the procedure by which any hierarchy is constructed from a given initial interpretation that the Liar sentence never becomes stably true. We cannot imagine an Archimedean point here. We cannot regard the vague "hierarchy of all hierarchies" as something that we can describe *from outside*, as it were. But *we can see from below* how things must go. That is, we can see that *no matter what we "get to"* in the way of reflection on the Liar, no matter what scheme of interpretation we arrive at, we can always use that scheme as the beginning of a new hierarchy, and we can see that, vague as the notion of a hierarchy is, at least *this* much is true of it: the Liar sentence will never become stably true. In fact, using Parsons' device of systematic ambiguity, I can say things like "If the Liar sentence has no truth value at a stage, it gets one to the next stage, and if it has a truth value at a stage, that truth value flips at the next stage." But now it seems to me that Alice may well become the victim of a Super-superliar paradox. The temptation now will be to think something like this:

> When we talk of all the hierarchies of interpretation we could produce, I know that we are not talking about something precise and well-defined, but nevertheless, as you have just shown, there is a sense – a LAST SENSE – in which the Liar sentence is not true: namely, it does not become stably true in *any* hierarchy, not even in, so to speak, the hierarchy of all hierarchies. But surely being eventually stably true in the hierarchy of all hierarchies is the last sense of being stably true, and so there is an absolute sense, namely the LAST SENSE, in which the Liar sentence is not true.

At this point, of course, she will have generated yet another interpretation – an ill-defined one, of course, but nonetheless an interpretation that can also be used as the basis of a hierarchy (even if we have to use intuitionist logic rather than classical logic to talk about it, in view of the fact that the only notion of "truth" that we appear to have in connection with it is some species of provability).

In short, the final temptation is the temptation to suppose that the notion of a LAST scheme of interpretation makes sense. What Parsons says, using a term of Herzberger's, in a sentence I quoted earlier, seems to be exactly right, namely, that when we talk of hierarchies in general, rather than of a specific hierarchy constructed in a specific set-theoretic way, we are necessarily talking schematically; and the schematic character of such talk is, it seems to me, just the difference between talking with systematic ambiguity and merely using Red Ink Language.

There is a further point that I want to make, one emphasized by Ulrich Blau. What is wrong with the temptation to which I said Alice might succumb is not that it is *impossible* to think of an interpretation of the language *L* under which one says of sentences such that we can prove that they will be unstable with respect to every hierarchy that they are "undecided" and of sentences such that we can prove that they will eventually become true in any hierarchy that goes far enough that they are "true" and of sentences such that we can prove that they will eventually become false in any hierarchy that goes far enough that they are "false" – although (because provability is not the same as classical truth) there will be sentences such that we cannot say that they are true, false, *or* undecided if we proceed in this way. As I already mentioned, there are logics that do not assume bivalence (e.g., intuitionist logic), which one might employ in this connection, but I shall not attempt a formal treatment. But if it is all right then, or possibly all right, to treat "the hierarchy of all hierarchies" as something we can reason about, at least in an intuitionist setting, then the mistake that Alice would be making if she gave in to the temptation that leads to what I called the Super-superliar paradox would not be that what I imagined her calling the LAST INTERPRETATION does not *exist*. The mistake is more subtle, it seems to me. The mistake, rather, lies in thinking of it as the "LAST." The phrase "LAST INTERPRETATION" assumes that limits have some kind of finality. But if we allow talk of the LAST INTERPRETATION, we must also allow that there is a *successor* to the LAST INTERPRETATION. That is, it is quite true that the Liar is undecided in the so-called LAST INTERPRETATION, but it is equally true that it becomes true again just AFTER the LAST interpretation. In short, the phrase "LAST INTERPRETATION" is a misnomer. The illusion is that, by this very act of looking up from below at what happens in our hierarchies, we can somehow generate an absolute sense of a "LAST INTERPRETATION," and the paradox itself shows this to be an illusion. Our desire to have a *final thing we can say about the Liar*, or an absolutely best thing to say about the Liar, is what always causes the Liar to spring back to life from the ashes of our previous reflections.

I am led back, in a way, to my own rejected solution in the unpublished James Thompson Memorial Lecture. If you want to say something about the Liar, in the sense of being able to finally answer the question "Is it meaningful or not? And if it is meaningful, is it true or false? Does it express a proposition or not? Does it have a truth value or not? And which one?" then you will always fail. And the paradox itself shows why this desire to be able to say one of these things must always fail.

In closing, let me say that even if Tarski was wrong (as I believe he was) in supposing that ordinary language is a theory, and hence can be described as "consistent" or "inconsistent," and even if Kripke and others have shown that it is possible to construct languages that contain their own truth predicates, the fact remains that the totality of our desires with respect to how a truth predicate *should* behave in a semantically closed language, in particular our desire to be able to say, without paradox, of an arbitrary sentence in such a language that it is true, or that it is false, or that it is neither true nor false, *cannot* be adequately satisfied. The very act of interpreting a language that contains a Liar sentence creates a hierarchy of interpretations, and the reflection that *that* generates does not terminate in an answer to the questions "Is the Liar meaningful or meaningless, and if it is meaningful, is it true or false?" On the other hand, Tarski's own suggestion of giving up on unrestricted truth predicates, and contenting ourselves with hierarchies of stronger and stronger languages, each with its own truth predicate, leaves us in much the same situation as does Parsons' hierarchy of interpretations of a single language. In the end, we are led to see that the things we say about formal languages must be (to use Herzberger's term) "schematic."

NOTES

The present essay and the one that follows it originated as my Alfred Tarski Lectures delivered at the University of California, Berkeley, April 24 and 25, 1995. I thank

Charles Chihara, Doug Edwards, Solomon Feferman, Charles Parsons, and especially Jamie Tappenden for helpful suggestions and criticisms.

1. Charles Parsons, "The Liar Paradox" in his *Mathematics in Philosophy* (Ithaca, NY: Cornell University Press, 1983).
2. "Realism with a Human Face," in H. Putnam, *Realism with a Human Face* (Cambridge, MA: Harvard University Press, 1990).
3. Haim Gaifman, in *Proceedings of the Second Conference on Theoretical Aspects of Reasoning About Knowledge*, M. Vardi (ed.), Los Altos, CA: Morgan Kaufman Publishers, 1988, 43–59. "Pointers to Truth," *Journal of Philosophy*, 85 (1992): 223–61.
4. Saul Kripke, "Outline of a Theory of Truth," *Journal of Philosophy*, 72 (1975): 690–716.
5. This may not be as implausible as it sounds. A speaker may well intend that all "paradoxical" sentences be left un-truth-valued, and the preceding might be regarded as one way of rendering that inexact intention precise.
6. Hans Herzberger, "Paradoxes of Grounding in Semantics," *Journal of Philosophy*, 67 (1970): 145–67.
7. In unpublished writing.

Paradox Revisited II: Sets – A Case of All or None?

HILARY PUTNAM

Through the years, I have noticed that set theorists tend to be either strong platonists or strong formalists (although my impression is that the formalists are in the minority). I have also observed that set theorists of both persuasions are quite sure that there are no problems remaining to be thought through! Nevertheless, some philosophers continue to seek an alternative to both formalism and platonism, and I know many mathematicians (and even an occasional set theorist) who are dissatisfied with extremes.

In this essay, I propose such an alternative; in addition, I present a reaction to two of Gödel's papers[1] and to some of Quine's highly influential doctrines, and I have to begin by saying what these are about if my reader and I are to be on the same wavelength.

The aspect of Gödel's views that concerns me is the attitude expressed toward the paradoxes in those papers. Gödel suggests that the paradoxes of set theory arise from a confusion of two ideas, a logical one (or, as I shall say, an *intensional* one), "extension of concepts," and another ("set of") that forms the basis of the cumulative hierarchy. He seems to have felt that we do not yet have a completely satisfactory view with respect to the intensional paradoxes (i.e., the Liar paradox, and the other semantical paradoxes), but that we do now have a clear and satisfactory view of the nature of sets, and not just a way of pushing paradoxes such as Russell's out of the formal language.[2]

Let me hasten to say that I do not intend to defend the view that the Russell paradox is irresolvable. Just as I believe that, in some sense, hierarchies are the answer to the Liar paradox, so I agree that, in a similar sense, hierarchies are the answer to the Russell paradox (and the Burali–Forti paradox, etc.).[3] But, as I argued in the preceding lecture, the hierarchy of meta-languages and the hierarchy of interpretations of the uses of truth in natural language are intrinsically puzzling hierarchies – they are "schematic" (to use Herzberger's term, mentioned at the close of the preceding lecture) and not perspicuously surveyable either from above or below. In the end, they are, as we say, "paradoxical" even if they are not formally contradictory. And I shall argue that the situation is set theory is not really very different.

The doctrines of Quine's that I referred to are (1) (in my words rather than Quine's) that "exists" is univocal – there is nothing at all to the idea that we are saying a very different sort of thing when we say "The power set of the countable ordinals *exists*" and when we say "The top quark *exists*"; and (2) one's use of the existential quantifier shows what *objects* one is committed to, in a sense of "object" that is also univocal.[4]

The "univocality" doctrine that I attribute to Quine takes account of Quine's insistence that there is "no fact of the matter" about translation. It is indeterminate, in Quine's view, just how a given idiom is to be "regimented" into the notion of quantification; but, once a given idiom has been regimented as an existential quantification, relative to that translation the idiom posits the existence of *objects* to serve as the values of the bound variables required to make the existential quantification true. In particular, Quine argues that the quantifiers we use must, in most cases, be interpreted as "objectual" and not "substitutional."[5] (Quine everywhere writes as if it were clear what it means to say that a quantification is objectual, although no criterion can be extracted from his writing.) If Quine were not treating the sign ∃ itself as univocal, then it would be strange to speak of it as implying the existence of *objects*. Thus I think it is fair to attribute both doctrines to Quine.

To the objection that "the logical notation of quantification is an arbitrary and parochial standard to adopt for ontological commitment," Quine's reply is that "the standard is transferable to any alternative language, insofar as we are agreed on how to translate quantification into it."[6]

I should mention that another doctrine of Quine's that both illustrates and supports his tendency to view mathematical objects and physical objects on a par is the epistemological claim that we accept the existence of mathematical objects (sets and numbers) for reasons perfectly analogous to the reasons for which we accept the existence of, say, electrons; this is a doctrine I have criticized elsewhere.[7]

These doctrines lead Quine to the view that sets are immaterial objects – which is just Gödel's view (though not Gödel's route to the view). Indeed, Quine himself writes, "I am a Predicate and Class Realist, now as of yore; a deep-dyed realist of abstract universals. Extensionalist yes, and for reasons unrelated to nominalism."[8]

I shall not attempt to *refute* Quine's view in the present lecture.[9] Instead, I state very briefly an opposing view and explore its implications for the controversies surrounding the notion of a set.

Conceptual Relativity

I have called my view "conceptual relativity."[10] Quite simply, I hold that, while "exist" (or, better, "∃") may be a single primitive in quantification theory, that is not a particularly interesting fact about the notion. I believe that the quantifiers themselves have many different uses, and that we are constantly extending those uses.[11] An example I have frequently used has to do with the controversy over the existence of "mereological sums" – supposed objects with arbitrary (and even disjoint) objects as parts – for example, a mereological sum of my nose and the Eiffel tower. In my view, the invention of mereology by Lesniewski was an *extension of language* – an extension of the *notion* of an object – and a new use of the quantifiers. To ask whether mereological sums "really exist" is to ask a pseudoquestion. Indeed, "object" is also a notion that we keep extending. Of course, some things are uncontroversially objects: as the etymology suggests, things that you can *throw* are. A brick and a cabbage are uncontroversially objects. But using set theory is not committing oneself to some objects (albeit immaterial ones); it is widening the notion of an object (if, for example, you refer to sets as mathematical objects). Rather than try to expound this conceptual relativist view once again, I have already indicated that what I am going to do is *apply it*, and the conclusions I shall reach will probably surprise a number of you.

"Set" as a Notion

A fact that figures surprisingly little in discussions of the foundations of mathematics is that the notion of a set is a modern one. It is true that people have spoken of "classes of things" from time immemorial; but in ancient and medieval writing, such talk was interchangeable with talk of predicates. Indeed, the entire medieval discussion of Abelard's question "What is a class?" was really concerned with the nature of predication. Talk of classes of classes of classes of classes, and so on, is the invention of Balzano, Cantor, and one or two other people of the nineteenth century. Set theory is not the theory of something we had always talked about or had mathematical intuitions about or had theories about. "Set," in the relevant sense, is a *neologism*.

Here it is important to resist the idea that mathematics is a seamless web. As Wittgenstein rightly said, mathematics is a "motley." It is not a question of "all or none." Big abstract philosophical positions such as platonism and formalism make it seem that classical number theory and classical set theory are really on a par; yet many mathematicians feel that, in some way, the notion of a number is much clearer than the notion of a set; as will emerge, I think that that feeling is exactly right.

One more preliminary point. Philosophers typically take the hard philosophical questions to be whether sentences about "abstract objects" (a notion that lumps together numbers, lines, and circles in geometry, functions in analysis, and Cantorian sets under the supposedly univocal label "object") are true or false even when undecidable, and if so, what "makes them" one or the other. My view with respect to that question is that it is not really a question that we understand. But I mention that issue only to set it aside for today. As far as I am concerned, *of course* statements of number theory are true or false – what is the third possibility? Speaking mathematically, there is none, and the supposed philosophical sense of the question is, I think, chimerical. The question I wish to talk about today is not "what makes sentences of set theory true or false," but *how do we understand what we are talking about here*?

Applying the Attitudes

I have spent some time on preliminaries, but I did not see any alternative. But now, I do what I said I would do, which is to apply this set of attitudes.

I begin, then, with the premise that "set" is *not* a notion we always have had, but a relatively new notion, a notion that still needs explanation. My procedure will be to ask what explanations have been given or could be given. Notably, as many have remarked, the explanations that were given in the nineteenth century were often confused, were psychologistic, and failed to distinguish some very different notions. Gödel himself mentioned two: the notion of the extension of a concept, and the notion of an arbitrary subset. The latter notion was supposed to be capable of indefinite iteration: thus it gives a natural motivition for both the *Aussonderung* (separation) axiom of Zermelo-type set theories instead of naive comprehension, and for the idea of a hierarchy. Thus Gödel writes:

> As far as sets occur in mathematics (at least in the mathematics of today, including all of Cantor's set theory), they are sets of integers or of rational numbers (i.e, of pairs of integers) or of real numbers (i.e., of sets of rational numbers).... This concept of set, however, according to which a set is something obtainable from the integers (or some other well defined objects) by iterated application of the operation "set of," not something obtained by dividing the totality of all existing things into two categories ... has so far proved completely self consistent.[12]

(Note that on the first interpretation – "the extension of a concept" – the Russell paradox is a carbon copy of a familiar semantical paradox: instead of talking about "the set of those sets that do not contain themselves," one could talk of "the predicate 'not self-applicable' ".) Gödel, like most contemporary set theorists, thought that set theory should now be thought of as the theory of what I just called *arbitrary* sets, and freed of all connection with the problematic "intensional" notion of an "extension of a concept." But let us examine both alternatives.

Set as the Extension of a Predicate (Concept)

This is the interpretation that guided both Frege and Russell. Frege's system, interestingly, did not permit predicates ("concept" variables or constants) themselves to be subjects of predicates of the same level, and so, the "not self-applicable" paradox (Is the predicate "not self-applicable" self-applicable or not self-applicable?) does not arise. However, sets are "objects" in the system, and "is a member of" and "is not a member of" and predicates defined in terms of them (including "is not a member of itself") are well-formed in the system, which is why Russell could derive the paradox that bears his name, but not its intensional counterpart, in Frege's system. In Russell's system, talk of "classes" (Russell's word for sets) is introduced as a *façon de parler*, and, as Gödel remarks, it is not quite clear what its status is. This is so because the status of defined expressions is unclear in *Principia*. If *Principia* were being put forward today, we would regard the theory not as set theory (theory of classes) at all, but as a theory of predicates. Moreover, what I am calling "predicates" – the intensional objects in *Principia* – are actually "propositional functions," but for our purposes the difference between a propositional function and a predicate is irrelevant. In addition, the definitions in use, by which class variables are supposed to be eliminated in favor of quantifications over predicates given by Russell and Whitehead, are defective, but it is possible to put this right by using the now-familiar technique of inductive definition over types. Applied to a simple type theory rather than to the complex ramified theory of *Principia*, we specifically define a predicate to be *hereditarily extensional* if

(1) It belongs to type 1 (i.e., is a predicate of individuals); or
(2) it belongs to type $n + 1$ and (a) all the predicates to which it applies are hereditarily extensional; and (b) whenever it applies to a predicate, it applies to all predicates that are coextensive with that predicate.

Then we reinterpret classes as hereditarily extensional predicates and reinterpret identity of classes as *coextensiveness*. However, it is clear, as I remarked, that insofar as what Russell presents is a theory of classes at all, it is a theory based on the notion that a class is *the extension of a predicate*.

But how should we understand that notion?

Let me imagine myself well before Russell, say back in 1870, encountering the new notion of a "set" or a "class" for the first time. Let me further imagine that instead of being offered the fuzzy explanation that a set is a lot of objects "gathered together in thought" into a single object, I had been offered the explanation that a set is the extension of a predicate. Although this obviously is *not* how Frege thought,[13] it seems to me that I would naturally have supposed that "predicate" meant predicate of some (perhaps vaguely specified) language *already* available. Certainly this is how we should think if we took the attitude I recommended, and saw ourselves as *extending* the notion of an object (and the uses of the quantifier, and the notion of a predicate) by quantifying over "extensions of predicates." Of course, the extension will introduce new "predicates" into the language, and so, it can be repeated and even iterated – but what one will obtain is *predicative set theory*, set theory that obeys Russell's Vicious Circle principle forbidding the definition of a set or predicate in terms of a totality that is supposed to already contain that set or predicate. For to explain a *novel* notion of a set in terms of a notion of a predicate that is itself dependent on the notion being defined is, literally, "viciously circular."

Notice that this argument for predicative set theory – for that is what it is, Charles Chihara will be pleased to note – is an argument for predicative set theory *as a theory of one conception of a "set."* It does not exclude the possibility that there can *also* be a legitimate theory of a different conception of a "set," a "theory of arbitrary sets," and, shortly, I shall turn to that subject.

Note also that this argument for predicative set theory does not depend upon any constructivist or intuitionist restrictions on mathematical reasoning. Those restrictions depend on complex metaphysical views on just the philosophical issues about "the nature of mathematical truth" that I said I regard as ultimately unintelligible.

The way that I just suggested we think about set-as-the-extension-of-a-predicate is, I think, close to the way Russell himself thought about it (if we prescind from his idea that class talk is itself dispensable in terms of propositional function talk), which is why Principia's own type theory, "ramified" type theory, is predicative. However, Russell could not derive all the mathematics he wanted in a strictly predicative way, and rather than either admit failure or seek a different procedure (possibly a different notion of a set), he chose to adopt a dubious axiom, the Axiom of Reducibility, which led Ramsey, in turn, to ruthlessly "simplify" Russell's system and to make it impredicative in the process.

I emphasize that the view that I see as lying behind the Vicious Circle principle treats the introduction of each new level as a *meaning bestowing operation.* On a platonist conception such as Gödel's, of course, there is a notion of a set – *one* notion – already there, laid up on Mount Olympus, as it were, and it only requires intuition to grasp it. But *that* view (as Gödel recognized) does not fit well with the idea that "set" means "extension of a predicate." For the idea that the extension of *any* predicate, in the Olympian sense of "predicate," is the extension of a set, in the Olympian sense of "set," is refuted by Russell's paradox. Thus Gödel takes the Olympian sense of set to be the other concept that I mentioned, the concept of an arbitrary set (i.e., an arbitrary subset of some domain of individuals, or an arbitrary subset of the arbitrary subsets of that domain, or an arbitrary subset of the arbitrary subsets of the arbitrary subsets of that domain . . . and so on through the "open series of extension").

Frege's way of thinking was more obscure. Michael Dummett sees Frege as a platonist. I see Frege's "realism" as informed by the idea that the ideal language that Frege constructed mirrors the structure of reason itself. The possibility of quantification over *all* concepts (predicates) of a given level is central to that conception of the ideal language. The ideal language cannot, on such a metaphysical conception, be capable of extension. But why Frege thought that the extensions of predicates should be treated as objects (in the sense of "object" that likewise is supposed to be part of the structure of reason itself, and likewise incapable of extension) is less clear to me. Perhaps because he was so impressed by Cantor's achievement?

Set as "Arbitrary" Set

It may seem obvious that one has to have a platonist attitude like Gödel's to accept the notion of an "arbitrary subset," for example, the notion of an arbitrary subset of the integers, or an arbitrary subset of the collection of all arbitrary subsets of the integers, or . . . . After all, the members of an arbitrary subset may not coincide with the extension of any predicate in any language we could understand. I want to call that "obviousness" into question. Here I shall draw on some remarkable observations by George Boolos concerning second-order logic.

Boolos[14] has connected those observation with a well-known problem concerning the notion of *validity* in first-order logic.[15] On the one hand, the quantifiers in first-order logic (for simplicity, I shall confine my attention to monadic quantifiers) are supposed to range over arbitrary bunches of individuals (in an arbitrary universe), and that has to mean arbitrary subsets of the universe, doesn't it? But first-order logic can be used to reason about totalities that are not *sets*, for example the universe of set theory itself. And the sub*sets* of that universe are not *all* the "bunches of individuals" in *that* universe. They are not even all the extensions of predicates. (For example, the von Neumann ordinals are the extension of a predicate of set theory, but not a set.) So, the standard *model-theoretic definition of validity* as validity in all models (in all sets with a certain structure) badly fails to capture the intended notion of validity, which is truth in all interpretations, including interpretations in which the extensions of some of the predicates are not sets.

Boolos' view is that there is a natural way of understanding second-order logic that does not involve the notion of a set at all. He employs, in fact, a linguistic device found in all natural languages – one as intuitive and probably more primitive than even talk of the numbers – namely, *plural quantification*.

Rather than try to define plural quantification, I will give an example:

(1) Some boys at the party talked only to one another.

Instead of hearing this as a disguised statement to the effect that:

(2) There is an object x (a set) whose "members" are all boys who were at the party, and such that no member of x spoke to anyone at the party who was not a member of x.

(Note that this reading of sentence (1) makes it part of the *content* of (1) that the boys form a *set*.) Boolos suggest that we hear sentence (1) as being related to the quantifier in

(3) There was a boy at the party who talked only to himself (which is expressible in first-order logic) as *plural* to singular.

Given a formula of first-order logic, we can say that it is valid *without* restricting the admissible universes of discourse to sets by using an appropriate formula of second-order logic and interpreting the quantifiers as plural quantifiers over the individuals in the universe in question. For example, to say that

(4) $(x)Fx \Rightarrow Fy$

is valid is to say that

(5) $(U)(y)_U(F)[(x)(Ux \Rightarrow Fx) \Rightarrow Fy]$.

Interpreted à la Boolos, this means that, *for any Us, for any y that is one of the Us, and for any Fs, if y is one of the Us, and if all of the Us are Fs, then y is a U* Stated in natural language, this means

For any things you please, and for any one of those things you please, and for any other things you please, if all of the first things are among the other things, then that one is one of the other things as well.

And this says that sentence (4) is valid over *all* universes, including the universe of "all sets" (if there is/were such a universe), provided that we do *not* construe the plural quantifiers as set quantifiers. Of course, this is only an illustration of Boolos's idea. His primary point is that *all* second-order quantification can and should be interpreted in this way, and that, so interpreted, there is no reason not to view second-order logic as "logic." But I wish to make a somewhat different application.

What Boolos's observations show is that, by virtue of our possession of the notion "some *A*s," we do have a way of speaking of arbitrary bunches of objects (at least we have a way of doing this *whenever* a notion of an "object" is already in place). However, that notion does not, of itself, automatically generate a new notion of an object. We may usefully think of the formation of the notion of an arbitrary set of *U*s (where *U* is some already-given universe of discourse) as split into two stages: (1) extending *plural* quantification to the universe *U*; and (2) reifying bunches of *U*s, by taking them to be "objects".[16] If, as is often the case, forming *U* already involved an extension of the notion of an object, then this second stage would constitute a *further* extension of the notion, and, in any case, it is this extension that constitutes the notion of an arbitrary set, just as the reification of extensions of predicates generated the Fregean and Russellian notions of a class. And like *those* extensions, this extension can be iterated indefinitely, and the result will again be a hierarchical structure. Indeed, if the iteration is just through the natural numbers, it will be the hierarchy of the simple theory of types (with the members of the original universe *U* as the individuals).

Of course, one can also iterate through infinite well-orderings that become available at any stage of the construction, and in this way (taking the types as cumulative) one can get models for Zermelo set theory. Note that, on the view I am proposing, there is no special problem at all about the power set axiom.[17]

But what of Zermelo–Frankel set theory? Quite frankly, I see no intuitive basis at all for the characteristic axiom of that theory, the axiom of replacement. Better put, I do not see that a *notion* of a set on which that axiom is clearly true has ever been *explained*. Instead, it seems to me, we have a formal maneuver that Gödel tried to justify as news from Mount Olympus.

Similarly, I see no basis at all at the present time for any of the proposed large cardinal axioms. On the last page of *Philosophical Investigations*, Wittgenstein famously wrote: "The confusion and barenness of psychology is not to be explained by calling it a 'young science'; its state is not comparable with that of physics, for instance, in its beginnings. (Rather with that of certain branches of mathematics. Set theory.) For in psychology there are experimental methods and *conceptual confusion*. (As in the other case conceptual confusion and methods of proof.)" The fact that axioms are proposed in set theory in a way that would certainly be regarded as questionable in arithmetic[18] (and yet the great set theorists see no problem about taking a platonist attitude toward the products of their opportunistic maneuvers!) is an illustration of just the phenomenon Wittgenstein had in mind.

To say that I see no basis for accepting any of the large cardinal axioms is not to say that there is some fixed place where the process of extending the hierarchy *must* stop. For any notion that we can *explain* of the "hierarchy of all sets," or of "the hierarchy of all ordinals," is one that we can transcend by taking the sets in that hierarchy as the basis of a new hierarchy iterated through all those ordinals and beyond. If we attempt to speak of "the hierarchy of all well-defined hierarchies," the result cannot be a well-defined hierarchy, on pain of inconsistency. The notion of a set, like the notion of a meta-language and the notion of a "scheme of interpretation," is inherently extendable – and the totality of possible extensions is not well defined. But for all that, if one wants to utter some truths about "all sets" without supposing that one has specified a *particular* stopping point, one can do so with the aid of systematic ambiguity, exactly as in the case of the Liar paradox. And this is what I meant by saying at the beginning of this lecture that I cannot agree with Gödel's view that the situation with respect to Russell's paradox is completely different from the situation with respect to the Liar.

NOTES

The present essay and the one that precedes it originated as my Alfred Tarski Lectures delivered at the University of California, Berkeley, April 24 and 25, 1995. I thank

Charles Chihara, Doug Edwards, Solomon Feferman, Charles Parsons, and especially Jamie Tappenden for helpful suggestions and criticisms.

1. Kurt Gödel, "Russell's Mathematical Logic" and "What Is Cantor's Continuum Problem," both in P. Benacerraf and H. Putnam (eds.), *Readings in the Philosophy of Mathematics*, 2nd Ed. (Cambridge, UK: Cambridge University Press, 1984).
2. The determinateness Gödel believed the cumulative notion of a set has – "the open series of extensions" as he puts it – reflects his own Platonism.
3. The Russell paradox, the reader will recall, arises from the observation that if *every* condition determines a set, then $x \notin x$ also determines a set, call it **R**, and, in virtue of the condition that determines membership in the set, $\mathbf{R} \in \mathbf{R} \leftrightarrow \mathbf{R} \notin \mathbf{R}$. The Burali–Forti paradox arises from the observation that if every well-ordering has an ordinal, then the well-ordering of the ordinals themselves has an ordinal, and this ordinal both is and cannot be the greatest ordinal.
4. Thus Quine writes, "The words 'five' and 'twelve' name two intangible objects, numbers, which are *sizes* of sets of apples and the like." Cf. "Success and Limits of Mathematization," in W. V. Quine's *Theories and Things*. Cambridge, Mass: Harvard University Press, 1981, 149.
5. W. V. Quine, *The Roots of Reference*, LaSalle, Illinois: Open Court, 1973, 108ff.
6. W. V. Quine, *The Pursuit of Truth*, Cambridge, MA: Harvard University Press, 1990, 27. Quine presumably would say that the univocality of the existential quantifier is trivial: what better standard do we have of univocality or ambiguity than the way in which a statement is "regimented"? To this I would reply that the assumption that symbolic logic has this sort of metaphysical significance implicitly views symbolic logic itself as the skeleton of an ideal language and not merely as a useful systematization of valid forms of deduction, and it is just that view of symbolic logic that I reject.
7. See "Rethinking Mathematical Necessity," in H. Putnam, *Words and Life* (Cambridge, MA: Harvard University Press, 1994).
8. Quoted from Quine's response to David Armstrong, in which Quine indignantly rejects the label "class nominalism." *Theories and Things*, p. 184.
9. But see my "The Dewey Lectures 1994: Sense, Nonsense, and the Senses; an Inquiry into the Powers of the Human Mind," *Journal of Philosophy*, 91(1994): 450 (footnote) for a brief indication of the reasons for my rejection of this view of Quine's, and "On Wittgenstein's Philosophy of Mathematics," in *Proceedings of the Aristotelian Society*, Supplement Vol. 70 (1996), 243–264 (the volume containing addresses to the Joint Session, Dublin, 1996) for a more extended discussion.
10. See *The Many Faces of Realism* (LaSalle, Il: Open Court, 1987), "The Dewey Lectures 1994," and, for a reply to criticisms of the doctrine, my "Reply to Simon Blackburn" in Peter Clark and Robert Hale (eds.), *Reading Putnam* (Oxford; Blackwell, 1994).
11. Philosophers of a certain bent will say that the notion of a "use" is too unclear to be of any value (a familiar objection to Wittgenstein's employment of the notion). "How can one tell what the *number* of uses of a word is?" they like to ask. I could reply that one can make the same objection to talk of "contexts," but nevertheless, it can be of great value to distinguish between the different contexts in which something is said.
12. Kurt Gödel, "What Is Cantor's Continuum Problem?" *Readings in the Philosophy of Mathematics*, pp. 474–5.
13. He would not have landed himself in the Russell paradox if he had!

14. George Boolos, "To Be Is To Be a Values of a Variable (or To Be Some Values of Some Variables)," *Journal of Philosophy*, 81(1984): 430–49; and "Nominalist Platonism," *Philosophical Review*, 94(1985): 327–44.
15. Cf. G. Kreisel, "Informal Rigour and Completeness Proofs," in J. Hintikka (ed.), *Philosophy of Mathematics* (Oxford: Oxford University Press, 1969); and J. Etchemendy, *The Concept of Logical Consequence* (Cambridge, MA: Harvard University Press, 1990).
16. Here I use "reify" as Quine does, applying it to cases in which a term that was not previously treated as eligible to instantiate a quantification becomes so treated. Of course, I do not view this as Quine does, as positing the existence of objects, in a univocal sense of "existence" and "object," but as extending both the notion of existence and the notion of an object, as explained earlier. Charles Parsons has pointed out (in conversation) that there is a grammatical device in English that facilitates the passage from plural quantification to reification: just reword "some *A*s" as "a plurality of *A*s."
17. One may object that I am assuming that plural quantification is *clearer* than talk of sets, and what right do I have to assume *that*? To this objection, I respond as follows: First (following Boolos), plural quantification is as familiar and ubiquitous a logical tool as singular quantification itself, and unlike the notion of a set, it does not immediately threaten us with contradiction. (Note that the threat of the Russell paradox arises on *both* the "naive" conception of a set as the extension of a predicate *and* the "naive" conception of a set as any collection of things you please). Second, when philosophers ask of a notion that has not caused us any problems whether it is really "clear," they unvariably import one or another *metaphysical* notion of "clarity." This second part of my response is elaborated in "Was Wittgenstein *Really* an Antirealist About Mathematics?" (forthcoming in *Wittgenstein in America*, edited by Timorthy McCarthy and Peter Winch, to be published by Blackwell, Oxford).
18. Of course, a formalist might argue that we should not regard it as questionable to add a new axiom to *either* arithmetic or set theory as long as we have a relative consistency proof. But not many mathematicians are formalists about *arithmetic*, and this does not explain why set theorists who profess to be "realists" feel so comfortable with such axioms as replacement.

Truthlike and Truthful Operators

ARNOLD KOSLOW

I. Introduction

There is a time-honored tradition, let's call it the implicational tradition, according to which truth and implication systematically mesh with each other. In outline, the idea is simple; in detail, adherence to the tradition can get snagged. There are, of course, the well-known challenges posed by the truth paradoxes. The responses to those challenges have been either to restrict the kinds of structures on which truth and implication interact or to modify the notion of truth, or implication, or various combinations. Central to all these deliberations has been the *Tarski Equivalence Condition*. In what follows, I call any sentential operator "T" (one that maps statements to statements) *truthful* if and only if it satisfies the Tarski Equivalence Condition: for all A, "$T(A)$" is equivalent to "A".[1] Once I describe some conditions that link truth with implication, it becomes evident that there is a natural class of operators (I call them *truthlike operators*) that satisfies those conditions, and that although the class contains operators that satisfy the Tarskian condition, it also contains a good deal more. Those operators that are truthlike but not truthful are called *definitely truthlike*.

The truthlike operators as a whole have some interesting features. They behave extremely well with respect to the logical operators, and no two of them are comparable (in a sense to be explained) to each other. In particular, none of them is comparable to any Tarskian operator. The truthful operators include some philosophically significant examples, which I discuss in some detail: the notion of truth under an interpretation, the notion of intuitionistic provability, and a coherentist account of truth. In contrast, the concept of classical provability is not even truthlike.

As noted earlier, the truthful (Tarskian) operators fall within the class of truthlike operators. Are they just one more example of being truthlike, or is there something special about them, something that distinguishes them from all the others? Sometimes there are no others. On certain structures (those that are syntactically complete), all truthlike operators have to be truthful (§VIII), but this special case aside, there are a number of ways of indicating just how special it is to be truthful. I discuss three quite different necessary and sufficient conditions for being truthful. The first (§VI) considers what happens when a truthlike operator continues to be truthful when there is a shift in the implication relation with respect to which it is characterized. The second introduces symmetry considerations and explores what happens when an operator is a symmetry of other operators on the structure (§XI). The third, and perhaps the most interesting, shows that a truthlike operator is Tarskian if and only if it is extensional (§XIII). Thus, any definitely truthlike operator has to be non-extensional. Each of these three results indicates that it takes something beyond simple truthlikeness to yield a truthful operator. That something extra, in each case, is far from obvious. For example, the arguments that support the claim that an operator is not just truthlike, but truthful, will also have to support the claim that the operator is (say) extensional – to use one of the results just described. That suggests that the use of a truthlike rather than a truthful operator on a structure involves far weaker assumptions than the use of a truthlike one, and that the use of operators that satisfy the Tarski condition is far from a minimal position. If, for example, all that we need in a given case is an operator that behaves well with respect to all the logical operators, then truthful is all we need. Why carry the excess baggage?[2]

My suggestion is that whether one needs one kind of operator or another is a matter that should be determined case by case, structure by structure, problem by problem. I do not see the need to opt for the truthful in all cases. In that sense, the proper attitude to take is the pragmatic, one in which the choice of operator is constrained by the particular structures and problems. For some structures, as noted earlier, there is no choice possible other than the truthful. For other structures, however, different choices may be possible.

The notion of a truthlike operator can be weakened in at least two ways. One way involves the use of symmetries for various sets of sentential operators. The other way exploits the use of the antecedents of certain types of conditionals. The latter suggests the notion of positive and negative *semitruthlike* operators. Semitruthlike operators, as we see (§X), share many, but not all, of the characteristics of truthlike operators. And their consideration widens the net of interesting operators beyond the truthlike, which can still be seen as close cousins of truth. For the present, however, we focus on the truthlike and truthful operators.

II. Truthlike Operators

Let us begin with a modest version of a connection between truth and implication. Let $I = \langle S, \Rightarrow \rangle$ be an implication structure, by which we mean that S is any non-empty set, and $\Rightarrow$ is an implication relation on it, where an implication relation on S is any relation on S that satisfies the Gentzen structural conditions (Koslow, 1992, p. 5).[3] Let T be any operator that maps S to S. We say that T

distributes over the implication relation "⇒" if and only if for every $A_1, \ldots, A_n$, and B in S, if $A_1, \ldots, A_n \Rightarrow B$, then $T(A_1), \ldots, T(A_n) \Rightarrow T(B)$. By a *thesis of I* (or a thesis with respect to "⇒"), we mean any A in S for which $C \Rightarrow A$, for all C in S. Let us say that *T is thesis preserving for* "⇒" if and only if $T(A)$ is a thesis if A is a thesis. This condition is the analogue of the necessitation condition on modal operators ($\Box A$ is a thesis if A is a thesis). By an *antithesis of I* (or an antithesis with respect to "⇒"), we mean any member A of S for which $A \Rightarrow C$, for all C in S. It follows easily (assuming that negation is classical on the structure), that A is a thesis (⇒) if and only if $\neg A$ is an antithesis (⇒), and where "$\Rightarrow^{\wedge}$" is the dual of "⇒" (defined later), A is a thesis (⇒) if and only if A is an antithesis ($\Rightarrow^{\wedge}$) (whether negation is classical or not).

Last, we also have the notion of the dual, $\Rightarrow^{\wedge}$, of the implication relation "⇒". If $I = \langle S, \Rightarrow \rangle$, then "$\Rightarrow^{\wedge}$" is the implication relation on S that extends all the implication relations $\Rightarrow^*$ on S such that, for all A and B, $A \Rightarrow^* B$ if and only if $B \Rightarrow A$. [An equivalent direct definition of the dual in terms of "⇒" can be found in Koslow (1992).] The dual of the implication relation on S is also an implication relation on S, and in the special case of implications with single antecedents, it follows that $A \Rightarrow^{\wedge} B$ if and only if $B \Rightarrow A$. It is easily seen that when there are several antecedents, and the disjunctions of members of S always exist, $A_1, \ldots, A_n \Rightarrow^{\wedge} B$ if and only if $B \Rightarrow (A_1 \vee \cdots \vee A_n)$.

We say that "T" is a *truthlike operator* (on the structure I) if and only if

(1a) T distributes over "⇒", and
(1b) T is thesis preserving with respect to "⇒", and the dual conditions that
(2a) T distributes over "$\Rightarrow^{\wedge}$" (the dual of "⇒"), and
(2b) T is thesis preserving with respect to "$\Rightarrow^{\wedge}$".[4]

Perfect symmetry prevails with conditions (1a) and (2a) for truth and implication, and is continued in conditions (1b) and (2b). It is simple enough to show that each of the operators mentioned above, and certainly any truthful operator, is truthlike. Unlike a truthful operator, however, their values on any A are not always equivalent to A itself; the Tarski Equivalence Condition does not hold. We now want to provide exact conditions for the truthlike operators, under which each half of the usual equivalence holds.

III. Trivial Exclusions

Note first that, although each of the operators mentioned earlier qualifies as truthlike, certain operators of a trivial character do not. Thus the operator T_C, which assigns to each A the conjunction of A with some fixed C that is not a thesis, is not truthlike. The reason is that it fails to satisfy condition (1b). If there is some C^* such that $T_C(C^*)$ is a thesis, then the conjunction of C and C^* is a thesis, and that is impossible. Of course, if C is a thesis, then T_C is just the truthful operator on the structure. Similarly, the operator that assigns to each A the disjunction of A with some fixed C that is not an antithesis, is not truthlike because, in that case, condition (2b) implies that C is an antithesis. Moreover, although the operator that assigns theses to everything (the verum operator) satisfies conditions (1a) and (2a), it fails to satisfy condition (2b). And the operator that assigns antitheses to everything (the falsum operator) also satisfies conditions (1a) and (2a), but fails to satisfy condition (1b).

IV. The Logical Behavior of Truthlikeness

There is an immediate, but surprising, dividend for truthlike operators. Despite the apparent weakness of the conditions for truthlike operators, they are still strong enough to yield some of what we would obviously obtain if the Tarski Equivalence Condition were satisfied.

Suppose that T is some truthlike operator on a structure, for which negation is classical. Then, for all A, and B, we have

(i) $T(A \wedge B) \Leftrightarrow T(A) \wedge T(B)$,
(ii) $T(A \vee B) \Leftrightarrow T(A) \vee T(B)$,
(iii) $\neg T(A) \Leftrightarrow T(\neg A)$,
(iv) $T(A \to B) \Leftrightarrow T(A) \to T(B)$.

Thus, not only do the truthlike operators mesh closely with the implication relation on the structure, they mesh perfectly with the logical operators on that structure. In fact it is hard to see how there could be a closer fit with the logical operators than that which is already provided by the truthlike operators. Of course, the truthful operators also fit as well, since they are also truthlike. However, as we shall see, there are truthlike operators that are not truthful, and still the close fit with the logical operators persists.

The proofs are of an elementary sort, and we assume here that negation is classical:

(i) Since $A, B \Rightarrow A \wedge B$, it follows from condition (1a) that $T(A), T(B) \Rightarrow T(A \wedge B)$, so that $T(A) \wedge T(B) \Rightarrow T(A \wedge B)$. Conversely, since $A \wedge B \Rightarrow A$ (and B), it follows by condition (1a) that $T(A \wedge B) \Rightarrow T(A)$ [and $T(B)$], and hence their conjunction.

(ii) Since $A \Rightarrow A \vee B$ (and similarly for B), it follows from condition (1a) that $T(A) \Rightarrow T(A \vee B)$ [and similarly for $T(B)$], so that $T(A) \vee T(B) \Rightarrow T(A \vee B)$. Conversely, $A, B \Rightarrow^{\wedge} A \vee B$ (since $A \vee B \Rightarrow A \vee B$). By condition (2a), it follows that $T(A), T(B) \Rightarrow^{\wedge} T(A \vee B)$. That is, $T(A \vee B) \Rightarrow T(A) \vee T(B)$. Thus (ii) holds.

(iii) Since for any A, $A \vee \neg A$ is a thesis ($\Rightarrow$), it follows by condition (1b) that $T(A \vee \neg A)$ is a thesis ($\Rightarrow$). By (ii), $T(A) \vee T(\neg A)$ is a thesis.

Consequently $\neg T(A) \Rightarrow T(\neg A)$. The converse goes this way: From condition (2b) we know that if A is a thesis with respect to the dual "$\Rightarrow^{\wedge}$" (i.e., $C \Rightarrow^{\wedge} A$ for all C), and therefore $A \Rightarrow C$, for all C, then $T(A)$ is a thesis with respect to "$\Rightarrow^{\wedge}$" – that is, $T(A) \Rightarrow C$, for all C. Now $A \wedge \neg A$ is a thesis with respect to the dual "$\Rightarrow^{\wedge}$". Therefore, $T(A \wedge \neg A)$ is also a thesis with respect to the dual. That is, $T(A \wedge \neg A) \Rightarrow C$, for all C, but by (i), $T(A) \wedge T(\neg A) \Rightarrow C$, for all C. Consequently, $T(\neg A) \Rightarrow \neg T(A)$.

(iv) From condition (1a), it follows that the value of T on equivalent members of a structure are equivalent. Consequently, $T(A \rightarrow B) \Leftrightarrow T(\neg A \vee B)$. By (ii), $T(\neg A \vee B) \Leftrightarrow T(\neg A) \vee T(B)$. By (iii), $T(\neg A) \Leftrightarrow \neg T(A)$. Therefore, $T(\neg A) \vee T(B) \Leftrightarrow [\neg T(A) \vee T(B)]$. However, $[\neg T(A) \vee T(B)] \Leftrightarrow T(A) \rightarrow T(B)$. Consequently, $T(A \rightarrow B) \Leftrightarrow T(A) \rightarrow T(B)$. If negation is not classical, then, as we might expect, only half of (iii) and (iv) hold: $T(\neg A) \Rightarrow \neg T(A)$, and $T(A \rightarrow B) \Rightarrow T(A) \rightarrow T(B)$.

V. Preliminaries

We need a few elementary notions. Let "$\Rightarrow$" be the implication relation on the structure I. We can say that, for any member C of the structure, "$\Rightarrow^{C}$", (C-implication) is the implication relation that is relativized to C. That is, $A_1, \ldots, A_n \Rightarrow^{C} B$ if and only if $C, A_1, \ldots, A_n \Rightarrow B$. C-implication is an extension of "$\Rightarrow$" (any $\Rightarrow$-implication is also a C-implication), and it is a conservative extension if and only if C is a thesis of I. In short, we can say that each $\Rightarrow^{C}$ is a *C-extension of* "$\Rightarrow$". As with the implication relation "$\Rightarrow$", we can also form the C-relativization of the dual. Thus, $A_1, \ldots, A_n \Rightarrow^{\wedge C} B$ if and only if $C, A_1, \ldots, A_n \Rightarrow^{\wedge} B$.

VI. From Truthlike to Truthful: Extensions of Implication

We can now state some rather simple, but still interesting results. The following four theorems make it clear that the exact condition for a truthlike operator T to be such that $A \Rightarrow T(A)$ for all A is that T distribute over all C-extensions of "$\Rightarrow$", and the exact condition for a truthlike operator to satisfy the condition that $T(A) \Rightarrow A$ for all A is that it distribute over all C-extensions of $\Rightarrow^{\wedge}$.

Theorem 1. If T is a truthlike operator on the structure $I = \langle S, \Rightarrow \rangle$, and T distributes over all C-extensions of "$\Rightarrow$", then $A \Rightarrow T(A)$ for all A in S.

Thus, one-half of the Tarski equivalence is obtained for any truthlike operator, provided that it distributes over all C-extensions of "$\Rightarrow$".

Proof: For any C in S, notice that, for all A in S, $A \Rightarrow^{C} C$ (because $C, A \Rightarrow C$). Since T distributes over all C-extensions of "$\Rightarrow$", $T(A) \Rightarrow^{C} T(C)$. Therefore,

$C, T(A) \Rightarrow T(C)$, for all A. Since T is a truthlike operator, there is some C^* in S such that $T(C^*)$ is a thesis of the structure, and $C, T(C^*) \Rightarrow T(C)$. Therefore, $C \Rightarrow T(C)$, and this holds for any C in S. This result is exact because there is a converse; as follows. ■

Theorem 2. Let T be a truthlike operator on $I = \langle S, \Rightarrow \rangle$, such that $A \Rightarrow T(A)$ for all A in S. Then T distributes over all C-extensions of "$\Rightarrow$".

Proof: We consider the case in which the implication has only one antecedent. The case of multiple antecedents is equally simple. Suppose that $A \Rightarrow^C B$. Then, $C, A \Rightarrow B$. Since T distributes over "$\Rightarrow$", $T(C), T(A) \Rightarrow T(B)$. However, $C \Rightarrow T(C)$. Therefore (by Cut), $C, T(A) \Rightarrow T(B)$. Consequently, $T(A) \Rightarrow^C T(B)$. That is, T distributes over all C-extensions of "$\Rightarrow$".

Note that only conditions (1) and (2) for truthlike operators were used in the proof. The other half of the Tarski equivalence [$T(A) \Rightarrow A$, for all A in the structure] is covered by the following theorem. ■

Theorem 3. If T is a truthlike operator on the structure $I = \langle S, \Rightarrow \rangle$, and T distributes over all C-extensions of "$\Rightarrow^{\wedge}$", then for all A in S, $T(A) \Rightarrow A$.

Proof: For any C in S, $A \Rightarrow^{\wedge C} C$ [since $C, A \Rightarrow^{\wedge} A$, which holds if and only if $A \Rightarrow (C \vee A)$]. Since T distributes over the C-extension "$\Rightarrow^{\wedge C}$", it follows that $T(A) \Rightarrow^{\wedge C} T(C)$. That is, $C, T(A) \Rightarrow^{\wedge} T(C)$, for all A. Consequently, $T(C) \Rightarrow [C \vee T(A)]$, for all A. Since T is truthlike, there is some C^* such that $T(C^*)$ is an antithesis; that is, $T(C) \Rightarrow [C \vee T(C^*)]$. Since $T(C^*)$ is an antithesis, it implies everything, and so implies C in particular. Consequently, $T(C) \Rightarrow C$ holds for all C in S. This result is exact, for there is a converse, as follows. ■

Theorem 4. Let T be a truthlike operator on the structure $I = \langle S, \Rightarrow \rangle$, such that $T(A) \Rightarrow A$ for all A in S. Then, T distributes over all C-extensions "$\Rightarrow^{\wedge C}$".

Proof: Again we consider the case in which the implication has only one antecedent. The multiple-antecedent case is similar. Suppose that $A \Rightarrow^{\wedge C} B$. Then, $C, A \Rightarrow^{\wedge} B$. Since T is truthlike, it distributes over the dual $\Rightarrow^{\wedge}$, so that $T(C), T(A) \Rightarrow^{\wedge} T(B)$. Consequently, $T(B) \Rightarrow [T(C) \vee T(A)]$, but $T(C) \Rightarrow C$, so that $T(B) \Rightarrow C \vee T(A)$. That is, $C, T(A) \Rightarrow^{\wedge} T(B)$. Therefore, $T(A) \Rightarrow^{\wedge C} T(B)$. That is, T distributes over the C-extension "$\Rightarrow^{\wedge C}$", and this for every C. Note that only conditions (1a) and (1b) of truthlike operators were used in proving Theorems 1 and 2, and conditions (2a) and (2b) in proving Theorems 3 and 4. ■

In the implicational tradition as we have described it, truthlike operators have certain features with respect to the implication on a structure. It does not follow that they will continue to have those features if the implication relation is changed. As we have just seen, if a truthlike operator with respect to "$\Rightarrow$" continues to be truthlike when there is a shift to every C-implication, then T has to be truthful (and conversely). This result distinguishes between truthlike and truthful; it does not favor the use of one over the other. Any argument for the truthful status of an operator would have to show that the operator is insensitive to C-extensions. It is not an obvious property; it requires an argument.[5]

VII. Mathematical Practice

We have suggested that various operators are truthlike. We have not given arguments for any specific operator that support either the claim that (1) it is a truthlike operator that distributes over all C-extensions of an implication relation, or (2) it distributes over all the C-extensions of the dual implication relation.

Here is a consideration drawn from informal mathematical practice, in which the use of "truth" seems to satisfy the condition that the operator distribute over all the C-extensions of the appropriate implication relation. Moreover, if the example is correct, it shows that the informal use of "true" satisfies the Tarski Equivalence Condition, without the imposition of a semantics on the language used in the mathematical argument.

In the presentations of proofs, we often find an informal use of "true." For example, it is pretty common that, after a deduction from certain assumptions, something like this is said: "So these results would be true, if the assumptions were true." This kind of informal talk, without much forcing, can be thought of as assuming that truth distributes over implication. On the other hand, any support for the claim that truth, in this mathematical kind of context, distributes over the dual of the implication relations ([as required by condition (2a)]) will be indirect. Although there are many examples of proofs that involve the dual implication, they are not directly expressed that way. Thus, the natural way of expressing that, given C, it follows that either A or B ($C \Rightarrow A \vee B$), is also exactly the case when $A, B \Rightarrow^{\wedge} C$ – though the connection is rarely expressed that way. In such examples, the exposition might easily continue: "So, if C is true, then either A is true, or else B is." But this is tantamount to assuming that "truth" distributes over the dual; that is, if $A, B \Rightarrow^{\wedge} C$, then $T(A), T(B) \Rightarrow^{\wedge} T(C)$ [i.e., $T(C) \Rightarrow T(A) \vee T(B)$]. Conditions (1b) and (2b) for truthlike operators can also be supported by these kinds of informal considerations. Thus, it seems reasonable to think that the natural use of some notion of truth in the running commentary on proofs is indeed the use of a truthlike operator.

The obvious question is whether this particular truthlike operator satisfies the Tarski Equivalence Condition. The following considerations, although not decisive, seem to suggest a positive answer. Consider the following part of a mathematical argument:

(1) Assume C. Now if in addition one assumes A, then it follows that B. So,
(2) we see that given C, if A is true, then so too is B.

One way to think about this is to note that, first, clearly according to (1), $C, A \Rightarrow B$, so that in our notation, $A \Rightarrow^C B$. According to (2), we have $C \Rightarrow [T(A) \rightarrow T(B)]$. That is, $C, T(A) \Rightarrow T(B)$, which in our notation is just $T(A) \Rightarrow^C T(B)$. Thus, it seems that, for any C, when $A \Rightarrow^C B$, then $T(A) \Rightarrow^C T(B)$. That is, this particular truthlike operator distributes over all C-extensions. For this kind of operator, as we have seen, $A \Rightarrow T(A)$ for all A (Theorem 1).

The second case, involving C-extensions of the dual $\Rightarrow^\wedge$ are, it seems, less direct. The chief reason is the one to which I already alluded. One simply does not find mathematical proofs that use the dual implication, so that any examples will have to be reconstructed. Here is a possible case: Suppose that there is a proof that argues that

(1) From B one can obtain the disjunction $A \vee C$. So, one might naturally say that
(2) given not C, it follows that if B is true, then so too is A.

Now (1) is simply $B \Rightarrow A \vee C$, which is just $C, A \Rightarrow^\wedge B$, and this can be rewritten as $A \Rightarrow^{\wedge C} B$. So much for the reconstruction of (1). As for (2), it says that $\neg C \Rightarrow [T(B) \rightarrow T(A)]$, which holds if and only if $T(B) \Rightarrow [\neg C \rightarrow T(A)]$, that is, $T(B) \Rightarrow [C \vee T(A)]$. That is $C, T(A) \Rightarrow^\wedge T(B)$, which is just $T(A) \Rightarrow^{\wedge C} T(B)$. Thus we have a case that can be described as distribution over C-extensions of the dual implication relation "$\Rightarrow^\wedge$", and so, $T(A) \Rightarrow A$, for all A (Theorem 3). We have to do some rewriting to see that this is so. It does not appear as naturally or as directly as in our first example. Nevertheless, when the rewriting is carried out, it seems that there are certain kinds of mathematical arguments that are, in fact, cases of what we have called distribution of some truthlike operator over all the C-extensions of the dual implication relations. If this is correct, then it would show that, given the theorems above, it follows that this truthlike operator satisfies the Tarski Equivalence.

VIII. Definitely Truthlike Operators

Before we turn to a consideration of specific truthlike operators, it is worth noting that not all truthlike operators are truthful. All operators that are truthlike but not truthful are called *definitely truthlike*. The distinctness of these two kinds of operators, however, may depend upon the type of implication structure on which they operate. There are structures for which there is no difference at all.

Let us say that an implication structure $I = \langle S, \Rightarrow \rangle$ is (syntactically) *complete* (sometimes called "complete with respect to negation") if and only if for every A in S, either A is a thesis of I or $\neg A$ is a thesis of I. Such structures are, of course, special, and on them, there is no distinction between truthlike and truthful operators. That is,

Theorem 5. If $I = \langle S, \Rightarrow \rangle$ is complete, and negation is classical, then every truthlike operator on I is truthful; that is, $T(A) \Leftrightarrow A$ for all A in S.

Proof: Let A be any element of S, and let T be any truthlike operator on I.

(1) If A is a thesis, then $C \Rightarrow A$ for all C in S. In particular, $T(A) \Rightarrow A$.
(2) If A is not a thesis, then, by completeness, $\neg A$ is a thesis. Then $C \Rightarrow \neg A$ for all C in S.

Since T is truthlike, it distributes over "$\Rightarrow$", so that $T(C) \Rightarrow T(\neg A)$ for all C. By (1b), there is a C^* in S such that $T(C^*)$ is a thesis. Since $T(C^*) \Rightarrow T(\neg A)$, it follows that $T(\neg A)$ is a thesis. By (iii), $T(\neg A) \Leftrightarrow \neg T(A)$. So, $\neg T(A)$ is a thesis. Consequently, $\neg C \Rightarrow \neg T(A)$ for all C. Therefore, $T(A) \Rightarrow C$ for all C and, in particular, $T(A) \Rightarrow A$. Thus, for all A in the structure, $T(A) \Rightarrow A$.

To show that T is truthful then, it remains to show that $A \Rightarrow T(A)$ for all A. If A is a thesis of I, then since T is truthlike, by condition (1b), so too is $T(A)$. So, $A \Rightarrow T(A)$. On the other hand, if $\neg A$ is a thesis of I, then $C \Rightarrow \neg A$ for all C. In particular, $\neg T(A) \Rightarrow \neg A$. Therefore, $A \Rightarrow T(A)$. So, $A \Rightarrow T(A)$ for all A. Consequently, for all A, $T(A) \Leftrightarrow A$; that is, T is truthful.

In other words, if T is a truthlike operator on a classical structure, which is not truthful, then I is incomplete. A simple example, which is a special case of Theorem 5, is this: Let I be the structure in which $S = \{t, f\}$, where each element implies itself, and $f \Rightarrow t$. The logical operators are the usual functions mapping S (and $S \times S$) to S, and negation is classical. I is clearly complete since, of the two elements, t is a thesis (t is implied by everything in S), and f is an antithesis (f implies everything in the structure). If T is any truthlike operator on I, then it has to satisfy the Tarski Equivalence Condition, and so, it is truthful. The reason is that, since t is a thesis, so is $T(t)$. However, t is the only thesis of the structure. Therefore, $T(t) \Leftrightarrow t$. Moreover, since T is truthlike, $T(f) \Leftrightarrow T(\neg t) \Leftrightarrow \neg T(t) \Leftrightarrow \neg t \Leftrightarrow f$. Therefore, $T(A) \Leftrightarrow A$ for all A in this structure.

It is also worth observing that some operators are definitely truthlike. Here is a simple structure on which there is a definitely truthlike operator T. Let I be an implication structure with $S = \{A, B, C, D\}$, and an implication relation with the implications indicated by single arrows, and the values of T indicated in square brackets. That is,

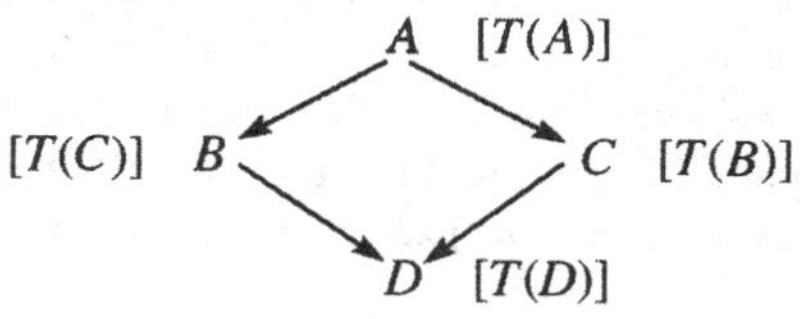

A direct computation shows that

(1) T distributes over this implication relation,
(2) T distributes over the dual [since T of any disjunction will be equivalent to the disjunction of the Ts, that is, for any E and F, $T(E \vee F) \Leftrightarrow T(E) \vee T(F)$],
(3) there is a C^* in S for which $T(C^*)$ is a thesis (namely D), and
(4) there is a $C^\#$ in S for which $T(C^\#)$ is an antithesis (namely, A). However, T is not truthful. In this structure, neither B nor C implies the other. Therefore $T(B)$ does not imply B (otherwise $C \Rightarrow B$, and that is impossible). Moreover, B does not imply $T(B)$ (otherwise $B \Rightarrow C$, and that is impossible). Thus T fails to satisfy the Tarski Equivalence Condition; in fact, each half of the condition fails.

Although T is not a particularly interesting truthlike operator, it does show that it takes something more than truthlikeness to obtain the Tarski Equivalence. Indeed, the point of Theorems 1–4 is to show exactly what more is needed. Nevertheless, note that the operator T in the preceding structure is not without some interest. It is a familiar result that if T satisfies the Tarski Equivalence Condition, then there cannot be a fixed point for the operator $\neg T$ (some C such that $\neg T(C) \Leftrightarrow C$). Otherwise, we get a contradiction. That is, $C \Leftrightarrow \neg T(C) \Leftrightarrow \neg C$. However, if T is a definitely truthlike operator (as in the preceding structure), then there need not be any contradiction. In fact, in the structure, both B and C are fixed points of the truthlike T. The reason is that, in that structure (in which negation is classical), $\neg B \Leftrightarrow C$ and $\neg C \Leftrightarrow B$. Consequently, $\neg T(B) \Leftrightarrow \neg C \Leftrightarrow B$, and $\neg T(C) \Leftrightarrow \neg B \Leftrightarrow C$. Thus, unlike the case for truthful operators, the negation of definitely truthlike operators can have fixed points.

Although the truth paradoxes are not the focus of this study, note that they can create trouble for truthlike operators, even if we assume only that they distribute over the implication and its dual (in §X, we call such operators *pretruthlike*). Suppose that A is *T-antinomic* on the structure $I = \langle S, \Rightarrow \rangle$. That is, $T(A) \Leftrightarrow T(\neg A)$ (following the terminology of G. H von Wright and H. Arlo-Costa[6]). Obviously, if T were truthful, a contradiction would follow. However, even if T is pretruthlike (but not truthlike) and negation is classical, then something less catastrophic than inconsistency, but still undesirable, happens: T is *trivialized*. That is, for all A and B, $T(A) \Leftrightarrow T(B)$, for suppose that A^* is T-antinomic. Then, $T(A^* \wedge \neg A^*) \Leftrightarrow T(A^*) \wedge T(\neg A^*)$ (by distribution over "$\Rightarrow$") $\Leftrightarrow T(A^*)$ (since A^* is T-antinomic). Now, since $A^* \wedge \neg A^* \Rightarrow B$, for all B, it follows, by distribution over "$\Rightarrow$" that $T(A^*) \Rightarrow T(B)$ for all B in the structure. On the other hand, since negation is assumed to be classical, $B \Rightarrow A^* \vee \neg A^*$ for all B. Therefore, $T(B) \Rightarrow T(A^* \vee \neg A^*) \Leftrightarrow T(A^*) \vee T(\neg A^*)$ by distribution over

"$\Rightarrow^\wedge$". Consequently, $T(B) \Rightarrow T(A^*)$ (since A^* is T-antinomic). Thus, for any B, $T(B) \Leftrightarrow T(A^*)$, and so for every A and B, $T(A) \Leftrightarrow T(B)$.[7] ■

IX. The Diversity of Truthlike Operators (Scattering)

We began by indicating that there were a number of operators that have sometimes have been viewed as involving truth. We suggested that they belong to a class of operators that we called truthlike. These, we noted, are not necessarily truthful. We listed several examples, which we consider below. Before we do so, there is an interesting observation worth noting about the family of truthlike operators. They are *noncomparable* with each other. By that we mean the following: Let T and T^* be two operators on some set S. We say that the operator T implies the operator T^* if and only if for all A, $T(A) \Rightarrow T^*(A)$, and that they are *equivalent* if and only if each implies the other. We say that T and T^* are *comparable* if and only if at least one of them implies the other. Then,

Theorem 6. Let $I = \langle S, \Rightarrow \rangle$ be an implication structure on which negation is classical. Let T and T^* be two non-equivalent truthlike operators on I. Then, T and T^* are not comparable.

Proof: If $T(A) \Rightarrow T^*(A)$ for all A, then $T(\neg A) \Rightarrow T^*(\neg A)$ for all A. Consequently, $\neg T^*(\neg A) \Rightarrow \neg T(\neg A)$. By (iii) of §IV, $T(A) \Rightarrow T^*(A)$ for all A. Therefore, T and T^* are equivalent, and that is impossible. Similarly, if for all A, $T^*(A) \Rightarrow T(A)$, then for all A, $T^*(A) \Leftrightarrow T(A)$. ■

Theorem 6 tells us that all the truthlike operators on a structure (including, of course, any truthful ones) are "scattered" from each other. No one of them can imply another, unless the two are equivalent. One immediate corollary is that the disjunction, conjunction, and so forth of truthlike operators will fail to be truthlike unless the operators are equivalent. Thus, there is no simple structure for the set of truthlike operators, although they are closed under functional composition.

Theorem 6 is surprising, but there is a more general result that explains why the truthlike operators scatter. It is that there is a larger class of operators on the structure that includes all the truthlike ones, and such that the truthlike operators are exactly the maximal operators of that class. Since the truthlike operators are maximal, it follows that they are not mutually comparable.[8]

X. Weaker than Truthlike: (1) Semitruthlike Operators, Negative and Positive

Truthlike operators (with respect to "$\Rightarrow$") distribute over "$\Rightarrow$" as well as its dual $\Rightarrow^\wedge$, and a bit more. It is the bit more that we now dispense with, in parts. We say that an operator is *semitruthlike* if and only if it satisfies conditions (1a) and (2a), and either condition (1b) (it is then positive semitruthlike) or condition (2b) (it is then negative semitruthlike). That is,

T is *pretruthlike* (with respect to "$\Rightarrow$") if and only if
 (1a) T distributes over "$\Rightarrow$", and
 (2a) T distributes over $\Rightarrow^{\wedge}$ (i.e., it distributes over disjunctions).

T is *positive semitruthlike* (with respect to "$\Rightarrow$") if and only if it is pretruthlike, and (1b) T is thesis preserving. That is, if A is a thesis, so too is $T(A)$.

T is *negative semitruthlike* (with respect to "$\Rightarrow$") if and only if it is pretruthlike, and (2b) T is antithesis preserving. That is, if A is an antithesis, then so too is $T(A)$.

T is *semitruthlike* if and only if it is either positive or negative semitruthlike.

The following results are immediate consequences:

Theorem 7. If T is pretruthlike then
 (i) $T(A \wedge B) \Leftrightarrow T(A) \wedge T(B)$,
 (ii) $T(A \vee B) \Leftrightarrow T(A) \vee T(B)$, and
 (iii) $T(A \to B) \Rightarrow T(A) \to T(B)$ all hold.

The proofs are just the ones that can be found in §IV.

This shows that, although the pretruthlike operators are weaker than those that are truthlike, they are nevertheless still quite well behaved with respect to the logical operators, and the "truthlike" terminology is appropriate.

The remaining theorems show how some further features of truthlike operators, which concern negation and conditionals, can be recovered. If T is truthlike, we have seen that $\neg T(A) \Leftrightarrow T(\neg A)$. Positive semitruthlike operators get back one-half, and the negative ones get back the other. That is,

Theorem 8. If T is a positive semitruthlike operator, then for all A, $\neg T(A) \Rightarrow T(\neg A)$.

Just note that, since T is positive, $T(A)$ is a thesis for all theses A. Since $(A \vee \neg A)$ is a thesis, so too is $T(A \vee \neg A)$. By Theorem 7(ii), then, $T(A) \vee T(\neg A)$ is a thesis. Therefore, $\neg T(A) \Rightarrow T(\neg A)$.

Theorem 9. If T is a negative semitruthlike operator, then for all A, $T(\neg A) \Rightarrow \neg T(A)$.

Just note that since T is negative, $T(A)$ is an antithesis for any antithesis A. Therefore $T(A \wedge \neg A)$ is an antithesis. By Theorem 7(i) then, $T(A) \wedge T(\neg A)$ is an antithesis. Therefore, $T(\neg A) \Rightarrow \neg T(A)$.

If we let T be any semitruthlike operator, and denote by $T^{\wedge}$ its dual, that is, $T^{\wedge}(A)$ is $\neg T(\neg A)$ for all A, then Theorem 8 tells us that, if T is positive, then it is implied by its dual, whereas Theorem 9 shows that, if T is negative, then it implies its dual. Of course, we have already noted ((iii) of §IV) that, if T is just truthlike, then it is self-dual; that is, T is equivalent to its dual.

The interesting thing about the semitruthlike operators is that they allow for a divergence between themselves and their duals, which disappears once one has moved to the level of the truthlike operators. Something even more subtle happens for conditionals:

Theorem 10. If T is a positive semitruthlike operator, then $T(A) \rightarrow T(B) \Rightarrow T(A \rightarrow B)$.

Just note that $T(A) \rightarrow T(B) \Rightarrow \neg T(A) \vee T(B)$, which by Theorem 8 and Theorem 7(ii) implies $T(\neg A \vee B)$. However, $(\neg A \vee B) \Leftrightarrow (A \rightarrow B)$, so that by condition (1a), we obtain $T(A \rightarrow B)$. Now, given that for pretruthlike operators (and so, for semitruthlike operators), we have $T(A \rightarrow B) \Rightarrow T(A) \rightarrow T(B)$, we see that, for the positive semitruthlike operators, we have equivalence: $T(A) \rightarrow T(B) \Leftrightarrow T(A \rightarrow B)$. That is, of course, a result that we have seen, which holds once we get to the truthlike operators.

The comparable result does not hold for the negative semitruthlike operators. In this case, the result that does hold depends upon the difference that there is between T and its dual. That is,

Theorem 11. If T is a negative semitruthlike operator, then $T\hat{}(A) \rightarrow T\hat{}(B) \Rightarrow T\hat{}(A \rightarrow B)$.

Thus for the negative case, the result of Theorem 10 does not hold for T, but it does hold for the dual, $T\hat{}$. By a straightforward calculation, we can show that, if T is a negative semitruthlike operator, then its dual $T\hat{}$ is a positive semitruthlike operator. Theorem 11 then follows from Theorem 10. Consequently, for any negative semitruthlike operator $T\hat{}$, we have the equivalence $T\hat{}(A \rightarrow B) \Leftrightarrow T\hat{}(A) \rightarrow T\hat{}(B)$.

There are interesting and well-known examples of positive and negative semitruthlike operators. Here we mention two that use the antecedents of different kinds of conditionals.

For any fixed C, let M_C be an operator that assigns to each statement A, the material conditional $C \rightarrow A$. Then M_C is a positive (but not negative) semitruthlike operator.

If we let S_C be the operator that assigns to each statement A, the Stalnaker (1968) conditional $C > A$; that is, for each C, let the value of S_C for any A be $C > A$. Then it is easy to check that S_C is a negative semitruthlike operator. With a slight change in the Stalnaker operator (to take care of the case in which C is impossible), for any C, let $S_{C'}(A)$ be the conjunction of $S_C A)$ and $\neg(C > \neg C)$ (in Stalnaker's study, this says that C is possible). The original operator and its variant are equivalent for all Cs that are possible. The variant operator is truthlike, though space does not allow for a fuller discussion of it. In contrast, the operators associated with the antecedents of intuitionistic conditionals are not, in general, even pretruthlike. They will always distribute over implication, but for some statements the operator will distribute over disjunctions, and for others, it will not. It is also interesting that the operator $\Box_C$ associated with the Lewis-Sprigge (1973) account of counterfactuals (for any A, $\Box_C(A)$ is $C\Box\!\!\rightarrow A$) is not semitruthlike because condition (2a) does not hold. The reason is essentially that, in the system VC, the law of conditional excluded middle fails to be a theorem. The relevant result is this: In the system VC, the operator $\Box_C(A)$ will fail to distribute over disjunctions if and only if C fails to satisfy the law of conditional excluded middle, where C satisfies *CEM* if and only if $\Box_C(B) \vee \Box_C(\neg B)$ is a theorem (of *VC*) for all B.

A direct computation shows that the verum operator (described in §III), which assigns a fixed thesis of the structure to all members, distributes over the implication relation as well as the dual implication relation (or equivalently, for structures in which disjunctions always exist, it distributes over all disjunctions). Since it is a thesis for every thesis A (it is a thesis for every member, thesis or not), it is a positive semitruthlike operator. Obviously, it can never be an antithesis, and so, it is not a negative semitruthlike operator.

On the other hand, the falsum operator (§III), which assigns a fixed antithesis to every member of the structure, distributes over the implication relation and its dual, and its value for any antithesis is, of course, an antithesis (since its value for every member is an antithesis). Therefore, it is a negative semitruthlike operator. Obviously, it can never be a thesis, and so, it is not a positive semitruthlike operator.

XI. Weaker than Truthlike: (2) Symmetries

We want to consider briefly a second way of generating operators that are even weaker than truthlike ones. The device that is used exploits certain symmetry properties that are obvious for truthful operators, and it yields a series of operators that pass from weaker than truthful to semitruthlike, truthlike, and truthful. The symmetry condition still falls within the general interest of the implicational tradition, since it concerns a special way in which a truth operator behaves with respect to implication.

Here is a possible motivation for considering truth as a symmetry of certain types of operators on a structure. Suppose that T is a truthful operator, and that P is any operator on the structure I, which maps S to S. It follows that

$$TP(A) \Leftrightarrow PT(A), \text{ for all } A \text{ in } S,$$

provided that P is an *equivalence operator* on S, by which we mean that, for any A and B, $P(A)$ and $P(B)$ are equivalent ($\Leftrightarrow$), provided that A and B are equivalent ($\Leftrightarrow$). This notion is easily generalized to the case of operators of more than one argument.[9] The argument is direct: $T(P(A))$ is equivalent to $P(A)$ (since T is truthful), and $P(T(A))$ is also equivalent to $P(A)$, since $T(A)$ is equivalent to A, and P is an equivalence operator. So, truthful operators play a symmetric role with respect to the equivalence operators on the structure. This is obvious for any truthful operator T and every equivalence operator P. What happens if we give this intuitively appealing idea of symmetry a more central role? Let U and P be equivalence operators on S. Then we say that

U is a symmetry of P if and only if for all A in S, $U(P(A)) \Leftrightarrow P(U(A))$.

If P is a mapping of pairs of S to S, and an equivalence operator, then the condition on symmetry will mean that U is a symmetry of P if and only if, for all A and B in S, $U(P(A, B)) \Leftrightarrow P(U(A), U(B))$.

There are many examples of equivalence operators. The verum operator (which assigns to each member some fixed thesis), the falsum operator (which assigns some fixed antithesis), and any constant operator P_A associated with some member A of the structure [i.e., for all B, $P_A(B) \Leftrightarrow A$] are all equivalence operators on the structure. The reason is that, for any members of the structure, these operators assign equivalent members. Moreover, all the logical operators are equivalence operators. By requiring that U be a symmetry of one or more equivalence operators, U will as a consequence have many of the features of the truthlike operators.[10] For example, by using symmetry conditions, we can obtain an operator on a structure that satisfies the Tarski Equivalence Condition *for any arbitrarily chosen member of a structure*. More exactly,

Theorem 12. Let C be a member of the structure, and P_C be the constant operator associated with C. Then, U is a symmetry of P_C if and only if $U(C) \Leftrightarrow C$.

The reason is that if U is a symmetry of P_C, then $UP_C(A) \Leftrightarrow P_CU(A)$ for all A. Since $P_C(A)$ is equivalent to C, and U is an equivalence operator, the left-hand side is equivalent to $U(C)$. The right-hand side is equivalent to C. Consequently, $U(C) \Leftrightarrow C$. The converse is obvious.

It follows that, if U is a symmetry of all the constant operators associated with the members of the structure, then U is a truthful operator. Actually, Theorem 12 is a special case of a more general phenomenon. Let us say that any operator P has a *fixed point* λ if and only if $P(\lambda) \Leftrightarrow \lambda$, and it has a unique fixed point if and only if all of its fixed points are equivalent (by "$\Leftrightarrow$"). Then we have the following result:

Theorem 13. If U is an equivalence operator and is a symmetry of an equivalence operator P that has a unique fixed point λ, then $U(\lambda) \Leftrightarrow \lambda$.

The reason is this: First note that, if λ is a fixed point of P, and U is a symmetry of P, then $U(\lambda)$ is also a fixed point of P. [Since $UP(A) \Leftrightarrow PU(A)$ for all A, we have $UP(\lambda) \Leftrightarrow PU(\lambda)$. Now, $P(\lambda) \Leftrightarrow \lambda$, and since U is an equivalence operator, $U(\lambda) \Leftrightarrow PU(\lambda)$; i.e., $U(\lambda)$ is a fixed point of P.] Since P has a unique fixed point, $U(\lambda) \Leftrightarrow \lambda$. Obviously, Theorem 13 is a consequence of Theorem 12, since any constant operator has exactly one fixed point.

The use of symmetries of various equivalence operators will generate a host of operators. Of course, by Theorem 12, if an operator is a symmetry of all the constant operators P_C, then it will be truthful. It is not an obvious condition, so that, in any particular case, its requirement needs some argument.[11] Nevertheless, there is an organizational power to symmetries: Sometimes the use of a restricted fragment of a theory of truth can be seen as a case of requiring that the operator be a symmetry of the members of a special set of equivalence operators.[12]

XII. Definitely Truthlike Operators: Examples

We have seen that truthlike operators need not be truthful. However, the proof of that fact (§VIII) used a truthlike operator that did the job, but had no special interest beyond that. It is worth looking at several more substantial examples of operators that are definitely truthlike. The first example, truth under an interpretation, is one in which both halves of the Tarski equivalence fail.

A. Truth-Value Interpretations

For the sake of the example, let i be any truth-value assignment on a structure consisting of the language of the classical sentential calculus (CSC), and let the implication relation on it ($\Rightarrow$) be given, say, by the usual deducibility relation ($\vdash$). In this case, by "$A_1, \ldots, A_n \Rightarrow B$", we mean that $\vdash A_1 \wedge \cdots \wedge A_n \rightarrow B$. To every truth-value assignment i, we can associate an operator on the language of CSC in a one-to-one fashion as follows. Let Ω be some fixed thesis of CSC, and ω be some fixed antithesis (contradiction). Then, set

$$T_i(A) = \begin{cases} \Omega, & \text{if } i(A) = t; \\ \omega, & \text{otherwise.} \end{cases}$$

It is easy to see that

(1) for any assignments i, and i^*, and for any A, $i(A) = i^*(A)$ if and only if $T_i(A) \Leftrightarrow T_{i^*}(A)$;
(2) for any i, T_i is a truthlike operator on the language of CSC; and
(3) each T_i fails to be truthful.[13]

We establish each of these as follows:

(1) Suppose that $i(A) = i^*(A)$. If the value of $i(A)$ is t, then both $T_i(A)$ and $T_{i^*}(A)$ are Ω, and if $i(A)$ is f, then they are both ω. In either case,

they are equivalent. Conversely, if $T_i(A) \Leftrightarrow T_{i^*}(A)$, then $i(A) = i^*(A)$, or else $\Omega \Leftrightarrow \omega$, which is impossible.

(2) This involves four conditions:

(1a) T_i distributes over "$\Rightarrow$", for suppose that $A_1, \ldots, A_n \Rightarrow B$. If $i(B)$ has the value t, then $T_i(A)$ is the thesis Ω, and is implied by everything. In particular, then, $T_i(A_1), \ldots, T_1(A_n) \Rightarrow T_1(B)$. If $i(B)$ has the value f, then since all the A_is imply B, it follows for some $r (1 \leq r \leq n)$ that $i(A_r)$ has the value f. Therefore, $T_i(A_r)$ is the antithesis ω, and so, $T_i(A_r) \Rightarrow T_i(B)$. Therefore, in this case as well, $T_i(A_1), \ldots T_1(A_n) \Rightarrow T_1(B)$.

(2a) The second condition is that each T_i distributes over the dual "$\Rightarrow^\wedge$". In this case, it involves showing for every A and B that $T_i(A \vee B) \Rightarrow T_i(A) \vee T_i(B)$. If $i(A \vee B)$ is f, then $T_i(A \vee B)$ is ω, which implies everything. In particular, then, $T_i(A \vee B) \Rightarrow T_i(A) \vee T_i(B)$. If $i(A \vee B)$ is t, then at least one of $i(A)$, $i(B)$ is t. Therefore, at least one of $T_i(A)$, $T_i(B)$ is Ω. Consequently, $T_i(A) \vee T_i(B)$ is Ω, which is implied by everything. In particular, then, $T_i(A \vee B) \Rightarrow T_i(A) \vee T_i(B)$. Thus, the second condition is satisfied.

(1b) The third condition is that T_i is thesis preserving ($\Rightarrow$). However, if A is a thesis, $i(A) = t$, and so, $T_i(A)$ is Ω, a thesis.

(2b) As for the fourth condition, that T_i is thesis preserving ($\Rightarrow^\wedge$), note that if A is a thesis ($\Rightarrow^\wedge$), then it is an antithesis ($\Rightarrow$). Therefore, $i(A) = f$, so that $T_i(A)$ is ω. Consequently, $T_i(A) \Rightarrow C$ for all C. Thus $C \Rightarrow^\wedge T_i(A)$ for all C. That is, $T_i(A)$ is a thesis ($\Rightarrow^\wedge$). Thus, all the T_i are truthlike operators. It only remains to show that:

(3) No T_i is truthful. That is, they all fail to satisfy the Tarski Equivalence Condition. In fact, each half of that equivalence fails. $T_i(A) \Rightarrow A$ fails for any A that is not a thesis, such that $i(A)$ has the value t (there always is such an A). In that case, $T_i(A)$ is Ω, so that $\Omega \Rightarrow A$. Consequently, A is a thesis. However, that is impossible. Moreover, $A \Rightarrow T_i(A)$ fails for any A that is not a contradiction, such that $i(A)$ has the value f (there always is such an A). For then, $T_i(A)$ is ω, and so, $A \Rightarrow \omega$. Therefore, A is a contradiction (antithesis), but that is impossible. So, each half of the Tarski equivalence will fail.

Thus even in this very simple case, each operator associated with a fixed interpretation i will be a truthlike operator that is not truthful. The same holds for the operators associated with "it is true in model M that ..." (for a fixed model M), and "it is true in world u, that ..." (for fixed world u). They are all truthlike, but fail to satisfy the Tarski condition.

B. Intuitionistic Provability

To focus the discussion, we consider a simple case of intuitionistic provability. Let I^{ISC} be a structure that consists of the sentences of the Heyting formulation of the intuitionistic sentential calculus (ISC), together with an implication relation "$\Rightarrow^{\text{ISC}}$" that is defined as follows:

For any sentences $A_1, \ldots, A_n$, and B, we say that $A_1, \ldots, A_n \Rightarrow^{\text{ISC}} B$, if and only if $\vdash_{\text{ISC}}[(A_1 \wedge \cdots \wedge A_n) \rightarrow B]$.

where $\vdash_{\text{ISC}} C$ means that C is a theorem of ISC. It follows immediately that A is a thesis of the intuitionistic structure; that is, $C \Rightarrow^{\text{ISC}} A$ for all C, if and only if A is a theorem of ISC($\vdash_{\text{ISC}} A$). Similarly, A is an antithesis of the structure; (that is, $A \Rightarrow^{\text{ISC}} C$ for all C, if and only if $\vdash_{\text{ISC}} \neg A$.

For the language of ISC, we now introduce an operator T_I, which represents the fact that a statement of the language is intuitionistically provable. We want such an operator ϕ to have the following features:

(1) that it map the structure to itself;
(2) that it be essentially a characteristic function on those sentences of ISC that are theorems, that is, that it have equivalent values on all the theorems and equivalent values on all the nontheorems, where the values in the two cases are non-equivalent; and
(3) it should behave like a provability operator:
 (i) for any thesis of ISC, $\phi(A)$ is a thesis;
 (ii) $\phi(A \rightarrow B) \Rightarrow [\phi(A) \rightarrow \phi(B)]$, and
 (iii) $\phi(A) \Rightarrow \phi\phi(A)$.

Accordingly, let Ω be some fixed thesis of the structure (i.e., $\vdash_{\text{ISC}} \Omega$), and let ω be some fixed antithesis of the structure (i.e., $\vdash_{\text{ISC}} \neg\omega$). Let

$$T_I(A) = \begin{cases} \Omega, & \text{if } A \text{ is a thesis of the structure;} \\ \omega, & \text{otherwise.} \end{cases}$$

T_I is a characteristic function on the set of all the intuitionistically provable members of the structure that satisfies 3(i)-(6). It is relatively straightforward to check that this operator, which corresponds to the prefix "It is intuitionistically provable (in ISC) that" is definitely truthlike.

First, we prove that T_I is truthlike. [It is assumed that ISC is consistent. It will be needed for the verification of condition (2b) on truthlike operators.] Condition (1a) requires that T_I distribute over $\Rightarrow^{\text{ISC}}$. Suppose that $A_1, \ldots A_n \Rightarrow^{\text{ISC}} B$. If all the A_is are theorems of ISC, then so too is B. So B is a thesis of the structure. Therefore, $T_I(B)$ is the thesis Ω. It is implied by all members of the structure. Consequently, $T_I(A_1), \ldots T_I(A_n) \Rightarrow^{\text{ISC}} T_I(B)$. If, however, some A_j is not a theorem, then it is not a thesis of the structure. Therefore $T_I(A_j)$ is ω.

Since ω is an antithesis of the structure, $\omega \Rightarrow^{\text{ISC}} C$ for all C. It follows that $T_I(A_1), \ldots T_I(A_n) \Rightarrow^{\text{ISC}} T_I(B)$.

Condition (1b) requires that, if A is a thesis, then so too is $T_I(A)$. However, if A is a thesis, then $T_I(A)$ is Ω, which is a thesis of the structure. Condition (2a) requires that T_I distribute over the dual implication $\Rightarrow^{\text{ISC}\wedge}$, which is equivalent to the condition that, for every A and B, $T_I(A \vee B) \Rightarrow^{\text{ISC}} T_I(A) \vee T_I(B)$. Now, if $(A \vee B)$ is a theorem of ISC, then it is a standard result of ISC that either A is a theorem of ISC, or B is a theorem of ISC. In either case, $T_I(A) \vee T_I(B)$ will be a disjunction at least one of whose disjuncts is Ω. So, $T_I(A) \vee T_I(B)$ is a thesis of the structure and, consequently, it is implied by every statement. In particular, then, $T_I(A \vee B) \Rightarrow^{\text{ISC}} T_I(A) \vee T_I(B)$. On the other hand, if $(A \vee B)$ is not a theorem of ISC, then $T_I(A \vee B)$ is the antithesis ω, which implies every statement in the structure. In particular, then, $T_I(A \vee B) \Rightarrow^{\text{ISC}} T_I(A) \vee T_I(B)$. Finally, Condition (2b) requires that, if A is a thesis with respect to $\Rightarrow^{\text{ISC}\wedge}$, then so too is $T_I(A)$. Now, if A is a thesis with respect to "$\Rightarrow^{\text{ISC}\wedge}$", then $C \Rightarrow^{\text{ISC}\wedge} A$ for all C. Then $A \Rightarrow^{\text{ISC}} C$ for all C, but this is equivalent to $\vdash_{\text{ISC}}(A \to C)$ for all C. In particular, then, $\vdash_{\text{ISC}}(A \to \neg A)$. From the fact that $\vdash_{\text{ISC}}[(A \to \neg A) \to \neg A]$, we conclude that $\vdash_{\text{ISC}} \neg A$. By the consistency of ISC, A is not a theorem of ISC. Consequently, $T_I(A)$ is ω. Since $\omega \Rightarrow^{\text{ISC}} C$, for all C, we have $C \Rightarrow^{\text{ISC}\wedge} \omega$, for all C. Therefore, $T_I(A)$ – that is, ω – is a thesis with respect to "$\Rightarrow^{\text{ISC}\wedge}$". Consequently, if A is a thesis with respect to "$\Rightarrow^{\text{ISC}\wedge}$", so too is $T_I(A)$. Thus, T_I is truthlike. It remains to show that T_I is not truthful.

Suppose that $A \Rightarrow^{\text{ISC}} T_I(A)$ for all A. Now, ISC is incomplete with respect to negation (syntactically incomplete). So, there is some statement A^* such that neither it nor its negation is a theorem of ISC. Consequently, A^* is neither a thesis nor an antithesis with respect to the implication relation $\Rightarrow^{\text{ISC}}$. Clearly, if $A^* \Rightarrow^{\text{ISC}} T_I(A^*)$; then, since $T_I(A^*)$ is ω (A^* is not a thesis), $A^* \Rightarrow^{\text{ISC}} \omega$. However, since $\omega \Rightarrow^{\text{ISC}} C$ for all C, it follows that $A^* \Rightarrow^{\text{ISC}} C$, for all C. Therefore A^* is an antithesis with respect to $\Rightarrow^{\text{ISC}}$, and that is impossible.[14]

It is clear that, even in the simple case of the sentential calculus, there is a significant difference between intuitionistic and classical provability. Whereas intuitionistic provability is definitely truthlike, classical provability, in sharp contrast, is not even truthlike. This can be seen by using a construction similar to that just used for the intuitionistic case. Let CSC be some formulation of the classical sentential calculus, and form the associated implication structure on its sentences, with the implication relation "$\Rightarrow^{\text{CSC}}$" defined so that $A_1, \ldots, A_n \Rightarrow^{\text{CSC}} B$ if and only if $\vdash_{\text{CSC}}[(A_1, \ldots, A_n) \to B]$. It follows that A is a theorem of CSC if and only if it is a thesis of the structure that we have associated with it. Now, define the operator T_C so that $T_C(A)$ is Ω (a fixed thesis of the structure), if A is a thesis of CSC, and it is ω otherwise. It satisfies conditions 3(i)–(iii) at the beginning of this subsection. Clearly, T_C distributes over $\Rightarrow^{\text{CSC}}$ by an argument similar to the one given for T_I. However,

T_C fails to distribute over the dual $\Rightarrow^{\text{CSC}\wedge}$. To see this, suppose otherwise. Then, for all A and B, $T_C(A \vee B) \Rightarrow^{\text{CSC}} T_C(A) \vee T_C(B)$. In particular, $T_C(A \vee \neg A) \Rightarrow^{\text{CSC}} T_C(A) \vee T_C(\neg A)$ for all A. Since CSC is incomplete with respect to negation (syntactically incomplete), there is some A^* such that neither A^* nor its negation is a theorem of CSC, so that neither is a thesis of the structure. Then, $T_C(A^* \vee \neg A^*) \Rightarrow^{\text{CSC}} T_C(A^*) \vee T_C(\neg A^*)$. Since $(A^* \vee \neg A^*)$ is a thesis of the structure, $T_C(A^* \vee \neg A^*)$ is Ω. Since neither A^* nor its denial are theses, both $T_C(A^*)$ and $T_C(\neg A^*)$ are ω. Consequently, $\Omega \Rightarrow^{\text{CSC}} \omega$. In that case, Ω would be an antithesis, but that is impossible. So, classical provability (on the sentential calculus) fails to satisfy condition (2a) and is not even truthlike.

C. *Coherentist Truth*

1. Truthlike but Not Truthful. We have in mind those theories of truth that are presently described (correctly we think) as coherentist theories of truth rather than belief. There is ample evidence that the classical writers in the Idealist tradition thought truth a metaphysical rather than an epistemological notion.[15] The following sketch is intended to be a typical idealist theory. Rough though it is, it suffices to show that the coherentist notion of truth is truthlike, but not truthful.[16] In fact, both halves of Tarski's Equivalence fail.

Let us suppose that there is some non-empty subset $\mathfrak{C}$ of statements, and some implication relation "$\Rightarrow$" on it such that

(1) $\mathfrak{C}$ is consistent (with respect to "$\Rightarrow$"), which we adopt in this form: for any A in the structure, $A \in \mathfrak{C} \Rightarrow \neg(\neg A \in \mathfrak{C})$; and
(2) $\mathfrak{C}$ is complete, in the sense that for any statement A : $[A \in \mathfrak{C} \vee \neg A \in \mathfrak{C}]$ is a thesis ($\Rightarrow$).[17]

In addition, the coherentist view usually includes some kind of closure condition. We say that it is *coherently true* that A if and only if $A \in \mathfrak{C}$. That is, the coherentist truths are just the members of the special set $\mathfrak{C}$. Then, the closure condition can be expressed this way:

(3) Closure. If $A_1, \ldots, A_n \Rightarrow B$, then $A_1 \in \mathfrak{C}, \ldots, A_n \in \mathfrak{C} \Rightarrow B \in \mathfrak{C}$.

We represent coherentist truth by the following operator: $T_C(A)$ is Ω if $A \in \mathfrak{C}$, and ω otherwise [where Ω is a fixed thesis (with respect to "$\Rightarrow$"), and ω is a fixed antithesis (with respect to "$\Rightarrow$")].

It follows that the coherentist truth operator T_C is a truthlike operator:

(i) T_C distributes over "$\Rightarrow$", for suppose that $A_1, \ldots, A_n \Rightarrow B$. If all the A_i are in $\mathfrak{C}$, then, by closure, $B \in \mathfrak{C}$. Therefore, $T_C(B)$ is Ω, so that $T_C(A_1), \ldots, T_C(A_n) \Rightarrow T_C(B)$. On the other hand, if some $A_j \notin \mathfrak{C}$, then $T_C(A_j)$ is the antithesis ω, and so, $T_C(A_1), \ldots, T_C(A_n) \Rightarrow T_C(B)$.

(ii) T_C distributes over disjunctions, for suppose that $(A \vee B) \in \mathfrak{C}$. Then it follows from the coherentist assumptions (1)–(3) that $(A \in \mathfrak{C})$ or $(B \in \mathfrak{C})$. Then, at least one of $T_C(A)$, $T_C(B)$ is Ω. Therefore, their disjunction is equivalent to Ω, so that $T_C(A \vee B) \Rightarrow T_C(A) \vee T_C(B)$. On the other hand, if $(A \vee B) \notin \mathfrak{C}$, then $T_C(A \vee B)$ is ω, and so, $T_C(A \vee B) \Rightarrow T_C(A) \vee T_C(B)$.

(iii) The coherentist truth operator is thesis preserving, for it follows from the closure condition, that every thesis A is in $\mathfrak{C}$, so that $T_C(A)$ is Ω (a thesis).

(iv) Finally, the operator is antithesis preserving, for if B is an antithesis, then (by the consistency condition), it is not a member of $\mathfrak{C}$. Therefore, $T_C(B)$ is ω (an antithesis). Thus, T_C is a truthlike operator.

Is it truthful? Does it satisfy the Tarski Equivalence Condition? The answer is surely negative. Suppose that E is such that neither it nor its negation is a thesis (with respect to "$\Rightarrow$").[18] In that case, both halves of the Tarski Equivalence Condition fail. Suppose that $T(A) \Rightarrow A$ for all A. Suppose that $E \in \mathfrak{C}$, and so, $T_C(\neg E)$ is ω. Then, since $T_C(E) \Rightarrow E$, and $T_C(E)$ is a thesis ($\Rightarrow$), it follows that E is a thesis, but that is impossible. Moreover, by the consistency condition, $\neg E \notin \mathfrak{C}$. Now, if $\neg E \Rightarrow T_C(\neg E)$, then $\neg E \Rightarrow \omega$. In that case, E would be a thesis ($\Rightarrow$), and that is impossible. So, A does not imply $T_C(A)$ for all A. So, if E is in $\mathfrak{C}$, then both halves of the Tarski Equivalence Condition fail. On the other hand, if E is not in $\mathfrak{C}$, then $\neg E \in \mathfrak{C}$, and an argument similar to the one just given shows that both halves of the Equivalence Condition fail [this time using the assumption that E is not an antithesis ($\Rightarrow$)].

Cohen (1978) stressed one condition that coherentists maintained, but which we have omitted. It is that any two truths are equivalent (*Interconnectedness*). For many implication relations, this claim is wildly implausible, but there are implication relations that are logically mainstream, for which the claim is correct. We suggest one possible way of explaining the interconnectedness claim.

2. *Truthfulness Regained?* We expressed the coherentist theory with the aid of an implication relation "$\Rightarrow$". Suppose that we now extend that implication to an implication relation, $\mathfrak{C}$*-implication* ("$\Rightarrow^{\mathfrak{C}}$"), that depends on the coherence set $\mathfrak{C}$. That is, let us say that, for any statements $A_1, \ldots, A_n$, and B,

$$A_1, \ldots, A_n \Rightarrow^{\mathfrak{C}} B \text{ if and only if } (A_1 \wedge \cdots \wedge A_n \wedge \neg B) \notin \mathfrak{C}.$$

Then the following observations are immediate consequences:

(a) "$\Rightarrow^{\mathfrak{C}}$" is an implication relation that extends "$\Rightarrow$": for any $A_1, \ldots, A_n$, and B, if $A_1, \ldots, A_n \Rightarrow B$, then $A_1, \ldots, A_n \Rightarrow^{\mathfrak{C}} B$.

(b) "$\Rightarrow^{\mathfrak{C}}$" is "truth" preserving in the coherentist sense: if $A_1, \ldots$ and A_n are in $\mathfrak{C}$, and $A_1, \ldots, A_n \Rightarrow^{\mathfrak{C}} B$, then B is in $\mathfrak{C}$.

(c) (*Interconnectedness*) If A and B are any members of $\mathfrak{C}$, then $A \Leftrightarrow^{\mathfrak{C}} B$. [Note that, if A fails to $\mathfrak{C}$-imply B, then $A \wedge \neg B$ is in $\mathfrak{C}$. But in that case, by closure, $\neg B$ is in $\mathfrak{C}$. That is impossible, since $\mathfrak{C}$ is consistent. The proof that B $\mathfrak{C}$-implies A is similar.]
(d) Any two statements that are not in $\mathfrak{C}$ are $\mathfrak{C}$-equivalent.
(e) Any statement in $\mathfrak{C}$ is $\mathfrak{C}$-implied by any statement not in $\mathfrak{C}$.
(f) Every member of $\mathfrak{C}$ is a $\mathfrak{C}$-thesis (every statement $\mathfrak{C}$-implies them), and every statement not in $\mathfrak{C}$ is an $\mathfrak{C}$-antithesis (it $\mathfrak{C}$-implies every statement).

So, the use of $\mathfrak{C}$-implication has some definite advantages for a coherentist. Since it is an extension of a standard implication relation it is not some weird deviant notion.[19] One other consequence is surprising: the Tarski Equivalence Condition holds if the equivalence is $\mathfrak{C}$-equivalence. That is,

(g) For all $A, T_C(A) \Leftrightarrow^{\mathfrak{C}} A$.

Proof: Suppose that $A \in \mathfrak{C}$. Then, $T_C(A)$ is Ω, a thesis ($\Rightarrow$), and so belongs to $\mathfrak{C}$. Therefore, A and $T_C(A)$ are $\mathfrak{C}$-equivalent, by (c), since they are both in $\mathfrak{C}$. If $A \notin \mathfrak{C}$, then $T_C(A)$ is ω, an antithesis ($\Rightarrow$), and so does not belong to $\mathfrak{C}$. Consequently, by (d), A and $T_C(A)$ are $\mathfrak{C}$-equivalent since they both fail to be in $\mathfrak{C}$.[20] ■

Of course, the shift to $\mathfrak{C}$-implication is adopted at an enormous cost to the coherentist. It saves some strange claims, but it does so by shifting away from implication as most would understand it to a $\mathfrak{C}$-implication that is tailored to the set of coherent truths. Perhaps in the end the coherentist account should be formulated as a theory that uses two kinds of implication, one being a complete extension of the other. It would be a kind of double implication theory. Nevertheless, it remains that the coherentist notion of truth (membership in the set $\mathfrak{C}$) yields a truthlike operator that is not truthful (with respect to "$\Rightarrow$"). Although it is correct that the shift to $\mathfrak{C}$-implication will yield the Tarski Equivalence, that shift means that we are given something far weaker and a good deal less interesting than what we thought we were promised.

XIII. The Place of Truthful Among the Truthlike (Extensionality)

Truthful operators are, as we described them, among the truthlike ones on a structure. We consider a third way of distinguishing them from their definitely truthlike cousins. The answer is simple: it is extensionality that makes the difference. That is,

Theorem 13. A truthlike operator satisifies the Tarski Equivalence Condition if and only if it is extensional.

Proof: What is needed is the appropriate account of the extensionality of an operator on a structure. Let $\Im$ be a set of interpretations, which are mappings from the structure to the set $\{t, f\}$ such that (1) every interpretation in $\Im$ assigns t to all the theses; (2) every interpretation in $\Im$ assigns f to all the antitheses (contradictions); and (3) for any A and B, $A \Rightarrow B$ if and only if there is no interpretation in $\Im$ that assigns t to A, but f to B.[21] Let us say that an operator T on the structure is *extensional*

If any interpretation that assigns the same truth-value to A and B, also assigns the same truth-value to $T(A)$ and $T(B)$.[22]

To prove Theorem 13, we first show that if $T(A) \Leftrightarrow A$ for all A, then T is extensional. Suppose that i is any interpretation in $\Im$ that assigns the same value to A and to B. If they are assigned t, then since $A \Rightarrow T(A)$ (similarly for B), it follows by (3) that both $T(A)$ and $T(B)$ are assigned the value t. If both are assigned the value f, then since $T(A) \Rightarrow A$ (similarly for B), it follows that both $T(A)$ and $T(B)$ are assigned the value f. Thus, if $T(A)$ is equivalent to A for all A, it follows that T is extensional. The second part of Theorem 13 goes this way: Let B be any thesis. Let i be any interpretation in $\Im$. If $i(A) = t$, then, since by (1), $i(B) = t$, it follows that $i(A) = i(B)$. Since T is extensional, i assigns the same value to $T(A)$ and $T(B)$. Since T is truthlike, $T(B)$ is a thesis, so that $i(T(B)) = t$. Therefore, $i(T(A)) = t$. Consequently, for any i, if $i(A) = t$, then $i(T(A)) = t$. By (3) then, $A \Rightarrow T(A)$, for all A. There remains only to show that, for all $A, T(A) \Rightarrow A$. Let i be any interpretation in $\Im$, and let C be any antithesis. If $i(A) = f$, then since $i(C) = f$ by (2), it follows that i assigns the same value to A and to C. By the extensionality of T, it follows that i assigns the same value to $T(A)$ and $T(C)$. However, since T is truthlike, $T(C)$ is an antithesis, so that i assigns it the value f. Therefore, i assigns f to $T(A)$. Consequently, for any i, if $i(A) = f$, then $i(T(A)) = f$. Therefore, for any i, if $i(T(A)) = t$, then $i(A) = t$. By (3) then, we obtain that $T(A) \Rightarrow A$ for all A. Thus $T(A) \Leftrightarrow A$ for all A.

This result gives us a necesary and sufficient way of distinguishing the (Tarskian) truthful from its truthlike cousins. It is similar in flavor to the familiar result that the identity relation, which is an equivalence relation, is special among all equivalence relations in that it is the only one that satisfies the additional requirement of substitutivity. Likewise, a truthlike operator is Tarskian if and only if it is extensional. The substitutivity condition has to be justified if, in a particular case, we claim that a relation is not just an equivalence relation, but the identity relation. Similarly, any argument that shows that a particular operator is not only truthlike, but truthful, will have to support the extensionality condition.

Extensionality, though familiar, is neverthelesss a very powerful assumption. Although we have seen that the extensionality of any truthlike T guarantees that $T(A) \Leftrightarrow A$ for all A, it is a result that has very little to do with the truthlike character of T. In fact, extensionality of *any* operator T that maps the structure to itself, together with some very weak assumptions, will guarantee that T satisfies the Tarski Equivalence Condition. We do not even have to assume that T distributes over implication, or over disjunction. That is, ■

Theorem 14. If T is any extensional operator that maps the structure to itself, and there is *some* thesis Ω such that $T(\Omega)$ is a thesis, and there is some antithesis ω such that $T(\omega)$ is an antithesis, then for all A, $T(A) \Leftrightarrow A$.[23]

There are certain operators for which it is clearly a mistake to require extensionality. Sentential probabilistic operators are one example and modal operators are another. In the former case, extensionality would imply that probabilities could have only two values, and in the latter, extensionality by Theorem 14 requires that every one of a huge class of modal operators would have to collapse.[24] We ought to proceed with some care in requiring an extensionality condition. In situations that call for a truthlike operator, it takes additional support to narrow the choice to the truthful.

NOTES

Thanks to Charles for his unflagging friendship over the decades. Thanks also to J. Azzouni, A. Lipeles, J. Kastin, E. Mendelson, and G. Priest. Versions of this paper were read to the Graduate Center Philosophy Colloquium and to the Cambridge Moral Sciences Club. For their helpful comments, I am beholden to my colleagues at the Graduate Center and to Jeremy Butterfield, Dorothy Emmet, Hugh Mellor, Michael Redhead, and Tim Smiley.

1. This is the operator version of the Tarski Equivalence Condition. Cf. Feferman (1984).
2. One type of response lies in an appeal to the demands of an adequate theory of reference. An adequate theory of reference (so the story would go) either requires the use of a Tarskian operator in its construction or the only truthlike operator that combines coherently with such a theory of reference has to be Tarskian. I do not believe either of these claims is correct, but the issue of how the demands of reference might make a difference is beyond the scope of this study.
3. For readers who may not be familiar with the Gentzen conditions, it suffices to note that they require

   (1) Refexivity: $A \Rightarrow A$, for all A in S;
   (2) Projection: $A_1, A_2, \ldots, A_n \Rightarrow A_k$, for any $k = 1, \ldots, n$;
   (3) Simplification (sometimes called Contraction): If $A_1, A_1, A_2, \ldots, A_n \Rightarrow B$, then $A_1, A_2, \ldots, A_n \Rightarrow B$, for all A_i and B in S;
   (4) Permutation: If $A_1, A_2, \ldots, A_n \Rightarrow B$, then $A_{f(1)}, A_{f(2)}, \ldots, A_{f(n)} \Rightarrow B$, for any permutation f of $\{1, 2, \ldots, n\}$;
   (5) Dilution: If $A_1, A_2, \ldots, A_n \Rightarrow B$, then $A_1, A_2, \ldots, A_n, \mathrm{C} \Rightarrow \mathrm{B}$, for all A_i, B, and C in S;

(6) Cut: If $A_1, A_2, \ldots, A_n \Rightarrow B$, and $B, B_1, B_2, \ldots, B_m \Rightarrow C$, then $A_1, A_2, \ldots, A_n, B_1, B_2, \ldots, B_m \Rightarrow C$.

4. Not all implication structures will have truthlike operators on them. The definition assumes that there are theses and antitheses [otherwise conditions (1b) and (2b) are vacuous]. A structure in which there are just three elements, A, B, and C, such that A implies B as well as C (but not conversely), and neither B nor C implies the other, has no theses. The use of the dual implication on this structure provides an example where there are no antitheses.
5. Here is a simple familiar case in which the assumption would be rejected. Suppose that $\Box$ is a T-modal with respect to the implication relation "$\Rightarrow$" (it distributes with respect to "$\Rightarrow$", $\Box A$ is a thesis if A is a thesis, and $\Box A \Rightarrow A$ for all A). If $\Box$ continues to be a T-modal with respect to all C-extensions of "$\Rightarrow$", then $\Box A \Leftrightarrow A$ for all A. That is, the modal collapses, and that is impossible for T-modals.
6. The helpful notion of T-anomaly was brought to my attention by Horacio Arlo Costa (private communication), who attributes it to von Wright.
7. Here is a simple example of a consistent structure, and a mapping T^* on it, which is pretruthlike (but not truthlike): Let $S = \{A, B\}$, where $A \Rightarrow B$ (but not conversely). Then, $A \Leftrightarrow \neg B$, and B is a thesis of the structure. Let $T^*(A)$ and $T^*(B)$ be equivalent to A. Since all the values of T^* are equivalent, T^* distributes over "$\Rightarrow$" and its dual. Therefore, it is pretruthlike, but B is a thesis and $T^*(B)$ is not. So, it is not truthlike. Notice that, in this structure, both A and B are T^*-antinomic [$T^*(\neg A) \Leftrightarrow T^*(B) \Leftrightarrow T^*(A)$, and $T^*(\neg B) \Leftrightarrow T^*(A) \Leftrightarrow T^*(B)$].
8. The result is this: Let Σ be the set of all operators on the structure $I = \langle S, \Rightarrow \rangle$ that distribute over "$\Rightarrow$", distribute over the dual $\Rightarrow^{\wedge}$ (or over disjunction), and preserve theses. Say that an operator ϕ is maximal in Σ if and only if it is in Σ, and for any operator ψ in Σ, if $\psi(A) \Rightarrow \phi(A)$ for all A, then $\psi(A) \Leftrightarrow \phi(A)$ for all A. Then, all the truthlike operators on the structure are maximal in Σ. The proof is essentially the same as that given for Theorem 6.
9. If P is an operator mapping, say, pairs of members of S to S (as are all of the logical operators aside from negation), then equivalence will mean that if $A \Leftrightarrow A^*$, and $B \Leftrightarrow B^*$, then $P(A, B) \Leftrightarrow P(A^*, B^*)$. The generalization to arbitrary n is obvious.
10. For example, if U is a symmetry of the logical operators, then the results (i)–(iv) will follow. If U is a symmetry of conjunction and negation, and a symmetry of the constant operators associated with each antithesis of the structure, then U will distribute over "$\Rightarrow$". If U is a symmetry of disjunction and negation, and a symmetry of all the constant operators associated with each antithesis of the structure, then U will distribute over "$\Rightarrow^{\wedge}$". And so, it follows that, if U is a symmetry of conjunction, disjunction, negation, and all of the constant operators associated with each antithesis, then T is a truthlike operator.
11. Here is an example in which its use would be rejected. Suppose that $\Box$ is a normal modal operator. Then the requirement that $\Box(P_C(A)) \Leftrightarrow P_C(\Box(A))$ for all the constant operators P_C leads to the collapse of the modal operator.
12. This seems to be true for the operators considered, for example, by von Wright (1984).
13. Thus, for any interpretation i, T_i is a truthlike operator mapping the structure to the set $\{\Omega, \omega\}$. It is also true that, for any truthlike operator T on the structure that maps it to $\{\Omega, \omega\}$, there is an interpretation j such that, for all A, $T_j(A)$ is equivalent to T.
14. However, it is easily seen that, for all A, $T_I(A) \Rightarrow A$. This would seem to contradict

our result about scattering, but there is no conflict. Scattering holds when negation is classical, and here it is intuitionistic.

15. See Cohen (1978) for an illuminating and lucid account of that tradition.
16. On this point, we differ from Cohen (1978), who claims that coherentist truth does satisfy the Tarski Equivalence Condition. The problem with Cohen's view is that the modal operator that he uses to explain their concept of truth can be shown to collapse.
17. The condition of completeness can be guaranteed by (i) introducing some notion of relative coherence that is a transitive, asymmetric relation on non-empty sets of statements; (ii) introducing a set K such that A is a member of K if and only if $\{A\} \cup K$ (the union of K with A) is more coherent than is $\{\neg A\} \cup K$; and (iii) requiring that K be decisive in the sense that, for any non-equivalent statements A and B, either the coherence of K with A is greater than the coherence of K with B, or conversely. However, this is a fine-grained feature of the coherentist position. What is important for our purposes is the inclusion of completeness (2), no matter how it might be supported.
18. This is generally true for many implication relations, though not for all. Of course, the assumption that the implication structure is syntactically incomplete may not be acceptable to some coherentists. We address that issue in §XII.C.2.
19. In fact, it is an example of a special class of implication relations that are called bisection implications by Koslow (1992). The latter are based on the notion of a bisection of a structure. It was Scott (1974) who first noted that these bisections are essentially the truth-value assignments on a structure.
20. The observation (g) is also a consequence of Theorem 5, since the coherentist structure with $\mathfrak{C}$-implication is classical and (syntactically) complete.
21. There is an easy construction showing how to obtain such a set of interpretations that satisfy (1)–(3) even in an abstract setting. See Koslow [1992, Theorems 8.11 and 8.12 (Strong and Weak Completeness)]. The set of strong bisections on the structure yields a set of interpretations on the structure by taking an interpretation to be the function i, such that $i(A) = t$ if A is in L, and $i(A) = f$ if A is in K. They satisfy the three conditions listed.
22. See Koslow (1992, Chap. 18), for a full account of the notion of extensionality, which is available in any implication structure that has at least two members. In particular, the notion of extensionality takes on a special richness, since it relies upon the appropriate notion of implication. If there is a shift in the implication relation, there is an associated shift in what the requirement of extensionality expresses. The cited chapter also contains a full discussion of the extensionality (or non-extensionality) of the various logical operators.
23. The proof is exactly the same as that given in Theorem 13 for showing that $T(A) \Leftrightarrow A$, given that T is truthlike. We have only to notice that, in that proof, only the fact that T is a thesis for some thesis, and that T is an antithesis for some antithesis, was used. In fact, the proof shows two things: (1) if T is a thesis for some thesis and is extensional, then $A \Rightarrow T(A)$ for all A; and (2) if T is a contradiction for some contradiction, then $T(A) \Rightarrow A$, for all A.
24. The case for probabilities is well known and raises a serious problem for someone like Quine, who seems to believe that all serious scientific idioms and ideology are extensional. The case of the modals is not as serious a problem since no one as far as I know has ever advocated the view that some modals are extensional. In fact, by Theorem 14, any normal modal for which necessity implies possibility would have

to collapse (i.e., $\Box A \Leftrightarrow A$ for all A). A more general proof of non-extensionality that covers most, if not all, of the modals can be found in Koslow (1992), pp. 261–264.

REFERENCES

Cohen, J. L. (1978). "The Coherence Theory of Truth," *Philosophical Studies*, 34: 351–60.

Feferman, S. (1984). "Toward Useful Type-Free Theories, I," *Journal of Symbolic Logic*, 49: 75–111; reprinted in R. L. Martin (ed.), *Recent Essays on Truth and the Liar Paradox* (Oxford: Oxford University Press, 1984), pp. 237–87.

Koslow, A. (1992). *A Structuralist Theory of Logic* (Cambridge, UK: Cambridge University Press).

Lewis, D. (1973). *Counterfactuals* (Cambridge, MA: Harvard University Press).

Scott, D. (1974). "Completeness and Axiomatizability in Many-Valued Logic," in L. Henkin et al. (eds.), *Proceedings of the Tarski Symposium* (Providence, RI: American Mathematical Society), pp. 411–35.

Stalnaker, R. (1968). "A Theory of Conditionals," in *Studies in Logical Theory, American Philosophical Quarterly*, Monogr. 2 (Oxford: Blackwell), pp. 98–112.

Von Wright, G. H. (1984). *Truth, Knowledge, and Modality, Philosophical Papers*, Vol. III (Oxford: Blackwell).

'Everything'

VANN McGEE

Ontology, as Aristotle conceived it,[1] is the most general of sciences. The other sciences are specialized, restricting their attention to some proper part of reality. It remains for ontology to take account of everything there is.

The question I would like us to investigate here is not a problem in ontology, but a problem in the meta-theory of ontology. I want to know whether, given the limits of logic and language, a science such as ontology is even possible, there being some reason to fear that we lack the conceptual resources to formulate a fully general science, because we have no any way reliably to ensure that our quantifiers range over everything there is. Even if we succeed in convincing ourselves that there is no logical impediment to the development of a fully general science, it will remain to be seen whether such a science has anything interesting to tell us. It could happen that the fully general truths are too abstract to be very important, and it could also happen that the fully general truths are beyond our epistemic reach, so that, while we are privileged to know rather a lot about our immediate surroundings, we are not in a position to know much of anything about the universe as a whole. But first things first.

The logical worry comes from the interplay of context and quantification. Often we use such words as 'each', 'any', 'every', 'all', 'anyone', 'everyone', and 'everything' without intending to talk about everyone and everything. When I say "Everyone who eats her broccoli will get ice cream," I do not promise to reward every broccoli-eater in the world, only those in my immediate family. I tacitly intend, and the people who hear me understand me tacitly to intend, to restrict my offer to a tiny subset of all the people there are, so that what we get is not a deep principle of universal karmic reward but a parent's pitiful attempt to bribe his children into eating their vegetables.

Restricted quantification is useful at mealtimes, but ontology requires unrestricted quantification. Without it, we shall have no assurance that the story we tell, despite its use of the words 'everyone', 'everywhere', and 'everything', is anything more than a parochial account of the people, places, and things in our immediate neighborhood. The logical worry is that such unrestricted quantification may not be reliably available.

I personally got interested in this question by reflecting upon the philosophy of mathematics, specifically the view, sketched by Ernst Zermelo[2] and developed by Charles Parsons,[3] that the phrase 'for any set *x*' is ambiguous. There are many different universes of set theory, and there is nothing we do, say, or think that singles out any one of them as the intended domain of our set-theoretic discourse. Quantificational indeterminacy produces indeterminacy of truth values: There is no fact of the matter whether the statement 'There is a largest inaccessible cardinal' is true. This is an important doctrine for the philosophy of mathematics, since it rules out the most extreme forms of mathematical realism. But even people who are not especially interested in the philosophy of mathematics are likely to be interested in the question whether unequivocally general judgments are possible, since questions of what there is are going to appeal to anyone with any metaphysical leanings whatever.

There are two worries that trouble the prospects for genuinely universal quantification, one more serious than the other. The lesser worry is that the universal quantifier is a monkey wrench that cannot be opened wide enough to fit reality. The range of our quantified variables varies according to our conversational purposes and interests, so that their range is wider in philosophical discourse than it is in typical mealtime chitchat. But the range is always limited, even when we are doing philosophy, so that even our bravest attempts at ontology fail to be truly universal in scope.[4]

The reason this is not a serious worry is that the thesis that, for any discussion, there are things that lie outside the universe of discourse of that discussion is a position that cannot be coherently maintained. Consider the discussion we are having right now. We cannot now coherently claim that there are things that lie outside the universe of our discussion, for any witness to the truth of that claim would have to lie outside the claim's universe of discourse.

Of course, the fact that a thesis cannot be coherently maintained does not strictly entail that the thesis is false. Even so, the fact that we cannot coherently hold the theory is surely reason enough not to attempt to embrace it.

The more serious worry is not that quantificational restriction is inevitable but that it is invisible. It seldom happens that we restrict our domain of quantification by an explicit stipulation. On the contrary, the processes by which quantifiers become restricted are subtle and poorly understood – so subtle, indeed, that we may not be consciously aware that our domain has been restricted. So subtle – this is our real worry – that there might be nothing we do, say, or think that is any different when we use unrestricted quantification and when we employ quantifiers that are restricted to some suitable, less inclusive domain. If this should happen, then our use of the quantifiers is hopelessly ambiguous. There is no fact of the matter of what our quantifiers refer to, and so, any scientific program, such as the founding of a science of ontology, that relies upon a distinction between universal and restricted quantification is sure to founder.

The cause of this worry is not idle speculation. Eminent thinkers, notably Quine and Putnam, have given influential arguments that would seem to show that quantificational ambiguity is unavoidable.

Quine and Putnam both talk about the quantifiers of the first-order predicate calculus, but their real target (as well as the target of the discussion here) is quantification in natural languages. We can regard their statements about first-order formulas as disguised claims about the English sentences that are standardly schematized by those formulas. More-or-less equivalently, we can understand their pronouncements as describing properties that the first-order languages would acquire if they were interpreted by explaining that, for example, '∀' means 'everything'.[5]

To pretend that standard formalizations adequately represent the logical structure of English oversimplifies on a vast scale, for the inadequacies of the formalization techniques are infamous. (One discrepancy that is particularly relevant to present concerns is that English permits and the formalized languages forbid nondenoting proper names.) The motive for nonetheless working with the formalized languages is a methodology of starting with the very simplest cases and advancing to the more complex. The formal languages are the linguistic equivalent of frictionless planes.

Quine's worry[6] is that, under special circumstances, it may be impossible to distinguish substitutional from objectual quantification. The special circumstances are these: For each true existential sentence, there is a name that designates an individual that bears witness to the truth of the sentence; as Quine puts it, the unnamed objects are "inseparable" from the named ones. Should such a circumstance obtain, there will be no ready way to distinguish ordinary objectual quantification from its substitutional counterfeit, which regards a sentence $(\exists x)\phi(x)$ as true if and only if every substitution instance of $\phi(x)$ is true.

The underdetermination of theory by available data is a familiar fact of scientific life; it is a disturbing fact, surely, but no more worrisome for the theory of meaning than it is for the other sciences, so that, if the only problem were that the physiological, psychological, and sociological factors that determine the range of our quantifiers were difficult or even impossible to identify experimentally, this would not be a cause for great alarm. What Quine has in mind, if I understand him properly, is the altogether more troubling possibility that there are no factors that determine the range of our quantifiers. The only thing that there is to fix the meaning of a word is the role of the word in determining the truth values of the sentences that contain it; so, if discovering the truth values of all sentences does not suffice to determine the range of the quantifiers, there is nothing else to do the job. So, our problem is not that we are unable to learn the relevant facts, but that the totality of relevant facts is insufficient to answer our question.

The doctrine that we have exhausted the relevant facts as soon as we have learned the truth values of sentences is derived from Frege,[7] who admonished us never to look for the meaning of a word in isolation, rather always to look within the context of a sentence. But while Frege's admonition seems altogether reasonable, Quine's application of it is entirely too severe. When a sentence has a determinate truth value, our linguistic practices determine the conditions under which the sentence is true, and the nonlinguistic facts[8] determine whether those conditions are met. In establishing the truth *conditions* for a sentence, our linguistic practices go beyond their role in determining the sentence's truth *value*. In addition to helping determine what truth value the sentence in fact has, our linguistic practices determine what truth value the sentence would have in various counterfactual circumstances. Our thoughts[9] and practices, together with the empirical facts, determine that the sentence "There are massage parlors in Antarctica" is false, and they go beyond this also to determine that, if there were massage parlors in Antarctica, "There are massage parlors in Antarctica" would be true.

Once we start looking at counterfactual as well as actual truth values, Quine's problem vanishes. Consider a world w in which there exist individuals who do not exist in the actual world. Suppose that there is some property – call it F – that no individual that exists in the actual world possesses in w, but that at least one individual that exists in w but not in the actual world possesses in w. If quantification is objectual, $(\exists x)F(x)$ will be true in w; if quantification is substitutional, it will be false. Inseparability, if it obtains, is an accidental feature of this world, and once we begin looking at other worlds, Quine's problem disappears.

This easy solution depends upon the assumptions that the "names" we talk about when we describe substitutional quantification are proper names, and that those names do not change their designation when we go from world to world. There are versions of substitutional quantification that do not satisfy these assumptions. For example, if our names include closed terms formed using Hilbert's ε-operator,[10] then the names will no longer be rigid designators, and named and unnamed individuals will be inseparable in every possible world. With this enlarged supply of names, we shall be unable to distinguish substitutional from objectual quantification by looking at truth conditions. But this inability is no failing, since the two forms of quantification are now logically equivalent. Indeed, Hilbert was able to treat objectual quantification as a defined operation:

$$(\exists x)\phi(x) =_{\text{Def}} \phi(\varepsilon x(\phi(x)))$$
$$(\forall x)\phi(x) =_{\text{Def}} \phi(\varepsilon x(\neg\phi(x))).$$

With Hilbert's ε-operator in the language, substitutional and objectual quantification are the same operation, differently described.

Numerous variations are possible. If our names do not include Hilbert's ε-operator but do include definite descriptions, then our names will not be rigid designators, but there will nonetheless be logically possible[11] worlds in which named and unnamed individuals are separable, so that the two forms of quantification give distinct truth conditions. On the other hand, taking our names to be definite descriptions, we can restore necessary inseparability by adjoining to the predicate calculus a new logical operator '$<$' with the property that it is necessary that '$<$' well-order the universe, whatever the universe be.

What we do not get – or, more guardedly, what the examples we have examined do not give us – is a case in which objectual and substitutional quantification are clearly distinct and yet the distinction is nowhere reflected in truth conditions.

In "Models and Reality,"[12] Putnam advances a different argument to much the same effect, using the Löwenheim-Skolem theorem to demonstrate the existence of a set S with the following properties:

- Every individual named by a proper name is a member of S.
- Every individual (if there are any) that we are able to name by direct ostension or to pick out via a causal connection is a member of S.
- Every n-tuple of members of S will satisfy precisely the same open sentences if the variables are taken to range over S or they are taken to range over everything. In particular, restricting the quantifiers to S makes precisely the same sentences true.
- S is countable.

What determines, Putnam wants to know, whether a speaker is speaking a language in which the variables range over S or a language in which the variables range over everything? The same sentences are true in both languages, and, indeed, the same sentences will be held true by a speaker, since the observation reports will be the same in both languages, as will the evidentiary relations among the sentences. Moreover, when the speaker uses a sentence to issue a request, promise, or warning, the expected effect on the listener will be the same, whichever language she employs.

There is, it appears, nothing in the behavior of a child trying to learn the language that determines whether the variables the child employs and takes her elders to be employing range over S or over everything. This will be so even if the child shows a precocious ability for meta-linguistic reflection. On either hypothesis, the precocious child will tell us without hesitation that her variables range over everything there is; she will say this because the quantifiers in the meta-language are subject to the same indeterminacy as those of the object language.

Not only is there nothing in the child's behavior to discriminate between the two hypotheses, there is nothing in her thoughts either. What she says to herself will be the same, whichever way she understands the variables, and the same words will provoke the same memories, images, and longings.

An antirealist conclusion quickly follows. If there is nothing the child does, says, or thinks that determines whether she is using universal quantification or

S-quantification, there is nothing you or I can do, say, or think to determine which mode of quantification we are using. There is no fact of the matter whether when we say 'everything' we refer to everything or we refer to S.

At this point, we could try replying to Putnam the same way that we did to Quine, asking him to consider not only which sentences are, in fact, true, but which sentences would be true in various counterfactual circumstances, but this time, Putnam has a ready rejoinder. Consider the whole structure of individuals, properties, and worlds (together with the similarity and accessibility relations on the worlds), as a first-order structure. This means representing n-ary relations by $(n+1)$-place predicates, the extra argument-place ranging over worlds. Now apply the Löwenheim–Skolem argument to this modal structure. In the original interpretation, there is an uncountable totality of individuals and an uncountable totality W of worlds. In the submodel, there is a countable collection S of individuals and a countable collection V of worlds. A complete determination of the semantic role of the logical operators would determine both what the range of the quantifiers is and what the logically possible worlds are. But, in fact, there does not appear to be anything we do, say, or think that determines whether the correct interpretation of the logic of the language is one according to which the quantifiers range over everything and the logically possible worlds are the members of W, or one according to which the quantifiers range over S and the logically possible worlds are the members of V. Whichever way we choose, the answers to questions of the form "If such-and-such circumstances obtained, what sentences would be true?" will all be the same.

It is possible to dismiss the whole problem simply by insisting on the prerogative of employing second-order logic, since this would block the application of the Löwenheim–Skolem theorem. However, suspicion of second-order logic remains sufficiently widespread that it is worthwhile to seek a first-order solution. (The status of second-order quantification will be discussed, albeit inconclusively, later.)

To forestall the antirealist conclusion, we need to find some reason to prefer the hypothesis that the child's variables range over everything to the hypothesis that they range over S. Such a reason is ready at hand if, instead of thinking of a language as a system of sentences in want of an interpretation, we think of a language as constituted, in whole or in part, of the rules and practices followed by its speakers. The rules of the language have to be rules that the speakers are capable of learning and following, and S-quantification is not learnable. To quantify over S, we would have to be able to distinguish the Ss from the non-Ss. Either the rule of universal specification would have to be restricted so that we could only infer $\phi(\tau)$ from $(\forall x)\phi(x)$ in the special case in which τ denotes a member of S or the grammatical rules would have to include a special provision that forbade closed terms that designated non-Ss. In either case, it would be necessary to distinguish the Ss from the non-Ss before we could learn and employ the rules. But we cannot tell the Ss from the non-Ss;

Putnam went to great pains to ensure that the boundary between them was entirely ineffable.

We can put the conclusion in the form of a dilemma. Either we hypothesize that a child learning the language is able to distinguish *S*s from non-*S*s, in which case there is a difference in her mental state when she refers to *S*s and when she refers to everything, a difference that can be made manifest by introducing a new predicate true of all and only the *S*s; or we hypothesize that she is unable to make such a distinction, in which case *S*-quantification will be unlearnable. In either case, the antirealist conclusion is thwarted.

Answering Putnam this way raises a provocative question: It is all very well to complain that *S*-quantification is not learnable, but the complaint is idle unless it marks a contrast between *S*-quantification and universal quantification. The difference, we want to say, between *S*-quantification and universal quantification is that the latter is learnable and the former is not. To say this credibly, we have to make a case that genuine universal quantification is, in fact, learnable. So, what is the story? How do we learn universal quantification?

The question I am raising is intended as a philosophical question, rather than a question about developmental psychology. What might a speaker, or a community of speakers, do, say, or think that would make their use of '∀' genuine universal quantification? One might try to answer this question by gathering information about patterns of speech dispositions, correlating the acquisition of speech dispositions with stages in a speaker's social, psychological, and neural development. But it is not immediately clear how such information is going to help us, because we still want to know what makes this particular pattern of speech dispositions universal quantification rather than *S*-quantification? Examination of developmental stages will show us when and how the child becomes able to use the quantifier the same way an adult does, but it will not directly answer the question that worries us here, which is, "What features of their thoughts and practices guarantee that adult and child alike are using genuine universal quantification?" As a preliminary strategy toward answering this question, I propose that we try to identify a scenario in which creatures like ourselves acquire a pattern of usage recognizable as universal quantification. Having done so, we can use the scenario as a starting point for developing a general understanding of what patterns of speech dispositions would constitute universal quantification, and a particular understanding of when and where such patterns are acquired by actual communities of human beings.[13]

The central idea of the scenario is simple enough: We learn to use the connectives and quantifiers by learning the rules of inference.[14] As developmental psychology, the story does not succeed, since abstract formal reasoning is something human beings do seldom and badly. But the story is only intended as rational reconstruction. The rules of natural deduction resemble closely enough the processes by which actual people reason that, if we understood how the formal rules of natural deduction determined the logical operations employed by idealized speakers of formal languages, we would be well along the way toward understanding how informal rules of reasoning fix the logical operations of flesh-and-blood speakers of natural languages.

How do the rules of inference succeed in pinning down the meaning of the connectives and quantifiers? Let me first trot out a plausible but unsuccessful answer, and then give another answer that I think works better. The unsuccessful answer is this: The natural deduction rules describe the inferential role of the logical symbols; their inferential role is their use, and to determine a word's use is to determine its meaning. Such an account is plausible enough as applied to the connectives, but wholly implausible as applied to the quantifiers. The natural deduction rules only describe the use of the quantifiers in deductive reasoning, and a central part of their use is their employment in inductive reasoning. It seems to me quite plausible to say that the meaning of the quantifiers is given by their conceptual role, or that it is given by their inferential role, if inferential role is understood widely enough to encompass use in inductive as well as deductive reasoning; but, conceived so widely, their inferential role is not described by the rules of natural deduction.

The better answer comes from a short paper by Nuel Belnap, entitled "Tonk, Plonk, and Plink."[15] Belnap provides two conditions on the basis of which a rule that purports to introduce a new logical connective into the language can be said to have been completely successful. The first is *conservativeness*. If, employing the new rule, a sentence ϕ that does not contain the new connective can be derived from a set of sentences Γ, none of which contain the new connective, then ϕ must be derivable from Γ without employing the new rule. The second, *uniqueness*, is more complicated. Let us say that the new connective we are introducing is '$\neg$' (the same criteria apply whether the new logical operations we are introducing are one or many), and let ϕ be a formula containing the new connective. Let ϕ_1 and ϕ_2 be the expressions obtained, respectively, by replacing each occurrence of '$\neg$' with '$\neg_1$' and by replacing each occurrence of '$\neg$' with '$\neg_2$'. The uniqueness condition requires that ϕ_1 and ϕ_2 be interderivable.

Belnap's criteria are proposed as sufficient conditions for a new connective to have been successfully introduced, not as necessary conditions. For example, it might happen that, whenever ϕ is derivable from Γ by means of the new rule, ϕ is a logical consequence of Γ (according to some model-theoretic conception of consequence), even though ϕ might not be derivable from Γ by the original rules; this could happen if the original rules were incomplete. If this happens, conservativeness fails, but we would not therefore want to say that the new rule is illegitimate.

The uniqueness criterion is intended to ensure that the new rule succeeds in unambiguously specifying the semantic role of the new connective. Belnap does not explain why he thinks that the criterion does this, but an explanation is not hard to supply. Where there is ambiguity, it is possible to disambiguate by replacing the old term with new terms that refer unambiguously. Use 'riverbank' and 'moneybank' in place of 'bank'. Use 'inertial mass' and 'rest mass' in place of 'mass'. But it is not possible to produce two logically inequivalent connectives to replace our current use of '¬', and so, our usage of '¬' must be unambiguous.

To fend off Putnam-style skepticism, it is not enough that it be psychologically impossible for us to learn and understand a language in which two logically inequivalent connectives supplant our current usage of '¬'. Indeed, Putnam's argument depends on the fact that universal quantification and S-quantification, though logically distinct, are indistinguishable by us. To show that negation is not subject to that sort of ambiguity, we need to show not merely that a language with two negations is psychologically unavailable to us, but that such a language is not possible even in principle. There is not even a mathematically possible language in which two distinct, logically inequivalent negation operations both obey the rules for '¬'.

To make sense of "mathematically possible languages," we have to conceive of a language abstractly, as perhaps an ordered pair consisting of a collection of expression types, appropriately arrayed in grammatical categories, and a function assigning to each expression a semantic value. This is opposed to the conception of a language as a concrete system of human conventions and practices.[16] I am not supposing that the abstract conception of language is the only one available, or that it is superior, only that it is better suited to present purposes. So, I am not making a significant assumption about the nature of language. I am, however, making a significant assumption about ontology. In treating languages as mathematical entities, I am presuming that mathematical entities exist, and whether mathematical entities exist is precisely the sort of question we look to ontology to answer. Thus, while trying to develop a logical framework in which to pursue an ontological inquiry, I am already making nontrivial ontological assumptions. Consequently, all the conclusions we reach must be regarded as provisional; ontological investigation could undermine their foundations.

In particular, in presuming classical mathematics, I am employing classical logic, even though the availability of classical logic remains very much in controversy. To those who deny that classical logic can be coherently used and understood, nothing we say here is going to be the least bit persuasive. This is a story about how classical logic is acquired, told from the viewpoint of classical logic. The rival intuitionist will tell an intuitionist story about how intuitionist logic is acquired. Nothing we say here will be any help in mediating between them.

The proof that the natural deduction rules for classical negation, and, indeed, for all the classical connectives and quantifiers, including second-order quantifiers, satisfy Belnap's uniqueness condition is easy. We treat negation, conjunction, universal quantification, and identity as primitive, and regard the other connectives as defined:

$$\bot =_{\text{Def}} (P \wedge \neg P),$$
$$(\phi \vee \psi) =_{\text{Def}} \neg(\neg\phi \wedge \neg\psi),$$

and so on. The rules we use are entirely standard; we put them down for definiteness:

(1) If ϕ is a member of Γ, one can infer ϕ from Γ.
(2) If one can infer ϕ from Δ and one can infer every member of Δ from Γ, one can infer ϕ from Γ.
(3) One can infer $(\phi \wedge \psi)$ from $\{\phi, \psi\}$.
(4) One can infer ϕ from $\{(\phi \wedge \psi)\}$.
(5) One can infer ψ from $\{(\phi \wedge \psi)\}$.
(6) One can infer ψ from $\{\phi, \neg\phi\}$.
(7) If one can infer ψ from $\Gamma \cup \{\phi\}$ and also from $\Gamma \cup \{\neg\phi\}$, one can infer ψ from Γ.
(8) One can infer $\tau = \tau$ from the empty set, for any closed term τ.
(9) One can infer $\phi(\rho)$ from $\{\phi(\tau), \tau = \rho\}$.
(10) One can infer $\phi(\tau)$ from $\{(\forall x)\phi(x)\}$.
(11) If the individual constant c does not appear in ϕ or in Γ and one can infer $\phi(c)$ from Γ, then one can infer $(\forall x)\phi(x)$ from Γ.
(12) One can infer $\phi(\psi)$ from $\{(\forall X)\phi(X)\}$.
(13) If the atomic predicate R does not appear in ϕ or in Γ and one can infer $\phi(R)$ from Γ, then one can infer $(\forall X)\phi(X)$ from Γ.

Rules 11–13 are subject to the proviso that there are no collisions of bound variables.

The proof is by a straightforward induction on the complexity of formulas. One case will be enough to convey the idea. let us prove that $\neg_2\phi_2$ is derivable from $\{\neg_1\phi_1\}$:

(i)	$\neg_2\phi_2$ is derivable from $\{\phi_1, \neg_1\phi_1\}$	[by rule 6, applied to '$\neg_1$']
(ii)	ϕ_1 is derivable from $\{\phi_2\}$	[by inductive hypothesis]
(iii)	$\neg_2\phi_2$ is derivable from $\{\phi_2, \neg_1\phi_1\}$	[from (i) and (ii), by rules 1 and 2]
(iv)	$\neg_2\phi_2$ is derivable from $\{\neg_2\phi_2, \neg_1\phi_1\}$	[by rule 1]
(v)	$\neg_2\phi_2$ is derivable from $\{\neg_1\phi_1\}$	[from (iii) and (iv), by rule 7]

This theorem, which was proved by J. H. Harris[17] (for the first-order predicate calculus), is quite a striking result. To see how powerful it is, compare the status it bestows on the connectives and quantifiers with the status of the phrase "the tallest mountain in Africa." The facts of geography determine that there is one and only one mountain in Africa taller than 19,000 feet; this, together with a standard Russellian analysis of the denotation conditions for definite descriptions[18] tells us that there is one and only one entity to which the phase "the tallest mountain in Africa" refers. On the other hand, the geographic facts, together with our practices in using definite descriptions, do not single out any particular land mass x such that the phrase "the tallest mountain in Africa" refers to x. There are innumerable collections of Tanzanian rock and dirt, each differing from the next by a pebble or so, such that each of them is an equally good candidate for what the phrase refers to. We have a case of indeterminacy of reference: The geographic facts and our linguistic practices determine that there is one and only one referent of the phrase (assuming the standard laws of logic and semantics apply to vague terms), yet there is not one and only one thing of which it is determined that it is the referent.[19] The tallest mountain in Africa has a status analogous to that of the winner of an indeterministic lottery: It is determined that there will be one and only one winner, but there is no one of whom it is determined that she will be the winner.[20]

By contrast, the rules of inference do single out a unique best candidate for the referent of '$\neg$'. If there are two, logically inequivalent candidates for what '$\neg$' refers to, one or both of them must fail to satisfy either rule 6 or rule 7. Whichever of them fails to satisfy one of the rules is thereby revealed as an unsatisfactory candidate for what '$\neg$' means.

What the theorem tells us is that, if we have two systems of candidates for what the logical operators mean, both satisfying the rules of inference, then a sentence containing operators from one set and the corresponding sentence with operators from the other set are interderivable. For this observation to be more than a curiosity, we need to know that when two sentences are interderivable, they are logically equivalent. We shall know this if we can be assured of the *soundness condition*: If ϕ is derivable from Γ, then ϕ is a logical consequence of Γ. The theorem does not purport to define the logical consequence relation; the soundness condition constrains the relation, but it does not pin it down. What the theorem does for us is this: Once we have a logical consequence relation and, hence, a logical equivalence relation,[21] satisfying the soundness condition, the theorem assures us that the rules succeed in pinning down the roles of the logical operations uniquely to within logical equivalence.

The proof of the theorem nowhere requires us to consider mixed sentences, sentences that contain connectives from both systems. It does, however, require us to consider mixed deductions. Where $\mathscr{L}_1$ is the set of sentences formed using the first set of operators and $\mathscr{L}_2$ is the set of sentences formed with the second set, we need to examine derivations that apply the rules of inference, both as they were originally formulated within $\mathscr{L}_1$ and as they were originally formulated within $\mathscr{L}_2$, to deduce a sentence of $\mathscr{L}_1$, say, from a set of premisses that includes members of both $\mathscr{L}_1$ and $\mathscr{L}_2$. We need to convince ourselves that, when the field of possible derivations is expanded in this way, the rules of inference are still sound.

The original soundness condition for $\mathscr{L}_1$ required that, whenever a sentence ϕ of $\mathscr{L}_1$ is derivable by the rules of $\mathscr{L}_1$ from a set Γ of sentences of $\mathscr{L}_1$, then ϕ is a logical consequence of Γ. To extend this condition to encompass derivations in $\mathscr{L}_1 \cup \mathscr{L}_2$, we require three things. We need to expand the notion of 'derivation' to encompass the larger language; we likewise need to expand the notion of 'logical consequence'; and we need to assure ourselves that any sentence derivable from a set of sentences Γ of the larger language will be a logical consequence of Γ.

In extending the notion of derivability, the relevant maxim is "Do the same things with sentences of the larger language that you did with the original language," but, as any reader of Wittgenstein will tell you, the application of this maxim is not fully automatic. For a rationally reconstructed account of language acquisition, there is first the problem, even within $\mathscr{L}_1$, of going from seeing which strings of sentences are correct derivations to recognizing the rules that govern the derivations; for it is possible that quite different rules should happen to permit the same derivations. Even after we get the rules for $\mathscr{L}_1$, their extension to the larger language can be problematic, because of such questions as "Which expressions of the enlarged language count as sentences?" and "When can a sentence of the enlarged language be said to contain the constant c?" There are treacherous shoals here, which need to be negotiated carefully.

Given a language $\mathscr{L}_1$, we form the language $\mathscr{L}_2$ by replacing members of one set of logical operations by members of a new set, and we ask when a sentence of the new language implies or is implied by a sentence of the original. If we think of the languages as concrete social institutions, we would not generally expect such a question to have a sensible answer. If one person speaks $\mathscr{L}_1$ and another person speaks $\mathscr{L}_2$, we have no reason to expect that their pronouncements would be commensurable. Stimulus implication[22] is the most we can hope for, in general, and that stops far short of logical implication. Even if both languages belong to a single speaker, we have no assurance that embedding $\mathscr{L}_1$ into the larger language has not altered the semantic roles of some of its words; after all, look at what happens to substitutional quantification when we enlarge the language.

So, it is just as well that we have chosen to think of languages as mathematical abstractions. For abstract languages, the notion of logical consequence will, presumably, follow the general lines of Tarski's definition[23]: ϕ is a logical consequence of Γ if and only if ϕ is true in every model of Γ, although exactly what one means by "model" is up for grabs. Assuming that changing the logical operators does not change the class of models,[24] it will make perfectly good sense to ask whether a sentence of $\mathscr{L}_1$ is true in every model of a given sentence of $\mathscr{L}_2$.

We start with a language $\mathscr{L}_1$, within which we know that, if a sentence ϕ is derivable from a set of sentences Γ, then ϕ is a logical consequence of Γ. We now expand the language to a larger language $\mathscr{L}_1 \cup \mathscr{L}_2$. Assuming that we are indeed able to make sense of the notion of applying the same rules of inference within the larger language that we applied within $\mathscr{L}_1$, and assuming that the notion of logical consequence can be sensibly extended to the larger language, how do we know that, when ϕ is derivable from Γ within the larger language, ϕ is also a logical consequence of Γ, in the sense of 'logical consequence' appropriate to the larger language? How do we know that, after we expand the language, the soundness condition will still hold? Conservativeness is no help; it is a principle about the internal structure of $\mathscr{L}_1$ that tells us nothing about how the language treats newcomers. What we require is that the rules governing the connectives of $\mathscr{L}_1$ be *open-ended*, meaning that the rules are valid (i.e., for each model, if the premises are true in the model, the conclusion is true in the model) not only within the language $\mathscr{L}_1$, but they will remain valid however the language may be enriched by the addition of new sentences.

Our acceptance of the classical rules of inference surely is open-ended. We accept *reductio ad absurdum* because its validity follows from the meaning of classical '$\neg$'. We do not accept it because we have surveyed the forms of expression found in English and found that its expressive power is circumscribed in such a way as to validate the rule. When we introduce a new predicate into the language – say, when we discover a new species or introduce a new product line – we do not have to inquire whether *reductio ad absurdum* remains valid for inferences involving the new predicate.

How to formulate an open-ended rule is a matter of some delicacy. No simple syntactic test is going to tell us when a new locution is to count as a new sentence,[25] for there is no syntactic rule that forbids us from introducing a new atomic sentence, the effect of which, within any sentence it contains, is to exchange the roles of '$\wedge$' and '$\vee$'. If the language is expanded in this way, no one will expect the classical rules of inference to be upheld.

Once again, our abstract conception of language saves us from perplexity. The semantic role of a sentence is determined, uniquely up to logical equivalence, by specifying the models in which the sentence is true. So, the effect of introducing a new sentence is to identify a new class of models, and the effect of open-endedness is to broaden the class of models with respect to which the rule is required to be truth preserving.

To illustrate the effect of open-endedness, let us examine rule 7, and let us say that a class of models $\mathscr{F}$ is *definable* if there is, within our original language, a set of sentences Γ such that $\mathscr{F} = \{\text{models of } \Gamma\}$. The usual way of thinking of rules of language is to apply them only within a single, fixed language, so that the thesis that rule 7 is truth preserving tells us this:

For any definable classes of models $\mathscr{E}$ and $\mathscr{F}$, if $\mathscr{E}$ includes every member of $\mathscr{F}$ in which ϕ is true and $\mathscr{E}$ likewise includes every member of $\mathscr{F}$ in which $\neg\phi$ is true, then $\mathscr{E}$ includes $\mathscr{F}$.

The open-ended acceptance of rule 7 requires that the rule remain truth preserving even when we enrich the language by adding new connectives and quantifiers. Model-theoretically, the effect of introducing new connectives and quantifiers is to permit the definition of hitherto undefinable classes of models. For example, once we introduce second-order quantifiers, we can define the class of models in which the arithmetical symbols describe a system isomorphic to the natural number system, a class not definable using only first-order quantifiers. There are surely limits – though no one knows much about them – on what classes of models it would be possible to define in some psychologically and practically feasible extension of English. Surely, however, there are no limits on which classes of models it is possible, in principle, to define in some mathematically possible language.[26] This being so, the open-ended acceptance of rule 7 gives us this:

For *any* classes of models $\mathscr{E}$ and $\mathscr{F}$, if $\mathscr{E}$ includes every member of $\mathscr{F}$ in which ϕ is true and $\mathscr{E}$ likewise includes every member of $\mathscr{F}$ in which $\neg\phi$ is true, then $\mathscr{E}$ includes $\mathscr{F}$,

which is equivalent to the principle of bivalence:

Either ϕ or $\neg\phi$ is true in each model.

The language-immanent version of rule 7 does not give us bivalence.[27]

We see that the Belnap criteria require quite a bit of machinery for their application, but once the machinery is in place, the criteria do exactly what we want: They assure us that the natural-deduction rules uniquely pin down the semantic role of the connectives and quantifiers. Indeed, one starts to worry that they are doing too much. They give us all we hoped for, a precise fixation of the logical operations of first-order logic; and they go on to give us more than we ever dreamed of, a determination of the logical operations of second-order logic. One starts to fret, recalling the maxim, "If something sounds too good to be true, it probably is." Let me enumerate six sources of concern:

First, to determine the semantic behavior of the first-order quantifiers, we have to fix the range of the variables. One possibility is that the variables range over everything, but the rules of inference will be the same even if their range is less inclusive. So, how are the rules of inference supposed to determine the range of application of the first-order quantifiers?

A similar question arises for the second-order quantifiers, even after we have fixed the range of the first-order quantifiers. For, the rules of inference are compatible with numerous nonstandard interpretations of the second-order quantifiers.

Third, to determine the semantic role of the connectives and quantifiers, we must determine the truth conditions of the sentences containing them. But the only way the rules of inference constrain the assignment of truth values is by way of the soundness condition, and that is not enough to determine the truth values. Even if we know that the rules of inference are truth preserving and we know which atomic sentences are true and which not, that will not be enough to determine the truth values of complex sentences.

Fourth, we know how the rules of inference can make a sentence valid. The rules serve to stipulate, in whole or in part, the meanings of the connectives and quantifiers, since the soundness condition constrains which sentences we are permitted by the rules of the language to count as true, while the conservativeness condition ensures that these stipulations do not come into conflict. Consequently, if a sentence is derivable from the empty set, then the meanings of its constituent logical words are such as to make it true. But, in second-order logic, there are valid sentences that are not derivable from the empty set. What makes them valid?

Fifth, natural deduction is computationally tractable, whereas second-order consequence is wildly intractable. It is startling to imagine that such a fiercely untamable set as the set of second-order validities should be defined by a system of recursive rules.

Finally, natural deduction is compact: If ϕ is derivable from Γ, then ϕ is derivable from a finite subset of Γ; second-order logic is not compact. Finitary rules of deduction somehow fix the meaning of the second-order quantifiers, and thereby fix an infinitary consequence relation. How?

I shall attempt to assuage these worries, as well as I can.

With regard to the first two worries, let me say this: The rules of inference do something quite amazing. They create a uniquely defined semantic role for each of the connectives and quantifiers, where before there was none. But they do not do something magical. They do not create this uniquely determined semantic role out of nothing. Instead, they presuppose the whole apparatus of naming and predication, for the bounds of quantification are fixed by determining to what things our names and predicates are able to refer.

Suppose that our first-order variables are restricted, so that a lies outside their range, and imagine that we are permitted to name anything we like by an individual constant, and we are permitted to refer to any things we like by a predicate. Then we can take c to be a constant referring to a, and we can take F to be a predicate referring to those things that are within the range of our first-order variables. Then, Fc will be false, and $(\forall x)Fx$ will be true, invalidating rule 10. Thus, the only way there can be things outside the range of our first-order variables will be for there to be restrictions on what individuals it is possible to name by an individual constant or on what collections of individuals it is possible to name by a predicate.

Are there such restrictions? We are certainly capable of adopting such restrictions if it suits our purposes. If, on occasion O, we have adopted a temporary rule ensuring that only the things that satisfy the predicate F can be named, then there will not be any models of the language, as it is used on O, in which Fc is false. Fc will be an O-logical consequence of the empty set, and so, in accordance with rule 11, $(\forall x)Fx$ will be an O-logical consequence of the empty set.

To get beyond O-logical consequence to the full notion of logical consequence, we need to look beyond the restrictions specific to a particular occasion. Now surely there are practical limitations on what individuals we are, as a matter of psychological and physical fact, able to name, but for questions of logical consequence, we are not interested in practical limitations. We want to know what restrictions, if any, on what things our names are able to denote are built into the rules of the language.

It is hard to believe there are any such restrictions, for to know how to use a language that had a rule that said that only members of class $\mathscr{C}$ could be named, we would have to be able to distinguish the $\mathscr{C}$s from the non-$\mathscr{C}$s, so we would have at least to be able to think about the non-$\mathscr{C}$s. But if we can think about them, why should we not be able to talk about them? We can make up stories according to which such a thing occurs, say, by proposing that the names and quantifiers of public language are restricted but the names and quantifiers of the language of thought are unrestricted, but no such story is the least bit credible. The only plausible account has it that the rules of language permit us to name anything at all.

It is possible, certainly, to adopt an occasion-specific restriction on what things can be named, but it is also possible to refrain from adopting any such restrictions. (Recall that the contrary thesis that there is always some restriction on what things can be named, and thereby a corresponding restriction on the range of quantification, is a thesis that cannot be coherently advanced.) However, if no occasion-specific restrictions are in effect, the only constraints on what things can be named will be those built into the rules of the language. But there are no such constraints built into the rules of the language. Consequently, except when some peculiarity of the context dictates otherwise, we are free to name anything at all. The rules of inference determine that the quantifiers we use on occasion O range over the things it is permissible to name on occasion O. So, the default value of '$\forall$', the value it takes when no occasion-specific restrictions are in effect, is quantification over everything.[28]

The rules of inference do not determine the range of quantification. What they ensure is that the domain of quantification in a given context includes everything that can be named within that context. This includes even contexts in which there are no restrictions on what can be named. In such contexts, the quantifiers range over everything.

To see how the rules determine Tarski-style truth conditions, let us look at the clauses for negation and universal quantification, which are typical.

Rules 1 through 7 give us a complete axiomatization of the classical sentential calculus. We get an equivalent system if rule 7 is broken into two rules:

(7a) If one can infer '$\bot$' from $\Gamma \cup \{\phi\}$, one can infer $\neg\phi$ from Γ.
(7b) One can infer ϕ from $\{\neg\neg\phi\}$.

If we look at the argument for the uniqueness condition, we find that rule 7a is all we need. The logical system consisting of rules 1 through 6 and 7a satisfies Belnap's criteria, but the system consisting of rules 1 through 6 and 7a is intuitionist logic, not classical logic; it is a system compatible with rule 7b, but it does not entail rule 7b. So, according to the Belnap test, the system of rules for the sentential calculus we get by replacing rule 7 by 7a uniquely pins down the semantic status of the connectives. But how can we say this, when the rules are incapable of determining whether rule 7b is valid, hence incapable of expressing a preference between intuitionist negation and classical negation?[29]

The solution to this problem, I would like to propose, lies in the thesis that the rules of deduction are open-ended. Open-endedness – the thesis that the rules of inference are truth preserving within any mathematically possible extension of the current language – is, you will recall, what we require for Belnap's criteria to give us any reason to suppose that the rules of inference determine the semantic values of the logical operators. What we shall find is that, within the current language, the rules are upheld both for classical negation and for intuitionist negation, but there is some mathematically possible extension of the current language in which intuitionist negation fails to satisfy either rule 6 or rule 7a. For suppose that $\neg\neg\phi$ does not imply ϕ. Then there are some models in which $\neg\neg\phi$ is true but in which ϕ is not true. For any class of models, there is a mathematically possible language in which there is a sentence true in just those models. So there is a mathematically possible language in which there is a sentence ψ true in just those models in which $\neg\neg\phi$ is true and in which ϕ is not true. There is no model in which ϕ and ψ are both true. Hence, '$\bot$' is true in every model in which ϕ and ψ are both true; that is, '$\bot$' is a logical consequence of $\{\phi, \psi\}$. It follows by rule 7a that $\neg\phi$ is a logical consequence of $\{\psi\}$. Also, by definition, $\neg\neg\phi$ is true in every model in which ψ is true; that is, $\neg\neg\phi$ is a logical consequence of $\{\psi\}$. It follows from rule 6 that '$\bot$' is a logical consequence of $\{\psi\}$. Assuming that there is no model in which every sentence is true, we conclude that there is no model in which ψ is true, and so ϕ is a logical consequence of $\{\neg\neg\phi\}$, after all.

The intuitionist will not believe a word of this, of course. The intuitionist will not accept the classical logician's characterization of logical consequence in terms of truth in every model; he will not condone the classical logician's faith in the existence of wholly mind-independent mathematical models; and he will not sanction the inference from "There does not exist a model in which $\neg\neg\phi$ is true and ϕ is not true" to "ϕ is true in every model in which $\neg\neg\phi$ is true." But our aim here is not to convert the intuitionist; it is to provide the classical mathematician with an internally coherent account of the acquisition of logical terms.[30]

The open-endedness of the rules gives us the classical truth condition for negation, namely,

$\neg\phi$ is true in a model if and only if ϕ is not true in that model.

That ϕ and $\neg\phi$ are not both true in any model follows immediately from rule 6, given our assumption that there is no model in which every sentence is true. To see that one or the other of them must be true in each model, introduce a new sentence θ, true in just those models in which neither ϕ nor $\neg\phi$ is true. Then, there are no models in which θ and ϕ are both true, so that '$\perp$' is a logical consequence of $\{\theta, \phi\}$. It follows by rule 7a that $\neg\phi$ is a logical consequence of $\{\psi\}$. Likewise, there are no models in which θ and $\neg\phi$ are both true, so that '$\perp$' is a logical consequence of $\{\theta, \neg\phi\}$ and, by rule 7a, $\neg\neg\phi$ is a logical consequence of $\{\theta\}$. Since both $\neg\phi$ and $\neg\neg\phi$ are logical consequences of $\{\theta\}$ and, by rule 6, '$\perp$' is a logical consequence of $\{\neg\phi, \neg\neg\phi\}$, '$\perp$' is a logical consequence of $\{\theta\}$. Thus there are no models in which θ is true, and so, in every model, either ϕ is true or $\neg\phi$ is true.

To extend this development to get the truth conditions for quantified sentences, we need to think about how to utilize the proviso "the individual constant c does not appear in ϕ or in Γ" in rule 11 and the analogous condition in rule 13. Let us say that two models $\mathfrak{A}$ and $\mathfrak{B}$ are *c-variants* if they are just alike, except perhaps in the value they assign to c. If c does not appear in the sentence ψ, then whether ψ is true in $\mathfrak{A}$ should not depend upon what value $\mathfrak{A}$ assigns to c, so that, if $\mathfrak{A}$ and $\mathfrak{B}$ are c-variants and ψ is true in one of them, it is true in both. Thus we assume that, if the constant c does not occur in ψ, then the class of models in which ψ is true is closed under c-variants, and, conversely, that, if a class of models is closed under c-variants, then there is, in some mathematically permissible language, a sentence not containing c that is true in all and only the members of the class.

Using this assumption, we get the classical truth condition for the first-order universal quantifier, namely,

If c does not appear in ϕ, then $(\forall x)\phi(x)$ is true in $\mathfrak{A}$ if and only if $\phi(c)$ is true in every c-variant of $\mathfrak{A}$.[31]

If $(\forall x)\phi(x)$ is true in $\mathfrak{A}$, then, since c does not appear in $(\forall x)\phi(x)$, $(\forall x)\phi(x)$ is true in every c-variant of $\mathfrak{A}$. Since $\phi(c)$ is a logical consequence of $\{(\forall x)\phi(x)\}$, it follows that $\phi(c)$ is true in every c-variant of $\mathfrak{A}$.

Conversely, suppose that $\phi(c)$ is true in every c-variant of $\mathfrak{A}$. Let ψ be a sentence, not containing c, that is true in all and only the c-variants of $\mathfrak{A}$. Hence $\phi(c)$ is true in all the models of ψ, so that $\phi(c)$ is a logical consequence of $\{\psi\}$, and hence, by rule 11, $(\forall x)\phi(x)$ is a logical consequence of $\{\psi\}$. Since ψ is true in $\mathfrak{A}$, it follows that $(\forall x)\phi(x)$ is true in $\mathfrak{A}$.

Once we have truth conditions, there is room for a disparity between derivability and logical consequence. For a garden-variety empirical statement, we do not expect the conventions of our language to determine whether the statement is true. We only expect linguistic conventions to determine the conditions under which the statement is true. To go on to determine the truth value of the statement, we have to go beyond language; we have to consult the empirical facts. There is no evident reason why the situation should be different with respect to statements about logic. The deductive rules determine the conditions under which a sentence is true in a model, but to determine whether the sentence is, in fact, true in a particular model, we have to look at the model. To determine whether the sentence is true in all the models – that is, valid – we have to look at all the models. There is the difference that, for a statement about logic, the relevant facts are mathematical facts, whereas, for an empirical statement, the relevant facts are empirical, but in both cases, the assessment of truth values requires us to look beyond the conventions of language.[32]

Within the first-order predicate calculus, deducibility and logical consequence coincide, but there is no reason to suppose that this comforting correlation will continue when we move to less familiar logical regions. Deducibility is one thing, logical consequence another. So, it is no great surprise that a sentence can be valid without being derivable from the empty set.

The rules of inference determine the truth conditions. The truth conditions, together with the mathematical facts, determine the logical consequence relation. The rules of inference are simple, but the mathematical facts are complex. So it should not be a cause of alarm that derivability is recursively enumerable but logical consequence is not.

As for compactness, we appeal again to open-endedness. For any set of sentences Γ, there is a mathematically possible language in which there is a sentence γ that is true in all and only the models in which all the members of Γ are true. Then ϕ is a logical consequence of Γ if and only if ϕ is a logical consequence of $\{\gamma\}$.

Returning briefly to my personal starting point, which is Parsons' discussion of the status of quantification in set theory, we find that, even if we have succeeded in specifying such a thing as no-holds-barred universal quantification, we have not yet succeeded in pinning down quantification in set theory. To pin down the truth values of set-theoretic statements, we have to determine two things: the extension of the predicate '$\in$', and the range of the quantifier 'for every set x'.[33] When doing set theory, we do not talk about everything, we only talk about sets. If we have unrestricted universal quantification, we can break the set-theoretic quantifier 'for every set x' down into simpler parts. Instead of saying 'For every set x, $\phi(x)$', we now can say 'For every x, [Set(x) $\rightarrow \phi(x)$]'; but we still have to fix the extension of the predicate 'Set'. Perhaps we have made some headway. Now we have to pin down two things of the same logical type, the predicates 'Set' and '$\in$', whereas before, we had to pin down a predicate and a quantifier. Our new problem is an instance of the question how a theory determines the meanings of its theoretical terms, which is perhaps less unwieldy than the question how an uninterpreted calculus determines the meanings of both its theoretical terms and it logical operators. So we may have made progress toward a solution, but we have certainly not achieved a solution.[34]

In the interest in truth in labeling, I emphasize how very unsteady the metaphysical underpinnings are for our proposed solution to the problem of defining unrestricted universal quantification. Of course, there is the presumption of classical mathematics. Also, I have presumed that second-order quantification is truly logic, rather than, as Quine[35] puts in, the theory of classes in sheep's clothing. If second-order quantification is nothing more than a disguised notation for first-order quantification over classes, then we cannot employ second-order quantification in contexts in which our first-order quantifiers range over everything – everything, that is, including all the classes – without falling foul of Russell's paradox. If we repudiate second-order logic, translating statements in which we hitherto employed second-order quantification into an explicitly first-order notation, we avoid all the puzzles that arose with the use of natural-deduction rules to define the operations of second-order logic. We are likely to find, however, than many of the same puzzles confront us, albeit in somewhat different guises, if we try in any other way to extend the reach of logic beyond the bare first-order predicate calculus.

There is a much more worrisome difficulty lurking in the shadows. We have talked about "truth in a model," but we have not said much of anything about what a model is or how truth in a model is to be defined. If we define truth in a model in the most natural way, the way Tarski defined it,[36] we find that, even for first-order sentences, truth in a model is a third-order notion.[37] If all this implied were the utilization of third-order quantifiers, there would be no cause for alarm. We could introduce third-order quantifiers via natural deduction rules, just as we did first- and second-order quantifiers. The real difficulty is that third-order logic requires a new grammatical category, predicates that take predicates as arguments.[38] One is naturally reluctant to make such a big conceptual leap.

There are three alternatives. We can accept higher-order logic at face value. We can renounce higher-order logic and all the ideas that go with it. Or we can try to economize, by using codings, reflection principles, and other such devices to attempt to achieve the effects of higher-order logic without unwelcome metaphysical commitments. Which of these paths we ought to take is a difficult question, requiring deep study.

Here is the curious thing: To evaluate the status of higher-order logic requires an *ontological* investigation. We set out to develop the meta-theory required for ontology, and we find that, to properly develop the meta-theory, we need to have already completed extensive ontological research. We are trapped.

Aristotle had it right. Ontology is first science. There is no prior science of meta-ontology. Instead, ontology and the meta-theory of ontology – which is, roughly speaking, logic – are both first science, equal partners in a complex relationship.

We start with tentative, unexamined ontological hypotheses. On the basis of these assumptions, we develop a meta-theory that we can use to go back and reexamine our ontological hypotheses. Having reassessed our initial ontological hypotheses, we have to reevaluate our meta-theory and, consequently, once again to revise our ontology. It is to be hoped that this dialectical process will, in the long run, bring us, however erratically, closer to truth.

NOTES

I read this paper at the Philosophy Colloquium at New York University, where I got a lot of useful comments. I thank George Boolos, Iris Einheuser, Hartry Field, Arnold Koslow, Shaughan Lavine, Brian Loar, Barry Loewer, Brian McLaughlin, and Peter Milne for their help.

1. *Metaphysics* Γ.
2. "Über Grenzzahlen und Mengenberreiche," *Fundamenta Mathematicae*, 16 (1930): 29–47; and "Über Stufen der Quantifikation und die Logik des Unendlichen," *Jahresbericht der Deutschen Mathematiker-Vereinigung (Angelegenheiten)*, 41(1932): 85–8. See also Gregory Moore, "Beyond First-Order Logic: The Historical Interplay Between Mathematical Logic and Axiomatic Set Theory," *History and Philosophy of Logic*, 1(1980): 95–137; and Shaughan Lavine, *Understanding the Infinite* (Cambridge, MA: Harvard University Press, 1994), pp. 134–41.
3. "Sets and Classes," *Nous* 8 (1974): 1–12, reprinted in Parsons' *Philosophy in Mathematics* (Ithaca, NY: Cornell University Press, 1983), pp. 209–20; "The Liar Paradox," *Journal of Philosophical Logic*, 3 (1974): 381–412, reprinted in Parsons' *Philosophy in Mathematics*, pp. 221–67, and in Robert L. Martin (ed.), *Recent Essays on Truth and the Liar Paradox* (Oxford and New York: Oxford University Press, 1984), pp. 9–45; and "What Is the Iterative Conception of Set?" in R. E. Butts and Jaakko Hintikka (eds.), *Logic, Foundations of Mathematics, and Computability Theory* (Dordrecht, The Netherlands: Reidel, 1977), pp. 335–67, reprinted in Parsons' *Philosophy in Mathematics*, pp. 268–97, and in Paul Benacerraf and Hilary Putnam (eds.), *Philosophy of Mathematics*, 2nd Ed. (Cambridge, UK: Cambridge University Press, 1983), pp. 503–29. A closely related view is developed by Tyler Burge, "Semantic Paradox," *Journal of Philosophy*, 76 (1979): 169–98, reprinted in Martin (ed.), *Recent Essays on Truth*, pp. 83–117.
4. Here is an argument that I have sometimes heard: We understand universal quantification by learning the truth conditions for univerally quantified statements, but to give the truth conditions requires a meta-language more extensive in what it talks about than the object language. So, the domain of the object language cannot include everything. The argument was answered satisfactorily by Quine, "Truth by Convention," in Otis H. Lee (ed.), *Philosophical Essays for A. N. Whitehead* (New York: Longmans, Green, 1936), pp. 90–124, reprinted in Benacerraf and Putnam (eds.), *Philosophy of Mathematics*, pp. 329–54). If learning to employ the logical operators indeed required us to be able to express the truth conditions for the sentences containing them, we would be caught in a vicious regress, since, before we

could learn the logical vocabulary of the object language, we would first have had to have learned the logical vocabulary of the meta-language.

5. Later, we shall expand our discussion to include second-order quantification. The method for discerning second-order quantification in English was discovered by George Boolos. See his "To Be Is to Be a Value of a Variable (Or to Be Some Values of Some Variable)," *Journal of Philosophy*, 81 (1984): 430–49, and "Nominalistic Platonism," *Philosophical Review*, 44 (1985): 327–44. Both articles are reprinted in Boolos, *Logic, Logic, and Logic* (Cambridge, MA, and London: Harvard University Press, 1998), pp. 54–72 and 73–87, respectively.
6. "Ontological Relativity," *Journal of Philosophy*, 65 (1968): 185–212, reprinted in Quine's *Ontological Relativity and Other Essays* (New York: Columbia University Press, 1969), pp. 26–68. See pp. 63–7 of the reprint.
7. *The Basic Laws of Arithmetic*, 2nd Ed. (Evanston, II: Northwestern University Press, 1978), German text with English translation by J. L. Austin, p. x.
8. Assuming that the sentence is not about language.
9. For a strict linguistic behaviorist such as Quine, the thoughts of the members of the speech community will not matter, or, rather, thoughts will not matter apart from their manifestation in behavior. For the present discussion, this does not matter. Something that does matter is that, in trying to determine meanings, even a strict behaviorist should consider not only actual verbal behavior but also dispositions to verbal behavior, dispositions that show us how members of the speech community would respond to various counterfactual circumstances. Determining those dispositions is a difficult matter, but it is a familiar difficulty; it is the problem of assessing the solubility of a thing that is never placed in water.
10. "Die Grundlagen der Mathematik," *Abhandlungen aus dem mathematischen Seminar der Hamburgischen Universität*, 6 (1928): 65–85; English translation by Stefan Bauer-Mengelberg and Dagfinn Føllesdal in Jean van Heijenoort (ed.), *From Frege to Gödel* (Cambridge, MA: Harvard University Press, 1967), pp. 464–79. The ε-operator is introduced on p. 466 of the translation.
11. In determinations of meaning, the relevant notion of possibility is surely logical possibility. Our concept of gold and our practices in using the word 'gold' are such that we can understand what it would be like for 'Gold is the element with atomic number seventy-nine' to be false, even if, as a matter of extraconceptual scientific fact, the sentence is metaphysically necessarily true. (The use of logical rather than metaphysical necessity might make a difference to the question whether talk of possible worlds is to be understood literally or metaphorically.)
12. *Journal of Symbolic Logic*, 45 (1980): 464–82; reprinted in Benacerraf and Putnam (eds.), *Philosophy of Mathematics*, pp. 421–44. Putnam's argument is built upon the foundation laid by Thoralf Skolem. See "Einige Bemerlungen zur axiomatischen Begründung der Mengenlehre," *Matematikerkongressen i Helsingfors den 4-7 Juli 1922, Den femte skandinaviska mathematikerkongressen, Redogörelse* (Helsinki: Akademiska Bokhandeln, 1923), pp. 217–32; English translation by Stefan Bauer-Mengelberg in van Heijenoort (ed.), *From Frege to Gödel*, pp. 290–301.
13. It could turn out that where the action is isn't in the acquisition of public language, but rather at the level of the language of thought, because we have something like the following situation: When we try to correlate public-language symbols with symbols of the language of thought, we find that, given our computational limitations, the only way we could use '$\forall$' as we do would be for '$\forall$' to correspond to a simple operation in the language of thought. Moreover, when we look at the simple operations in the

language of thought, we find only one such operation whose action at all resembles our usage of '∀'. In such a situation, it is no wonder that youthful speakers all settle on the same usage of '∀'; nature has provided them with such a limited menu of choices that they have no real alternative. In this situation, the philosophically interesting question is this: What is it about the neural computational role of this simple symbol of the language of thought that makes it universal quantification, rather than *S*-quantification or something else?

14. The general thesis that the meanings of the logical terms are given by their position in the rules of inference is developed by Gerhard Gentzen; see his *Collected Papers* (Amsterdam: North–Holland, 1969), *passim*. The program is given an especially graceful systematic development in Arnold Koslow's *A Structuralist Theory of Logic* (Cambridge, UK: Cambridge University Press, 1992).
15. *Analysis*, 22 (1962): 130–4, reprinted in P. F. Strawson (ed.), *Philosophical Logic* (Oxford: Oxford University Press, 1967), pp. 132–7. Belnap's paper is a response to a puzzle posed by Arthur Prior, "The Runaway Inference Ticket," *Analysis*, 21 (1961): 38–9, reprinted in Strawson (ed.), pp. 129–31.
16. This distinction is developed in David Lewis's "Language and Languages" in Keith Gunderson (ed.), *Minnesota Studies in the Philosophy of Science*, Vol. 7 (Minneapolis: University of Minnesota Press, 1975), pp. 3–35, reprinted in Lewis's *Philosophical Papers*, Vol. 1 (New York and Oxford: Oxford University Press, 1983), pp. 163–88.
17. "What's So Logical About the 'Logical' Axioms?" *Studia Logica*, 41 (1982): 159–71.
18. "On Denoting," *Mind* 14(1905): 479–93, reprinted in Russell's *Logic and Knowledge* (London: Allen and Unwin, 1956), pp. 39–56.
19. The example is discussed in some detail in my "'Kilimanjaro'," in A. Kazmi (ed.), "Meaning and Reference," *Canadian Journal of Philosophy* supplementary 23 (1997): 141–198.
20. This analogy is discussed on pp. 221f. of Vann McGee and Brian McLaughlin, "Distinctions Without a Difference," *Southern Journal of Philosophy*, 33 (Suppl., 1995): 203–51.
21. That is, ϕ and ψ are logically equivalent if and only if ψ is a logical consequence of $\{\phi\}$ and ϕ is a logical consequence of $\{\psi\}$.
22. See Quine, *Word and Object* (Cambridge, MA: MIT Press, 1960), Ch. 2.
23. "Über den Begriff der logischen Folgerung," *Actes du Congrés International de Philosopie Scientifique*, 7 (1936): 1–11; English translation by J. H. Woodger in Tarski's *Logic, Semantics, Metamathematics*, 2nd Ed. (Indianapolis: Hackett, 1983), pp. 409–20.
24. Most, but not all, of the connectives and quantifiers we encounter in practice meet this condition. One that does not is H. J. Keisler's probability quantifier; see his "Probability Quantifiers" in Jon Barwise and Solomon Feferman (eds.) *Model Theoretic Logics* (New York and Berlin: Springer–Verlag, 1985), pp. 509–56. To introduce it, one needs to graft a measure onto the model. It is not altogether clear that this ought to count as a *logical* operator. Tarski, in "What Are Logical Notions?" *History and Philosophy of Logic*, 7 (1986): 143–54 (this is the posthumously published text of a 1966 lecture, edited by John Corcoran) and F. I. Mautner, in "An Extension of Klein's Erlanger Program: Logic as Invariant Theory," *American Journal of Mathematics*, 68 (1946): 345–84, recommend that an operator be counted as logical only if it is invariant under arbitrary permutations of the universe. By this standard, the probability quantifier would not count as logical. Cf. Gila Sher, *The Bounds of Logic*

(Cambridge, MA: MIT Press, 1993); and McGee, "Logical Operations," *Journal of Philosophical Logic*, 25 (1996): 567–80.

25. Following logicians' somewhat sloppy custom, I use 'sentence' to refer to those sentences used to make assertions; so, for present purposes, orders and questions do not count as sentences.
26. One way to define a class of models is to associate with each model a state description that completely describes the model, and then to define $\mathscr{C}$ as the class of models of the disjunction of the state descriptions of members of $\mathscr{C}$. (If it should happen that there are two models that satisfy the same state description, this procedure will permit us to define only classes that group them together; for present purposes; this limitation does no harm.)
27. The easiest way to see this is in terms of supervaluations, as described in Bas van Fraassen, "Singular Terms, Truth Value Gaps, and Free Logic," *Journal of Philosophy*, 63 (1966): 464–95. If a sentence is counted ture if it is true in every member of a class of acceptable models (which do not all make precisely the same sentences true), then the laws of second-order logic will be truth preserving, yet bivalence will fail. The reason is that the set of models in which a given sentence is either true or false need not be definable.
28. The picture we have in mind is that restricted quantification is gotten by taking unrestricted quantifiers and restricting them, so that, if, on occasion O, '$\forall$' means 'for all human beings', then on occasion O the sentence '$(\forall x)x$ is mortal' expresses the thought that would be expressed in neutral contexts by saying '$(\forall x)(x$ is a human being $\rightarrow x$ is mortal)'. An alternative picture has it that relative quantification is logically primitive, so that the locution 'All____s are____s' cannot be broken down into anything simpler. The advantage of the alternative account is that it generalizes straightforwardly to other forms of quantification in English. See Jon Barwise and Robin Cooper, "Generalized Quantifiers and Natural Language," *Linguistics and Philosophy*, 4 (1981): 159–219. For present purposes, it does not matter which picture we prefer, since on the alternative picture, unrestricted quantification is what we get when we let the first argument be something vacuous, like 'x is self-identical'.
29. This problem has been widely discussed in the literature, starting with Harris's "What's So Logical?" (see note 17), though generally from a somewhat different perspective from the one we are developing here. See, among others, Kosta Došen and Peter Schroeder-Heister, "Conservativeness and Uniqueness," *Theoria*, 51 (1985): 159–73; Michael A. E. Dummett, *The Logical Basis of Metaphysics* (Cambridge, MA: Harvard University Press, 1991), pp. 245–64; Michael Hand, "Negations in Conflict," *Erkenntnis*, 38 (1993): 115–29; Arnold Koslow, *Structuralist Theory of Knowledge*; Peter Milne, "Classical Harmony: Rules of Inference and the Meaning of the Logical Constants," *Synthese*, 100 (1994): 49–94; Peter Milne, "The Uniqueness of Negation," forthcoming; and Timothy Williamson, "Equivocation and Existence," *Proceedings of the Aristotelian Society*, 87 (1987–88): 109–27.

    Here I am taking classical mathematics for granted, and using classical mathematics to give an account of the acquisition of classical logic. More ambitious than I, the other authors assume only such mathematics as is common ground to classical and intuitionist mathematicians, and they use the Belnap criteria to try to arbitrate between the classical and intuitionist positions.
30. As I see it, the principal conclusion about the connection between classical and intuitionistic logic to be drawn from the "Tonk, Plonk, and Plink" literature is that the two schools not only have conflicting ways of understanding the connectives and quantifiers but conflicting notions of logical consequence. We cannot graft the

intuitionistic way of employing the connectives onto the classical notion of consequence or graft the classical way of employing the connectives onto the intuitionistic notion of consequence; if we attempt to do so, the differences between them completely collapse.

31. This device for giving truth conditions, which enables us to provide the truth conditions without making a detour through satisfaction conditions, is due to Benson Mates. See his *Elementary Logic*, 2nd Ed. (New York: Oxford University Press, 1982).
32. For the logicist, this is a disturbing conclusion, since the logicist does not acknowledge any basis for mathematical truth other than the conventions of languages. The mathematical realist, however, is unperturbed.
33. Or, perhaps, 'for every pure set x'.
34. A little more progress is made, or so, at least, I would like to hope, in my "How We Learn Mathematical Language," *Philosophical Review*, 106 (1997): 35–68.
35. *Philosophy of Logic*, 2nd Ed. (Cambridge, MA: Harvard University Press, 1986), p. 66.
36. "What are Logical Notions?" (see note 24).
37. The simplest way to think about models is to treat them as functions assigning an entity of appropriate type to each nonlogical symbol of the language. Assuming that the nonlogical symbols are of the sorts familiar from the predicate calculus – individual constants, predicates, and function signs – we can encode a model as a relation pairing each nonlogical symbol with whatever individual, individuals, or finite sequences of individuals it designates. Truth in a model is thus a binary relation between first-order objects (sentences) and second-order objects (models).

    In treating the domain of quantification as something that remains constant from one model to another, we are following the proposal of Tarski's 1936 paper. We can get the effect of the more familiar approach, which allows for a variable domain of discourse, by explicitly relativizing all quantifiers.
38. The grammatical category is not wholly new. As Frege realized, quantifiers can naturally be regarded as predicates of predicates. See §24 of *Grundgesetze der Arithmetik*, Vol. 1 (Jena: Verlag Hermann Pohle, 1893); English translation by Montgomery Furth, *The Basic Laws of Arithmetic* (Berkeley and Los Angeles: University of California Press, 1967), pp. 78f. Even so, it requires a substantial renovation to take a grammatical position previously occupied only by the quantifiers and open it up for occupancy by quantifiable variables.

On Second-Order Logic and Natural Language

JAMES HIGGINBOTHAM

I. Introduction

I explore some considerations on behalf of, and against, second-order logic that take for at least part of their motivation the properties of natural languages. A couple of preliminary remarks are appropriate here.

First, there is the question of whether second-order logic really is logic. Suppose that logic is understood in a traditional way, as the most general theory of the true and the false, abstracting from the subject matter of the special sciences, but applicable to all of them. The traditional conception certainly includes all of the logic of the truth functions – that much of logic arises as soon as we have distinguished truth from falsehood – but it is not altogether trivial to arrive at the conclusion that even first-order logic is logic. The reason is that, besides truth and falsehood, first-order logic requires the additional notion of satisfaction, or a predicate's being true *of* an object, hence two additional concepts, that of 'truth of' and that of an object. However, if all first-order logic is admitted as logic, then second-order logic appears at first to require, at most, only one further additional concept, namely, that of the value of a predicate variable. Supposing this concept acceptable, second-order logic would have the required degree of generality and topic neutrality: it is not biased in such a way as to exclude (except, perhaps, by failing to contain enough logical resources) any special science from being expounded in a language of which it is the logic. Thus, I will suppose that, if the notions additional to second-order logic can be motivated, then second-order logic qualifies as logic in at least as strict a sense as first-order logic with identity.

Preliminary notice must also be taken of the concept of "natural language" that will be at stake in what follows. A natural language, as I understand it, is not an historically given language, as opposed to an artfully or artificially created one, but rather is a language that is natively available to us for acquisition and use as a first language, under normal environmental conditions. Many natural languages in this sense of the term are no longer spoken, some have not yet been spoken but will be, and yet others will never be spoken by anyone. To look at the matter platonistically, natural languages are a proper subset of possible languages, and a proper superset of existing human first languages.

If natural languages are identified with historically given modes of human communication, then they will include mechanisms that do not figure in human first languages at all, an obvious example being the use of small Roman and Greek letters to mark places of quantification. Of course, these and similar pieces of mathematical and other general scientific idioms are in part artificial, having been, like musical notation, self-consciously created. However, it would be wrong to regard them as wholly artificial to *us*, who stand downstream from refinements and adjustments by many people over many years. If we think of our language as taking these historical elements on board, on the ground that we are educated to use them freely, then we shall have obliterated the distinction that I wish to make, between motivations for second-order logic that can be recovered from the design of human first languages and motivations that stem from our general practices of thinking, speaking, and writing. From the perspective that I am taking, however, it is possible that a good case for second-order logic can be made out by appealing to the additional practice, even if it cannot be launched entirely from our clarification of the semantics of the most basic parts of language.

Much of the literature on the problem of defending or advancing second-order logic concentrates first on the case for adding monadic predicate variables in predicate position, thus not including predicate variables of more than one place or even monadic predicates as arguments. I will follow this practice for the most part, noting, however, that if extensions beyond the monadic case are more difficult to motivate, the reasons may be essential or adventitious; that is, they may reflect limitations on the expressive power of natural languages, or what we might call syntactic accidents that preclude the well-formedness of what would otherwise be the most intuitive constructions. Our questions are naturally related to questions about the motivation for property- and relation-abstraction from within natural language, and whether properties and relations should be taken as essentially predicational entities or as objects of some sort; but I will not consider these issues here.[1]

II. Elementary Linguistic Motivations

Contemporary native speakers of the language in which this article is written will have been introduced to the explicit treatment of generality by means of the systematic discussion of the expressions 'all' and 'some' and related words, appearing in construction with possibly complex nominals. Not only these, but also the quantificational adverbs 'always' and 'sometimes' and related words, can be used for the purpose, and in some human languages these are far more common than the nominal quantifiers. Abstracting from the grammatical difference between 'All men are mortal' and Russell's 'Men are always mortal', we have the restricted quantification 'for every x such that x is a man, x is mortal' or the equivalent unrestricted version 'for every x, if x is a man, then x is mortal.' Now we find these quantifiers, as well as (analogues of) the nonstandard quantifiers, only some of which are first-order definable, also in predicate position, as in, for example

(1) John is everything we wanted him to be.

We also find inferences with predicational quantifiers paralleling inferences with objectual quantifiers, as in

(2) John is mostly what we expected him to be. The only things we expected John to be are: honest, polite, and scholarly. Therefore, John is either honest or polite.

A systematic treatment of these inferences will call for explicit variable-binding, but this time, of predicate positions, on a par with ordinary quantification over the argument positions of predicates.

Examples such as (1) and (2), however, do not show much. The argument (2) would readily be graspable, and correct, whether the quantification were truly second-order or merely projected by analogy onto the category of adjective phrases from that of the argument categories, and so in particular if it were substitutional. The quantifier words are generally restricted to certain constructions (as we had to use the dummy noun 'thing' in the second premise of (2)), but the '*wh*' expressions, which function at least in a quasi-quantificational manner in questions, range over more categories, including prepositional phrases of manner ("How did John fix the car?" "With a wrench"), quantificational adverbs ("How often does John walk to work?" "Rarely"), and others. These will correspond to quantifications using dummy nouns when the questions are embedded, and numerical words are added. So, we have

(3) I know three ways John could fix the car,
(4) Mary wondered with what frequency John walked to work,

and so forth. Once we are launched into a second-order semantic framework for natural language, we may (and many do, following especially the lead of Richard Montague) take these constructions as exhibiting the possibility of ascending to higher simple types. Thus, if prepositional phrases are adverbials in the sense of categorial grammar, belonging to a category that makes one-place predicates from one-place predicates, then in the extensional setting their model-theoretic semantics will have them taking as values functions from sets of individuals to sets of individuals; and if expressions such as 'rarely', 'more often than Bill does', and the like are adverbs of quantification, having for their values sets of individuals, then abstraction over such adverbial prepositional phrases as 'with what frequency' in example (4) will give predicates true of sets of sets of individuals. However, since we are concerned with the question of whether any ostensible form of quantification over other than argument positions of ordinary predicates (nouns, verbs, adjectives, and prepositions) can motivate the higher-order perspective in the first place in anything like the serious way reflected in the type-theoretic hierarchy, we have as yet no reason to subscribe to the extension.

It is a natural conjecture that, in the first instance, quantification over other than ordinary argument positions is substitutional. Of course, we intend these quantifications, for example, over positions of one-place predicates, not to be restricted to the substitution class that we actually have available at this stage of our language, but also to include all predicates possible for us, and perhaps coextensive with actual predicates for speakers of other languages. The point would count against the substitutional treatment if we were thinking of *defining* truth for statements of the class.[2] If what we want to do, however, is simply to use the truth conditions of statements as part of the clarification of their meaning, then nothing prevents us from saying that we intend, as new predicates are added to our language, to enlarge the substitution class so as to include them. We think of ourselves as being prepared to "go on as before," and we can even explain what going on as before means to us. A similar response is available to the worry that, if logical laws are stated by means of "semantic ascent" rather than higher-order quantification, then the theorist is prisoner of whatever language is in question at the time, though in this case one may note that the laws that hold in our language will provably hold also in extensions of it. Hilary Putnam once argued that, when one asserted the validity of, say, $\ulcorner(\mathbf{A}x)[F(x)\vee\neg F(x)]\urcorner$ and the like, one was "implicitly making second-order assertions" (Putnam 1971, p. 31). However, a demonstration is lacking that we must intend by the assertion more than, "No matter what may be put for 'F', $\ulcorner(\mathbf{A}x)[F(x)\vee\neg F(x)]\urcorner$ is true." On the other hand, to have escaped the problem of the values of predicate variables and their intended range by going substitutional yields only a very weak theory.[3]

If there are linguistic considerations in favor of second-order logic – such as examples (1) and (2), and the suspicion that the substitutional interpretation may not do justice to what we intend – there are also linguistic considerations against it. Whereas it is easy to construct ordinary English (or other natural language) sentences instantiating quite arbitrary schemata of first-order logic, the second-order instances that present themselves seem to be very restricted in their nature, generally speaking, taking the form

$$[QX]\ldots X\ldots,$$

where Q is a quantifier, and replacement of X by a predicate constant F would give merely a first-order schema. A particular difficulty that presents itself is that of giving second-order definite descriptions

(5) the X such that ... X ...

in subject position, where the description is rounded off by a common noun in construction with a relative clause. The problem is partly that significant common nouns are true of objects, and so, one must use a dummy noun such as 'thing' in place of a substantive head, and partly that ordinary predicates are likewise true of objects. Notoriously, Frege, having stated that the reference of a predicate-expression was a concept, found that one could not felicitously use this noun as a predicate of a definite description of a concept. Frege laid the problem to a difficulty "in which language finds itself"; urged that, at any rate, the problem did not apply within *Begriffschrift*; and laid the difficulty to the definite article *the* (or *der*), which "points to an object." As I have argued elsewhere (Higginbotham and Schein 1989, Higginbotham 1990), Frege appears to have been mistaken in the last assertion, since the definite article (like any quantifier) can function perfectly well in construction with second-order variables. The difficulty, rather, is that ordinary predicates demand *saturated* arguments, whereas the definite description schema (5) is unsaturated. Being unsaturated, it has instances that can function as predicates, as in

(6) John is [the very thing we expected him to become].

The trouble is that these descriptions generally are not acceptable as arguments. One can, however, agree with Frege otherwise: there is no difficulty in the predicating of concepts in *Begriffschrift*, and Frege would be justified from his own point of view in taking as adventitious the design feature of natural language that forces arguments generally to be saturated. I conclude, therefore, that second-order logic cannot just be written off on the score of such limitations.

The preceding discussion, if correct, leaves the problem of motivating or dismissing the promotion of second-order logic within natural language almost exactly where it was. To promote second-order logic requires showing that a substitutional interpretation of apparent quantification over predicate positions is too weak, and that the limitations in constructing instances of second-order schemata are adventitious. On the other hand, second-order logic cannot be dismissed out of hand.

III. Values of Variables

I said above that second-order logic would qualify as logic if one could get over the hump of identifying the values of predicate variables in an appropriate way, legitimating the additional concept. It has often been suggested that the additional concept puts second-order logic at least out of logic if not out of court. The latter conclusion would follow from a harsh reading of W. V. Quine's notorious slogan, that second-order logic is "set theory in sheep's clothing" (Quine 1970, p. 66). A classic rebuttal, from George Boolos (1975), is that there is simply no necessity to see the values of predicate variables (or constants) as sets or classes in disguise; no necessity, that is, to see them as surrogate names. Quine's objection can recur, however, in derived form. Thus Jody Azzouni (1994) asks in recent work how, if at all, second-order theories with standard models differ from certain first-order translations of them in terms of classes and membership, where the translation is accompanied by a simultaneous restriction on the class of models. He writes that

> the notation '*Pa*', which is a one-place predicate symbol syntactically concatenated with a constant symbol, is not taken to contain an (implicit) representation of ϵ.

> However, as soon as we allow ourselves to quantify (standardly) into the predicate position, this is precisely how syntactic concatenation *must* be understood. Furthermore, . . . , syntactic concatenation in these contexts is *not* open to reinterpretation – it is an (implicit) logical constant. (Azzouni 1994, p. 17)

The concept of concatenation figures in syntax, since the clauses that build formulas use it. If it has any semantic interpretation, then that interpretation is uniform across models. However, it does not follow that concatenation has any semantic interpretation. Azzouni's objection, however, is that the notion of a standard model *for* a second-order language stipulates that the monadic predicate variables are to range over *all* families of objects of the individual universe of discourse.[4] The stipulation is essential, because second-order logic is not recursively axiomatizable, and for recursively axiomatized fragments of it there will be a class of "generalized" models of the type due to Leon Henkin (1950), such that exactly the theorems of the fragment are true in every model of the class; these models are not standard. Confined as we may be said to be to recursively axiomatized systems, we cannot implicate as it were within the second-order language that the models to be considered for evaluating logical truth are just the standard ones. Now, we could in a way implicate this if we took concatenation as suppressing a silent ϵ, viewed as a logical constant, and took the predicate variable as ranging, for each universe of discourse, over all of its subsets. However, when with this understanding we write $(\exists F)F(a)$, we invisibly posit a uniform relation between the values that the predicate variable F may take on and the reference of a. The consequence that Azzouni draws is a version of Quine's thesis that second-order logic is set theory in sheep's clothing (with the modification that the wolf revealed when the clothing is stripped away is not set theory, but a two-sorted first-order theory of objects and their classes).

Thus far, I have been following Azzouni's text, and to some extent interpreting it. In connection with our present concerns, the conclusion would be that insofar as we implicate the understanding that we are supposed to have of the range of our predicate variables, the value of a predicate variable is an object. Of course, the implication cannot always be forthcoming. Even if the values of the ostensible predicate variables are classes, we do not implicate the standard interpretation of our discourse except where the universe of individuals constitutes a set. However, if we can take it as a set, then it falls short of what we intend for set theory, viz., that the quantifiers range over all sets. In any case, Azzouni's argument assumes that the set-theoretic models for second-order languages on the one hand and the interpretations of those languages on the other assign the same sorts of things as values of predicate variables, or at least that there is no significant difference between them. However, from within at least one perspective on second-order languages, we need not look at matters this way. We can instead regard the model theory as merely giving a picture, within a first-order language, of what is intended, and deny that the values of predicate variables are objects, construing them instead, as Frege did, as essentially predicational.

To specify an interpretation for a quantificational language, we need to specify the range of quantification over individuals (which may, indeed, be everything). Then, however, for second-order languages as well as first-order ones, it seems that there is nothing *more* to be specified about quantification; that is, the range of the monadic predicate variables is itself determined by the range of the objectual variables. Consider this point from a Fregean perspective. As you know, Frege held that predicate-expressions, what was left over when one or more occurrences of a proper name were identified in a complete sentence and the sentence was viewed as being constructed from that name and the residue left upon removal of those occurrences, referred to "unsaturated" things, which he called concepts. Frege took it as a matter of importance that the concept, being unsaturated, did not, in order to deliver a truth value when given an object, require any relation between the concept and the object at all; rather, the concept was itself of such a nature as to have this property. Because of his doctrine that concepts were a special case of functions, Frege's way of setting things up does not exactly correspond to ours. Still, the unsaturatedness of the values of predicate variables does play at least a negative role in responding to the suspicion that second-order quantification is really quantification over classes of individuals, in that we need not explain why the concatenation of a monadic predicate variable with a proper name yields a syntactic object that has a truth value on assignment of a value to the variable. Moreover, the realm of concepts will comprise all ways of discriminating among objects; that is, the range is simply all concepts.

The last paragraph expresses a point of view toward second-order quantification that Stewart Shapiro (1990) calls *neutral realism*. Neutral realism is not undermined by Azzouni's objection, because it is by no means committed to the thesis that we understand second-order quantification in terms of membership in classes (though that is not to say that we can refute someone who chooses to understand us in this way). The objection invites one to defend the thesis that the second-order quantifiers range over all – really, all – concepts, by specifying intended models in terms of set theory. However, that is an adventure that need not be undertaken. There is no question of interpreting, or reinterpreting, concatenation. Concatenation expresses nothing, serving only to indicate what is being predicated of what.

It is intrinsic to Frege's conception of second-order logic, however, that it at once invites us to ascend further. Once we have, say, $F(a)$, with proper name a and variable F, we have the notion of a concept under which a falls, something that discriminates among concepts: some concepts are concepts under which a falls, others are not. The way is then open to concepts of Frege's second level, and so on up. Now, the unrestricted quantifiers, as Frege showed, refer in this scheme to concepts of the second level. In natural-language quantifiers are generally, and perhaps without exception, restricted (but Frege noticed this point too, remarking that, in language, quantifiers referred to relations between concepts (Frege 1892, p. 48)). Beyond the level of relations between concepts, it requires argument to show that natural language has any serious means of expression at all. This or any other limitation on natural language, however, does not limit the hierarchy, which will ascend inevitably to a logic of order ω. Why then does our language not do the same, if the theory is correct?

Boolos (1984) proposes to define truth for second-order languages within second-order meta-languages without making classes, or anything else, count as the values of predicate variables. The definition uses quantification over relations $R(W,x)$, where x ranges over objects and W over monadic second-order variables, or, given a pairing function, predicates $R(\langle W,x\rangle)$ over ordered pairs of variables and objects. On either conception, if V is such a variable, then each object o such that $R(V,o)$ is *a* value of V according to R. The class of such o is the extension of V according to R, but the extension need not be brought in for the purpose of defining truth, and so, there is no need to think of anything as being *the* value of V according to R. Thus one might conclude that, after all, the notions of objecthood and truth-of suffice for the semantics of at least second-order languages with only monadic predicate variables. Charles Parsons (1990, pp. 327–8) proposes a difficulty, which I interpret as follows. What is defined inductively is a relation

$$R \text{ and } s \text{ satisfy } \Phi,$$

where Φ is a formula, s is an ordinary infinite sequence of objects, and R is as defined earlier. However, *that* notion is expressed by a triadic predicate with a second-order argument, namely R. If such predicates are admitted, we may consider in particular the predicate true of X if and only if X is $(\lambda x)[R(V,x)]$ for some R and some V. That predicate is X is a value of a second-order variable. Moreover, that X is *the* value of V according to R is: $X =$ (the Y)[$Y(x) \longleftrightarrow R(V,x)$]. The values of second-order variables are in this way reconstructible within the system proposed.

By virtue of the preceding considerations, I henceforth assume that to motivate second-order logic it must be shown that apparent second-order variables have values, and that these values are not objects, but rather things of an essentially predicational character. The latter statement responds to Azzuoni's version of the sheep's-clothing charge, whereas acquiescence in some conception of the value of a predicate variable appears, in view of the difficulty Parsons brings to bear for Boolos's construction, to be required for second-order semantics. The issues in the following section concern the extent to which the Fregean conception can be supported within natural language.

IV. Plural Reference and Quantification

Boolos (1984) and, more recently, Lewis (1991) have taken up plural quantification as a means of interpreting second-order theories (Boolos), or articulating non-first-order theories to serve as foundations for set theory (Lewis). Inversely, Schein (1986), Higginbotham and Schein (1989), and Schein (1993) have discussed interpreting plurals in terms of second-order logic. All of these discussions commence by arguing against a popular alternative, namely, the treatment of plural terms as referring to something like sets, classes, or properties. In this section, I review and to some degree refine these arguments, turning afterward to Boolos's discussion.

By a *plural term*, I mean either a definite description in the plural, or a conjunction of singular or plural terms. Thus 'Peter and Paul', 'the boys', 'the books on my shelf and the magazines on my shelf', and 'Peter and the other boys' are all plural terms. There are plural terms that are only revealed as such after taking account of quantifying into them: thus, 'every man and his dog' contains a plural term 'x and x's dog', which will be the subject of predication in a sentence like 'every man and his dog went hunting together'; but, I do not consider these cases further here. The problem of plural reference is the problem of the reference of plural terms, and also of plural quantifications, as in 'some (of the) boys', 'all (of the) books on my shelf and magazines on my shelf,' and the like. These quantifications are not exhaustive, since besides quantifiers such as 'many' and the numerals we have conjunctions as in 'some of the boys and Peter' and the like; but again I restrict the domain to the simplest cases.

In the literature both in philosophy and in formal semantics, plural reference has often been taken to be reference to sets, classes, or properties, and plural quantification as quantification over sets, classes, or properties. What matters is not so much that the reference is to, for example, sets exactly, but rather that plural reference, on the views in question, is taken to be singular reference in grammatical disguise. Naturally, something must mediate between the reference of a definite plural, such as 'the books', and certain individual books. In the influential work of Link (1983), each individual book on my shelf is an "atomic i-part" of the reference of the phrase 'the books on my shelf', and it is such a part just by virtue of being a book on my shelf; and nothing other than a book on my shelf is such a part. However, as pointed out by Higginbotham and Schein (1989), it follows that, for Link, the schema (C) holds unrestrictedly:

(C) x is an atomic i-part of the Fs $\longleftrightarrow$ $F(x)$

and therefore that, on pain of Russell's Paradox, the predicate 'is an atomic i-part of' cannot figure in the object language. The argument can be strengthened dialectically, as by Lewis (1991), or more straightforwardly, as follows: Consider all plural terms 'the Fs' constructible in our language; such a term exists for every nominal expression F with a count noun head and, by hypothesis, has a reference, provided that there are at least two Fs. Given a universe of discourse for our language, there will be some objects in it that are the reference of plural terms; call these the *plural objects*. Then every plural term refers to a plural object, if to anything at all. A logic for plurals will not be complete unless it allows distributive quantification and the use of plural terms as parts of predicates, as in

(7) Each of the Fs is G.
(8) x is one of the Fs/x is among the Fs.

Assuming that the phrases 'the Fs' function in examples (7) and (8) just as they would function as the arguments of other predicates, these locutions will be allowed for by positing a relation, call it R, such that, if $R(x,y)$, then y is a plural object, and (7) and (8) are construed as using this relation tacitly so that schematically, they become, respectively,

(9) For all x such that $R(x$, the Fs), $G(x)$,
(10) $R(x$, the Fs).

Moreover, provided that the presupposition that there are at least two Fs is satisfied, (7) is equivalent to (11), and (8) is equivalent to (12):

(11) For all x such that $F(x)$, $G(x)$
(12) $F(x)$.

(The equivalence of (8) to (12) of course implies the equivalence of (7) to (11).) Because (13) is an instance of (10) and (14) is the corresponding instance of (12), (13) and (14) are equivalent, provided that, for at least two y, $\neg R(y,y)$:

(13) R [the objects y such that $\neg R(y,y)$, the objects y such that $\neg R(y,y)$].
(14) $\neg R$ [the objects y such that $\neg R(y,y)$, the objects y such that $\neg R(y,y)$].

However, everything that is not a plural object is a y such that $\neg R(y,y)$, and so, the presupposition is satisfied. And so we have Russell's Paradox.

There are several points where the above deduction may be questioned, but none of them appear very promising:

(i) Perhaps the plural term 'the objects y such that $\neg R(y,y)$' has no reference. However, the books on my shelf (for instance) are each of them things that are not plural objects, and so do not bear R to themselves. If the plural term has no reference, it must be that a meaningful predicate, satisfied by some objects, may fail to determine a plural object.

(ii) Expression (8) is not to be taken as in (10), but rather as a somewhat roundabout way of saying 'x is an F', schematically (12) itself; similarly, (7) is just a roundabout way of saying, 'Each F is a G'. The definite article has a function, in that it delimits the universe of discourse – 'three of the boys are here' is felicitous only if the boys in question are antecedently identifiable, whereas 'three boys are here' is at best neutral – but this difference is a matter of proper assertion rather than logical form.

This way out amounts to denying that plural reference exists at all. Now, what motivated plural reference in the first place were cases in which the plural appeared not to refer to a single object, whose parts in some appropriate sense were the Fs, but rather to the Fs taken somehow collectively, as could be seen from the fact that the distributional interpretations were intuitively false. These cases are matched by those in which the form is just as in (8). For example, we do not regard

(15) x is one of the boys who built the boat

as equivalent to

(16) x is a boy who built the boat.

Similarly for forms (7), since

(17) Each of the boys who built the boat got a merit badge

is not regarded as equivalent to

(18) Each boy who built the boat got a merit badge.

The point may be put more sharply by considering simple uses of plural pronouns or demonstratives, without antecedents in a discourse. If I say, indicating some boys, 'three of them built a boat', then even if my utterance is understood as requiring and intending completion by a sortal, so that I am understood as talking, say, about *those boys*, there need be no particular complete description 'the boys such that F' that I intend to communicate, so that the predicate that applies to each of them is just: 'x is among those boys'. If 'those boys' refers to a plural object, then some relation must mediate between this term and x.

(iii) The equivalence of (10) to (12) is not unrestricted. On the contrary, it seems to be completely without exception, unimpeded by the question of whether 'the Fs', on the construal suggested, form a set or something like a proper class. In fact, 'x is one of the proper classes' strikes us as equivalent to 'x is a proper class', and 'x is one of the plural objects' as equivalent to 'x is a plural object'.[5]

The paradox of plurality just sketched seems to have a different status from the paradoxes of set theory, on the one hand, and the semantic paradoxes on the other. In set theory, we are not working within a given linguistic and theoretical scheme, so that the paradoxes can arguably be regarded as solved if there is some way out that allows theory to proceed. The semantic paradoxes arise naively, inasmuch as the concept of truth is not a technical one advanced for some theoretical purpose but a part of everyday vocabulary, and the general disquotational principles that give rise to the Liar and related paradoxes belong to common linguistic practice. Plurality is like truth in appearing to be antecedently given, but the extra material that we brought in to explain the behavior of plurals, namely, the hypothesis of plural objects and the relation R, are theoretical. All of this suggests that it is not the principles, whatever they are, governing the transition from 'x is among the Fs' to 'x is an F' and conversely that are to be faulted, but rather the hypothesis of plural objects that required positing R.

Boolos (1984) uses plural quantification to interpret monadic second-order formulas in set theory. He does not provide a theory of the reference of plural terms such as 'the Cheerios in the bowl' (his example), but he does hold that their use in making ordinary assertions does not augment the ontological commitments of our theory of the world. It is not easy to evaluate his thesis with respect to examples like this, however, because the Cheerios in the bowl are finite, indeed not many, so that it is hard, at least for those of us who admit small finite sets, to see what objection there could be to an ontology that admitted the Cheerios in the bowl, and because ontological commitment, if borne by plurals, would apply first to existential statements with plural quantificational prefixes. Putting the question of ontological commitment to one side, let me note by way of fixing ideas that Boolos's translations take as primitive the notion

(19) it (or: x; or: *who*, *which*, etc.) is one of them,

where the singular and plural pronominals are both variables, indicating positions related to antecedents. Consider a second-order sentence such as

(20) $(\exists F)((\exists y)(\exists w)\{[F(y)\ \&\ F(w)\ \&\ y \neq w]\ \&\ (\forall x)[F(x) \rightarrow (\exists! z)[z$ is a parent of $x\ \&\ \neg F(z)]]\})$.

Let the universe of discourse be people. In standard terms, sentence (20) is true if a non-empty F can be chosen that is true of people of exactly one of whose parents it is not true. (Naturally, there are many such F.) Under Boolos's translation, this becomes

(21) There are some people such that every person that is one of them has exactly one parent who is not one of them.

The truth of sentence (20) can be defended in second-orderese by letting F be true of me and my female ancestors, and no one else. Likewise, sentence (21) can be verified by letting the quantifier 'some people' have (as one might say) for its value just me and my female ancestors.

There are some limitations on Boolos's construction, which I do not consider here. The point of the construction, that monadic second-order logic is vindicated insofar as the translations into pretty ordinary English are well understood, is a powerful consideration in favor of second-order logic, whatever the peculiarities of natural language might otherwise be. With this much in mind, I consider plurals and plural quantification in themselves, so apart from the particular expressions that may be used to paraphrase second-order schemata.

V. Interpreting Plural Quantification

The instantiation of second-order schemata by means of plural quantification raises the question of how plural quantification itself should be understood, and this in turn leads to the more fundamental question of how ordinary definite plurals ('the books'), indefinite plurals ('some books') and bare plurals ('books') are interpreted in the sentences in which they occur. Schein (1986), Higginbotham and Schein (1989), and, far more elaborately, Schein (1993) defend the view that plurals in fact involve second-order quantification. That does not of itself undermine the interpretation of second-order quantification by means of plurals, because it may be that the sentences of the meta-language, unpacking as they do the truth conditions of plural sentences, can in turn be instanced in English by means of plural quantification, again interpreted by means of second-order quantification, provided anyway that the interpretations are transparently equivalent. It could, however, be taken to mean that basing second-order logic on plural quantification is not fundamentally distinct from basing it on predication.

The fundamental problem of plural interpretation is the interpretation of undistributed plurals, as in the most salient interpretation of

(22) The boys built a boat.

(23) Peter and Paul built a boat.

We may be prepared to assert sentence (22) without being prepared to assert, of any particular boy, that he built a boat, or sentence (23) without being prepared to assert that Peter built a boat. Indeed, we may correct a person's assertion that Peter built a boat by saying, "No, Peter and Paul built a boat." Rejecting as we have plural objects (whether thought of as constituted, like sets, by their members, or in some other way), and assuming that the case is not one in which a plural is being used to refer to a single thing of which the various elements mentioned (Peter and Paul), or the things satisfying the predicate (those in the range of 'boy' in the context of utterance), this feature of our behavior is not trivially explained.

I think that there are locutions in which definite plural NPs have singular reference. As discussed by Moltmann (1995) following Simons (1987), there are cases in which the reference is to a single thing, of which the objects falling under the plural predicate are taken to be parts, in a contextually determined sense of that notion. To say, for example, that the boxes together are heavy is to say that a certain single thing, the boxes together (or: taken together), of which the individual boxes are the parts, is heavy; to say that the books on the shelf are arranged alphabetically is to say that a certain thing, the family of books, of which the individual books are the parts, lies in a certain arrangement. However, many cases are not plausibly treated in this way: when I hear the children crying, I do not hear the crying of a single object, of which the children are the parts; rather, I hear an episode of crying to which various of the children make their contributions, or so it would appear. Furthermore, there are predicates that cannot be true of individual objects at all, such as 'clump' (Schein's example), 'rain down' (Boolos), 'cluster', and perhaps the reciprocal intransitives 'fight', 'meet', 'collide', and others.

If plural reference and quantification are taken as primitive, then, as we might say, we can have predicates true of the many, as well as of the one, and of course relations between the many and the one, as the many boys and the one boat they built. However, if we consider the logical form of plural sentences from a more elaborated point of view as Schein's (1986, 1993), then plural arguments undistributed with respect to the surface predicates may be reconstrued as distributed with respect to other predicates analytically posited.

To this end, consider the widely known thesis of Donald Davidson that action verbs contain a position for events, and suppose this thesis extended (as in, e.g., Higginbotham (1985), Terence Parsons (1989), and others) to all predicates whatsoever. Assume furthermore that the primitive predicates (nouns, adjectives, verbs, and prepositions) are related to their arguments through a family of specific relations, including the familiar relations of grammatical theory: "agent," "patient," "beneficiary," and so on. These relations (known in contemporary generative grammar as "thematic roles," though not always thought of as having the semantic content associated with them here) are borne by the arguments of the primitive predicate to the events (or, more generally, events and states) over which they range. The result is a picture of the simplest predication that natural language makes available as containing a complex structure, so that the skeleton for 'John walked', for example, would be as in

(24) walk(e) & Θ(John, e),

where 'walk' is a predicate of events, and Θ relates the event to its participant John, the reference of the subject. Sentences like 'John walked' are completed by adding tense, not considered here, and existential quantification over the position marked by e, so that we have the familiar Davidsonian paraphrase

(25) ($\exists e$) walk(e) & Θ(John, e).

More generally, the picture is what might be described as a planetary theory of thematic roles: the head verb, or other primitive, is true of events whose participants are grouped around it like planets around a sun, attached by the various relations Θ. In a typical open sentence, 'x saw y,' for example, we have

(26) $$x \xleftarrow{\Theta_1} E = \text{see} \xrightarrow{\Theta_2} y.$$

Could not one and the same relation Θ relate each of several participants to the same event? Nothing in the picture prevents that, and so, alongside the simple (26) we could also have

(27) $$\begin{array}{l} x_1 \xleftarrow{\Theta_1} E = \text{see} \xrightarrow{\Theta_2} y. \\ x_2 \swarrow \end{array}$$

On this view, when we say that Peter and Paul built a boat, we do not say anything about a complex agent, but rather aim to report an event of boat building, the agents of which were Peter and Paul. We thus have for (23) the structure

(28) built a boat(e) & Θ(Peter, e) & Θ(Paul, e).

(I ignore the further structure that would come from the analysis of the predicate 'built a boat'.)

It remains to complete the structure by binding the event variable; but here a logical point must also be addressed. We do not want (23) to imply that Peter built a boat, but of course it will imply that if the latter is just (29), and the structure for (23) is completed by existential closure applied to e:

(29) ($\exists e$) [(built a boat(e) & Θ(Peter, e) & Θ(Paul, e)].

What we require is that to say that Peter built a boat is not merely to say that there was an event e of boat building of which Peter was an agent (to which he was related as in Θ), but that there was a boat building of which he was the *sole* agent, as in

(30) $(\exists e)$ {built a boat(e) & $(x)\,[\Theta(x, e) \longleftrightarrow x = \text{Peter}]$}

Similarly, (23) is understood ultimately as in

(31) $(\exists e)$ {built a boat(e) & $(\forall x)[\Theta(x, e) \longleftrightarrow x = \text{Peter} \vee x = \text{Paul}]$}

With this much to hand, we can proceed to the more complex case, exemplified by

(32) Some boys built a boat.

In full, it will be as in

(33) $(\exists e)$(built a boat(e) & $(\exists X)\{(\exists x)(\exists y)[X(x)\ \&\ X(y)\ \&\ x \neq y]\ \&(\forall z)[X(z) \rightarrow \text{boy}(z)]\ \&\ (\forall w)[\Theta(w, e) \longleftrightarrow X(w)]\}$)

Thus far I have been following Schein, who shows a number of further applications of his basic idea. Let us introduce the prefix

there are some things whose number is at least 2, and ...

for

$$(\exists X)\{(\exists x)(\exists y)\ [X(x)\ \&\ X(y)\ \&\ x \neq y]\ \&\ \ldots\}$$

Then we may, following Boolos, translate (33) back into English syntactic forms

(34) There was a boat building and some things whose number is at least 2, all of whom were boys, such that everything is an agent of it if and only if it is one of them.

Such harmony between the interpretation of second-order logic using plural constructions on the one hand, and the interpretation of plural constructions in terms of second-order quantification on the other, shows, pending examples to the contrary, that so far as we leave in place Boolos's locution 'it is one of them', we will not have a conflict between the two perspectives. I now argue, however, that plurals and plural quantification should be taken in the much older terms of the "class as many," in the sense adumbrated by Russell (1903), rather than in terms of second-order logic.[6]

VI. Classes as Many

There are two rather obvious difficulties with the interpretation of plurals in second-order terms. The first difficulty is that the second-order logical forms considered to this point do not really allow primitive predicates with plural arguments; rather, any undistributed plural is taken to be distributed with respect to some relation, to one term of which the plural corresponds. In a statement such as 'these are a few of my favorite things', we do not appear to rely upon any mediating situations or relations to say that it is true if 'these' are (at least) a few, and each of "them" is a favorite thing of mine. In the case of 'these are at least a few', it appears that the numerical predicate is true of the reference of 'these', whatever it may be. And in the case of 'each of them is a favorite thing of mine', the question is how the quantification works – it appears to be universal quantification over objects x standing in some relation to the reference of the necessarily undistributed plural 'them'.

The second difficulty impresses itself with respect to demonstrative undistributed plurals. Suppose I wave my hand at some boys, saying, "They built a boat yesterday." I then make a definite reference of some sort – but to what? Adhering strictly to the second-order theory, one would take "they" as answering to a predicate demonstrative, not a bound variable as in

(35) $(\exists e)\ \{\text{built a boat}(e)\ \&\ (\forall x)[\Theta(x, e) \longleftrightarrow A(x)]\}$,

where A just *is* what I referred to with the demonstrative. However, there seems to be nothing predicational about the plural demonstrative.

The above difficulties are overcome if we regard plurals as referring to classes as many. We can enlarge standard semantics for English in such a way as not to disturb the insight that some undistributed plurals express, not the relation of a single object to an event, but rather the multiplicity of the objects that stand in that relation, for example, not group agency of some sort but multiple agency. However, the axioms governing plural reference will themselves use plurals, so that undistributed plurals will appear in the specification of the semantics. An initial axiomatization for a relatively untendentious fragment of English would, I think, include the following:

(36) for any singular nominal F, 'the Fs' refers to the Fs if there are at least two x such that F is true of x.

(37) For any singular or plural terms α and β, 'α and β' refers to α and β.

In (37), the plural 'α and β' that is the object of 'refers to' is irreducible; that is, 'α and β' is said to refer, not to α and to β, but rather to (α and β). We can formulate laws for terms such as Boolos's 'is one of', which takes one singular and one plural argument, for instance as in

(38) x is one of the Fs $\longleftrightarrow$ $F(x)$.

(39) If α is a singular term and $x = \alpha$, or β is a singular term and $x = \beta$, or α is a plural term "τ and δ" and x is one of τ and δ, or β is a plural term 'ϵ and μ' and x is one of ϵ and μ, or α is a plural term 'the Fs' and x is one of the Fs, or β is a plural term 'the Gs' and x is one of the Gs, then x is one of α and β.

To develop full semantics for plurals as referring to the class as many, the assignments of values to variables will be construed plurally; that is, an assignment to a variable occupying a position where plural terms could go would itself have to be expressed plurally as, for instance, $s(\text{`}x\text{'}) =$ Peter and Paul (or Paul and Peter), and not as the unordered pair {Peter, Paul}. Similarly for the ranges of quantifiers, the quantifier 'some people' for a universe of discourse containing just Peter, Paul, and Mary ranges over Peter and Paul; Peter and Mary; Paul and Mary; Peter, Paul, and Mary; and nothing (no things) else.

The term 'is one of' is the plural ersatz for membership, and in the presence of plural quantification and plural variables it is not going to be eliminable. However, we do not lose the advantages of Schein's fundamental thesis, that undistributed plural reference may involve, not group agency, but multiple agency of single events. Using capital Greek letters for variables ranging over classes as many, examples such as (40), reproduced here, will be as in (41):

(40) Some boys built a boat.

(41) $(\exists e)$ (built a boat(e) & $(\exists\Gamma)$ $\{(\exists x)(\exists y)$ (x is one of Γ & y is one of Γ & $x \neq y$) & $(\forall z)$ [z is one of $\Gamma \rightarrow$ boy(z)] & $(\forall w)$ [$\Theta(w, e) \longleftrightarrow w$ is one of Γ]$\}$)

The language is essentially two-sorted; that is, there are contexts in which only plural, or only singular, terms (including variables) are admitted, one of them being the argument positions of 'is one of' itself. For this reason, attempts to reproduce Russell's paradox should founder on ungrammaticality. Thus,

*the things that are one of themselves.

The ersatz subset relation, call it 'Γ are included in Σ', is definable in terms of 'is one of' in the expected way: some things are included in some other things if everything that is one of the former is one of the latter, or

(42) Γ are included in $\Sigma \longleftrightarrow (\forall x)$ (x is one of $\Gamma \rightarrow x$ is one of Σ).

There are a number of points to be clarified, and even decisions to be made, for a full two-sorted language of singulars and plurals. For one thing, I have omitted discussion of apparently plural terms such as 'Bill Clinton and the President' that fail to be in order because their singular components are coreferential, or 'the boys and the members of the band', where some of the members of the band are boys. Also, our language is not *rigidly* two-sorted with respect to singulars and undistributed plurals, since many univocal predicates can accommodate both. Thus one can say

(43) The soldiers surrounded the palace.

(44) The picket fence surrounded the palace.

We could think of the reference of 'the soldiers' in (43) as to a single thing whose parts are the soldiers, much as 'the picket fence' refers to a single thing whose parts are the pickets and the pieces connecting them. However it may be preferable to think of the reference as to the class as many, which in virtue of having its elements spatially arranged in a certain way, surrounds the palace.

David Lewis (1991) has formulated a basis for set theory using mereology and plural quantification. The interpretation of plurals that I have suggested here is not, I believe, one that would suit his purposes. He takes plural quantification to be (as he says) "*sui generis*" and ontologically "innocent." Plural quantification is a special mode of quantification over what there is, namely objects. Mereology is also ontologically innocent, in his view, so that, as he puts it:

> The fusion [of cats] is nothing over and above the cats that compose it. It just *is* them. They just *are* it. Take them together or take them separately, the cats are the same portion of Reality either way. (Lewis 1991, p. 81).

If Felix and Possum are two cats, and Felix + Possum is their fusion, then, on the class-as-many view of plurals, the words 'Felix and Possum' refer to Felix and Possum (and so, Felix and Possum are two), whereas 'Felix + Possum' is a singular term (and so, Felix + Possum is one). So, Felix and Possum ≠ Felix + Possum. So *they* are not *it*, contrary to Lewis's declaration. One may, of course, add that the fusion of Felix + Possum is nothing "over and above" Felix and Possum, or vice versa, but the question would then be whether that statement expresses anything more than one's conviction that there is nothing ontologically committing about fusions, or plurals.

In sum, I have argued that plural reference and quantification are subject to analysis that deprives them of ontological innocence, and also of being a satisfactory basis for motivating second-order logic. But is the distinction between quantification over predicate positions and quantification over classes as many a "shadow of a grammatical distinction," as Charles Parsons once put it, or is it, as it was for Frege, "written deeply into the nature of things?" Considerations such as I have given here are not decisive.

NOTES

A first draft of a paper from which this article is excerpted was presented at the British Academy symposium on philosophical logic, March 1996, David Bostock commenting. A revised version of that paper appears in the Proceedings. Besides my obvious debt to Charles Parsons, who has been my teacher now for some twenty-eight years, I would like to record in this place my indebtedness to George Boolos, who brought me to see many things I would not have seen otherwise, and likewise to express my sorrow that I cannot now receive his criticism.

1. Here and later, I use 'predicational' for 'having the value that a predicate has' as opposed to 'having the value that a singular term has' (of course, in some views, these cannot be distinguished). I also distinguish 'predicational' from 'predicative,' reserving the latter for definitions, or things definable in terms of definitions, that

are mathematically predicative, that is, do not involve quantification over domains that include the thing being defined.

2. It is analogous to the objection that if reference, satisfaction, and truth are defined in the familiar Tarskian manner, then we are deprived of saying the natural things about language change. In particular, we lose the distinction between replacing one language with another, and merely extending a language. See Field (1972).
3. Parsons (1971) shows how to construct a predicative theory of classes along substitutional lines.
4. Here and later, I use the words 'family of objects' as neutral among various conceptions of what kinds of values predicate variables may be said to have.
5. This point, originally due to Boolos, is made by Lewis (1991) and by Schein (1993, Ch. 2).
6. Parsons (1990) notes the affinity between Boolos's interpretation and the class as many.

REFERENCES

Azzouni, Jody (1994). *Metaphysical Myths, Mathematical Practice: The Ontology and Epistemology of the Exact Sciences* (Cambridge, UK: Cambridge University Press).

Boolos, George (1975). "On Second-Order Logic," *Journal of Philosophy*, 72: 509–27.

Boolos, George (1984). "To Be Is to Be a Value of a Variable (or to Be Some Values of Some Variables)," *Journal of Philosophy*, 81: 430–48.

Field, Hartry (1972). "Tarski's Theory of Truth," *Journal of Philosophy*, 64: 347–75.

Frege, Gottlob (1892). "Über Begriff und Gegenstand," translated as "On Concept and Object" in P. Geach and M. Black (eds.) *Translations from the Philosophical Writings of Gottlob Frege* (Oxford: Blackwell, 1952), pp. 42–55.

Henkin, Leon (1950). "Completeness in the Theory of Types," *Journal of Symbolic Logic*, 15: 81–91.

Higginbotham, James (1985). "On Semantics," *Linguistic Inquiry*, 16: 547–93.

Higginbotham, James (1989). "Frege and Grammar," (abstract). *Proceedings and Addresses of the American Philosophical Association*, 63(2): 52.

Higginbotham, James (1990). "Frege, Concepts, and the Design of Language," in *Information, Semantics and Epistemology*, ed. Enrique Villanueva (Oxford: Basil Blackwell), pp. 153–71.

Higginbotham, James (1998). "On Higher-Order Logic and Natural Language," in T. Smiley (ed.). *Proceedings of the British Academy 95: Philosophical Logic* (Oxford: Oxford University Press), pp. 1–27.

Higginbotham, James, and Schein, Barry (1989). "Plurals." *North East Linguistics Society XIX Proceedings*. Graduate Linguistics Students' Association (Amherst: University of Massachusetts), pp. 161–75.

Lewis, David (1991). *Parts of Classes* (Oxford: Basil Blackwell).

Link, Godehard (1983). "The Logical Analysis of Plurals and Mass Terms: A Lattice-Theoretic Approach," in *Meaning, Use, and Interpretation of Language*, eds. Rainer Bäuerle, Christoph Schwarze, and Arnim von Stechow (Berlin: de Gruyter), pp. 302–23.

Moltmann, Friederike (1995). "Part-Structure Modification," Manuscript, The Graduate Center, City University of New York.

Parsons, Charles (1971). "A Plea for Substitutional Quantification," *Journal of Philosophy*, 68: 231–37; reprinted in Parsons, Charles (1983). *Mathematics in Philosophy: Selected Essays* (Ithaca, NY: Cornell University Press), pp. 63–70.

Parsons, Charles (1990). "The Structuralist View of Mathematical Objects," *Synthese*, 84: 303–46.

Parsons, Terence (1989). *Events in the Semantics of English* (Cambridge, MA: MIT Press).

Putnam, Hilary (1971). *Philosophy of Logic* (New York: Harper).

Quine, W. V. (1970). *Philosophy of Logic* (Englewood Cliffs, NJ: Prentice–Hall).

Russell, Bertrand (1903). *Principles of Mathematics* (New York: W.W. Norton); 2nd Ed., 1938.

Schein, Barry (1986). "Event Logic and the Interpretation of Plurals," Ph.D. dissertation, MIT, Cambridge, MA.

Schein, Barry (1993). *Plurals and Events* (Cambridge, MA: MIT Press).

Shapiro, Stewart (1990). "Second-order Logic, Foundations, and Rules," *Journal of Philosophy*, 87: 234–61.

Simons, Paul (1987). *Parts: A Study in Ontology* (Oxford: Clarendon Press).

The Logical Roots of Indeterminacy

GILA SHER

I. Indeterminacy as Relativity to Logical Frameworks

In 1915, Leopold Löwenheim proved a remarkable theorem:

(L) If the domain is at least denumerably infinite, it is no longer the case that a first-order fleeing equation is satisfied for arbitrary values of the relative coefficients. (Löwenheim 1915, p. 235)

In contemporary terminology the theorem says that if a formula Φ of first-order logic with identity is finitely valid but not valid, then for every cardinal $\lambda \geq \aleph_0$, Φ is not λ-valid (i.e., if $\neg\Phi$ is satisfiable in an infinite model, then for every infinite cardinal λ, $\neg\Phi$ is satisfiable in a model of cardinality λ).[1] It follows from this theorem, Löwenheim pointed out, that "[a]ll questions concerning the dependence or independence of Schröder's, Müller's, or Huntington's class axioms are decidable (if at all) already in a denumerable domain." (1915, p. 240).

In a series of articles, Thoralf Skolem (1920, 1922, 1929, 1941, 1958) presented a new version of Löwenheim's theorem and offered a new kind of proof for it. We can formulate Skolem's result as

(LS) Let T be a countable 1st-order theory (where a theory is a set of 1st-order sentences). Then, if T has a model, T has a countable model; in particular: (i) T has a model in the natural numbers; (ii) If $\mathfrak{A}$ is a model of T, then there is a countable submodel $\mathfrak{A}'$ of $\mathfrak{A}$, such that $\mathfrak{A}'$ is a model of T.

Skolem's theorem was extended by Tarski to

(LST) Let T be a set of sentences in a language L of cardinality $\kappa \geq \aleph_0$. Then, if T has an infinite model (a model with an infinite universe), T has a model of cardinality λ for every $\lambda \geq \kappa$.

Skolem regarded LS as signaling the unavoidable relativity of mathematical notions to *logical frameworks*. Skolem's view is sometimes referred to as Skolem's paradox on the basis of passages such as this:

So far as I know, no one has called attention to this peculiar and apparently paradoxical state of affairs. By virtue of the [set-theoretical] axioms we can prove the existence of higher cardinalities, of higher number classes, and so forth. How can it be, then, that the entire domain B [the universe of an "LS model" of set theory] can already be enumerated by means of the finite positive integers? (Skolem 1922, p. 295)

However, the "paradox" is swiftly explained away:

The explanation is not difficult to find. In the axiomatization "set" does not mean an arbitrarily defined collection; the sets are nothing but objects that are connected with one another through certain relations expressed by the axioms. Hence there is no contradiction at all if a set M of the domain B is nondenumerable in the sense of the axiomatization; for this means merely that *within* B there occurs no one-to-one mapping Φ of M onto Z_0 (Zermelo's number sequence). Nevertheless there exists the possibility of numbering all objects in B, and therefore also the elements of M, by means of the positive integers; of course, such an enumeration too is a collection of certain pairs, but this collection is not a "set" (that is, it does not occur in the domain B). (Ibid.)

LS (*LST*) is often viewed as setting a limit to the axiomatic method, and this view is, according to Skolem, correct in a sense:

[A]xiomatizing set theory leads to a relativity of set-theoretic notions, and this relativity is inseparably bound up with every thoroughgoing axiomatization. (Ibid.; p. 296)

[By means of the axiomatic method] the theorems of set theory can be made to hold in a merely *verbal* sense. (Ibid. See also Skolem 1941, p. 468.)

[T]here is no possibility of introducing something absolutely uncountable except by means of a pure dogma. (Skolem 1929, p. 272, translated by Wang 1970, p. 38)

This limitation, however, is nothing more than a price paid for a precise formulation of intuitive notions: In the course of formalizing absolute, yet imprecise, mathematical notions within a formal, i.e., logical, calculus, these notions are inevitably relativized to the logical calculus itself. Thus, Skolem says:

[L]a distinction est essentielle entre la notion simple ou absolue d'ensemble et la notion telle qu'elle découle d'une méthode déterminée de la préciser. La notion étant, dans ce second sens, plus précise et en même temps relative seulement à la façon de la délimiter.[2] (Skolem 1941, p. 480)

The expectation that a formalization will preserve the absolute nature of the intuitive notions is, according to Skolem, unfounded:

Que l'axiomatique conduise au relativisme, c'est un fait parfois considéré comme le point faible de la méthode axiomatique. Mais sans aucune raison. Une analyse de la pensée mathématique, une fixation des hypothèses fondamentales et des modes de raisonements ne peut être qu'un avantage pour la science. Ce n'est pas une faiblesse d'une méthode scientifique, qu'elle ne puisse donner l'impossible.[3] (Ibid., p. 470)

And this impossibility is intimately connected with the idea of a *formal system*:

Mon point de vue est donc qu'on doit utiliser les systèmes formels pur le développement des idées mathématiques. On peut ainsi préciser les notions et les méthodes mathématiques ....

Si donc nous désirons avoire une théorie générale des ensembles, cette théorie aussi doit être développée comme un système formel .... Je ne comprends pas pourquoi la plupart des mathématiciens et logiciens ne semblent pas être satisfaits de cette notion d'ensemble définie par un système formel, mais au contraire parlent de l'insuffisance de la méthode axiomatique. Naturellement cette notion d'ensemble a un caractère relatif; car elle dépend du système formel choisi.[4] (Skolem 1958, pp. 634–5)

Tarski adds a twist to Skolem's analysis:

Le théorème de Löwenheim–Skolem lui–même n'est vrai que dans une certaine interprétation des symboles. En particulier si on interprète le symbole ∈ d'une théorie des ensembles formalisée comme un prédicat à deux arguments analogue à tout autre prédicat, le théorème de Löwenheim–Skolem s'applique, et il existe un modèle dénombrable. Mais par contre si l'on traite ∈ comme les symboles logiques (quantificateurs, etc.) et qu'on l'interprète comme signifiant appartenance, on n'aura plus en général de modèle dénombrable.[5] (Ibid., p. 638)

Indeterminacy, thus, according to Skolem and Tarski, is relativity to logical frameworks. For Skolem, this relativity concerns the "order" of the logical system involved (*LS* does not hold in full second-order logic) as well as the choice of formalization (axiomatization) within a given logic. For Tarski, this relativity concerns the choice of logical terms: not only does *LST* fail in second-order logic, but it also fails in first-order logic with the membership relation as a logical constant.[6]

II. Indeterminacy as Relativity to Background Language

In a series of influential works, Quine (1958, 1960, 1969a), Putnam (1977, 1978, 1981, 1983a), and others have argued that relativity and indeterminacy are characteristic of language in general, not just of language formalized within a logical framework. Meaning, reference, and ontology are relative to background theory (background language), and a direct, unique, absolute correspondence between words and objects is impossible. A unique correspondence, Putnam says, would require something that goes well beyond our critical resources: something on the order of a *metaphysical dogma*. [Compare with Skolem (1929).]

Indeterminacy as relativity to an external "system of coordinates" was illustrated by a number of suggestive parables. One of the best known of these is Quine's parable of the linguist–explorer: A linguist reaching an unknown land is trying to decipher its inhabitant's discourse. Meaning, however, is underdetermined by linguistic behavior, and it is only by imposing his own scheme of reference (individuation, ontology) upon the native speakers that the linguist is able to create a serviceable manual of translation. In principle, the linguist could use a different conceptual scheme in interpreting the native's discourse, arriving at an empirically equivalent but theoretically divergent manual of translation. The two manuals would agree with the same observational data, but whereas one, for example, would construe 'Gavagai' as a sortal term (or as a statement about discrete objects), the other would construe it as a mass term (or as a statement about undifferentiated "stuff").

Whereas in Quine's parable the background coordinate system is embodied in a member of our own culture (a linguist), in Putnam's the background framework is embodied in a transcendent metaphysical force. Consider the following excerpt from Putnam's parable of "God and the Indeterminacy of Reference":

> [*God and the Indeterminacy of Reference.*] [A]t the time of the Tower of Babel episode, God became bored.... Not only did He cause us to start speaking different languages, but He started to play around with the satisfaction relations, the 'correspondences', upon which the words–world connection depends.
>
> To understand what He did, pretend that English was one of the languages in existence back then. Imagine that C_1 and C_2 are two admissible 'correspondences' (satisfaction relations), i.e., that C_1 (respectively, C_2) is the satisfaction relation that one gets if M_1 (respectively, M_2) is the model that one uses to interpret English, where M_1 and M_2 are both models which satisfy all the operational and theoretical constraints that our practice imposes. Then what He did ... was to specify that when a *man* used a word, the word would stand for its image or images under the correspondence C_1, and that when a *woman* used a word, the word would stand for its image or images under the correspondence C_2.
>
> This situation continues to the present day. Thus, there is one set of things – call it the set of *cats* – such that, when a man uses the word 'cat' it stands for that set (in a God's eye view), and a different set of things – call it the set of *cats** – such that when a woman uses the word 'cat' it stands for *that* set (in a God's eye view)....
>
> Notice that the same sentences are true under both of His reference-assignments....
>
> It amused God ... to see men and women talking to each other, never noticing that they were almost never referring to the same objects, properties and relations.... (Putnam 1983a, pp. ix–x)

Indeterminacy, in this parable, is an external feature of language (discourse, theories). Men and women do not realize the indeterminacy of their discourse; God does. God hears Man say: ‡Some cats are black‡, [7] and Woman nodding, ‡Yes, some dogs are white‡; God hears Woman say: ‡I have a white dog‡, and Man echoing: ‡You have a black cat‡. Men and women do not sense anything strange in their dialogue. It is only from God's external, absolute point of view that the meaning (reference, ontology) of cross-gender discourse is indeterminate.

III. Indeterminacy as Loss of Information

Why is relativity to an external framework troubling for philosophers? After all, such relativity does not affect the interaction between language users: from a perspective internal to discourse, meaning, reference, and ontology are perfectly determinate.[8]

Indeterminacy is a meta-theoretical phenomenon, but philosophical theories in general are meta-theoretical in nature. Philosophy seeks to understand the relation between language and the world "from above," so to speak, and it is just at this level of understanding that the impact of indeterminacy is felt. Any discussion requiring both (i) an external point of view on language or theory, and (ii) an unequivocal determination of meaning, reference, or ontology, is made impossible by indeterminacy: the correspondence theory of truth, realistic epistemology, and so forth. From another perspective, indeterminacy adds a new weapon to the skeptic's arsenal: If it is not determined what the statement 'There are indenumerably many stars' says about the world, how can we trust it to tell us "the truth" about it? How can we rely on our theories to give us accurate information about the world if we cannot determine what they say about it? We can view indeterminacy as a barrier to the transmission of information: in theories formulated within the framework of standard first-order logic, a considerable amount of information (e.g., information about the size of large collections) is lost.[9] The idea that indeterminacy is *loss of information* can be illustrated by a sequel to Putnam's parable.

God and the Indeterminacy of Reference – Part II: Many years have passed. God is getting old. God's powers are leaving Him, His perception is deteriorating, His memory is not what it was. At some point God becomes dependent on humankind to provide Him with information about the world. God listens to humans' utterances about the world and updates His ledgers. God hears a man's voice saying: "Some cats are black", and God notes that some cats are black; God hears a woman's voice saying: "Some dogs are black", and God notes that some cats are white. . . .

One day God realizes that He no longer distinguishes between men's and women's voices. God hears a human's voice saying: "Some cats are black", but God does not know whether it is a man saying that some cats are black or a woman saying that some dogs are white. God is now paying a price for His youthful acts. Had He not tampered with Man – Woman communication, He would have known whether in the world some cats are black or some dogs are white. As things stand, He knows that in the world either some cats are black or some dogs are white, but He does not know either that in the world some cats are black or that in the world some dogs are white.

This is Field's 1974 rendition of Quine's indeterminacy thesis. The existence of men – models and women – models amounts to loss of information. God can obtain disjunctive information about the world, but not categorical information. Had God muddled also with children's and adults' use of language, he would have suffered a greater loss of information. We who obtain our information from theories formulated within logical frameworks are worse off than Him.

Discourse formulated within a logical framework is deeply indeterminate. Not only is the meaning of physical terms ('cat', 'black', 'atom', ...) highly indeterminate in such a discourse, but the meaning of mathematical terms ('number', 'set', 'uncountably many' in the standard logical framework) is also indeterminate. Indeterminacy in logical frameworks can be characterized in terms of "nonstandard" models and denotations: indeterminacy of terms is the existence of *nonstandard denotations*, indeterminacy of theories is the existence of *nonstandard models*. The notions of nonstandard model and nonstandard denotation are relativistic notions. Relative to one criterion of meaning ("intended meaning," "preformalized meaning," "standard meaning," etc.), a given denotation is nonstandard, relative to another – standard. I will call indeterminacy relative to an external standard of meaning 'relative indeterminacy'. Relative indeterminacy within a framework is determined by two things: (i) an external standard of meaning and (ii) the framework's expressive resources. A notion (or a theory) is relatively indeterminate within a given framework if and only if the distinctions required for capturing its "intended" meaning cannot be drawn within that framework. I will call a term's inability to distinguish between referents (or a theory's inability to distinguish between models) 'absolute indeterminacy'. 'Absolute indeterminacy' is a nonrelativistic notion underlying 'relative indeterminacy'.[10]

The difference between relative indeterminacy and absolute indeterminacy can be seen by the following example: Consider the notion 'exactly one', defined in a language L of some logical framework $\mathscr{L}$. Let $\mathfrak{A}_1$ and $\mathfrak{A}_2$ be two models for L in $\mathscr{L}$, the first with a universe $A_1 = \{1, 2\}$ and the second with a universe $A_2 =$ {Bill Clinton, George Bush}. The denotation of 'exactly one' in $\mathfrak{A}_1$ is $\{\{1\}, \{2\}\}$, in $\mathfrak{A}_2$ – {{Bill Clinton}, {George Bush}}. Clearly, these two denotations differ one from the other, and in this sense 'exactly one' is absolutely indeterminate in $\mathscr{L}$. 'Exactly one', however, is not relatively indeterminate in $\mathscr{L}$ since its absolute indeterminacy is perfectly compatible with its common meaning. (Relative to that meaning, both denotations are "standard.")

Now consider the concept 'x is president'. Let 'president' be defined by some theory T of L, and let the denotation of 'president' in $\mathfrak{A}_2$ be {Bill Clinton, George Bush}. Then 'president' has a standard denotation in $\mathfrak{A}_2$. Assume $\mathfrak{A}_2$ is isomorphic to $\mathfrak{A}_1$. Then the denotation of 'president' in $\mathfrak{A}_1$ – $\{1, 2\}$ – is nonstandard. The absolute indeterminacy of 'president' leads to relative indeterminacy. The difference between 'president' and 'exactly one' amounts to this: the distinctions required by our external standard of meaning for 'exactly one' *can* be drawn within logical frameworks, whereas those required by our external standard for 'president' *cannot*. We can sum up the distinction between absolute and relative indeterminacy by saying that absolute indeterminacy is the existence of a multiplicity of models (denotations), whereas relative indeterminacy is the existence of nonstandard models (denotations).

To appreciate the depth of indeterminacy in logical frameworks, I will now offer a brief account of absolute indeterminacy in such frameworks.

(Absolute) Indeterminacy in Logical Frameworks

Background Notions

1. *Logical framework.* We regard a logical framework, $\mathscr{L}$, as a pair, $\langle \mathcal{L}, \mathscr{M} \rangle$, or a triple, $\langle \mathcal{L}, \mathscr{M}, \mathscr{P} \rangle$, where $\mathcal{L}$ is a class of formalized languages, $\mathscr{M}$ is a class of apparati of models for languages in $\mathcal{L}$, and $\mathscr{P}$ is a proof system for languages in $\mathcal{L}$. To fix the notion of logical framework, we will restrict ourselves to languages of three kinds: standard first-order languages, generalized first-order languages, and standard higher-order languages, and the corresponding systems of models. In this paper, we disregard $\mathscr{P}$.[11]
2. *Meta-theoretical notions.* The meta-theoretical notions are defined in the usual way. Note in particular the notions of *model for L*, *model of T* (where T is a theory, i.e., set of sentences, in L), and *logical/nonlogical constant of L*.
3. *Term of L.* ξ is *term of L* iff ξ is either a nonlogical constant of L or an open formula of L.
4. *Reference of a term ξ of L in $\mathfrak{A}$.*
   (a) if ξ is a nonlogical constant of L, its reference in $\mathfrak{A}$ is the extension (denotation) assigned to it by $\mathfrak{A}$;
   (b) if ξ is an open formula of L, its reference in $\mathfrak{A}$ is its extension in $\mathfrak{A}$, based on the Tarskian definition of satisfaction-in-a-model for formulae of L.

We will symbolize the reference of the term ξ of L in $\mathfrak{A}$ by $R_{L,\mathfrak{A}}(\xi)$.

5. *Ontology of T.* There are two kinds of theories: (i) theories with a characteristic one-place predicate (primitive or defined) that determines their ontology, (ii) theories without such a predicate. We will say that the *ontological predicate of T* – O_T – is the characteristic predicate of T, if it has one; the self-identity predicate ($x \approx x$), otherwise. The *ontology of T in* $\mathfrak{A}$ is the *referent of O_T in* $\mathfrak{A}$.

Indeterminacy in a logical framework has to do with *variability of reference and ontology under models*. We distinguish four *modes* of indeterminacy and four *types* of indeterminacy.

Modes of indeterminacy. The following four modes have to do with *what is said to be indeterminate*:

(1) *Indeterminacy of terms in L*: variability of the reference of terms of L under models for L.

(2) *Indeterminacy of terms in T*: variability of the reference of terms of L under models of T.
(3) *Indeterminacy of the ontology of T*: variability of the ontology of T under models of T.
(4) *Indeterminacy of T*: indeterminacy of terms in T or indeterminacy of the ontology of T.

Types of Indeterminacy. For each mode, we distinguish four types of indeterminacy, which have to do with *what kind of variations under models* are involved:

(a) *NE-indeterminacy:* variability under non-equivalent models.
(b) *NI-indeterminacy:* variability under nonisomorphic models.
(c) *I-indeterminacy:* variability under isomorphic models.
(d) *A-indeterminacy:* variability under automorphic models.[12]

Combining the mode-type distinctions, we define:

(A) *The* term ξ of L is *NE-/NI-/I-/A-indeterminate in L* iff there are at least two models, $\mathfrak{A}_1$, $\mathfrak{A}_2$ for L such that $\mathfrak{A}_1$, $\mathfrak{A}_2$ are non-equivalent/nonisomorphic/isomorphic/automorphic, and the referent of ξ in $\mathfrak{A}_1$ is different from its referent in $\mathfrak{A}_2$.
(B) *The* term ξ of L is *NE-/NI-/I-/A-indeterminate in T* iff there are at least two models, $\mathfrak{A}_1$, $\mathfrak{A}_2$ of T such that $\mathfrak{A}_1$, $\mathfrak{A}_2$ are non-equivalent/nonisomorphic/isomorphic/automorphic, and the referent of ξ in $\mathfrak{A}_1$ is different from its referent in $\mathfrak{A}_2$.
(C) *The ontology of the theory T* is *NE-/NI-/I-/A-indeterminate* iff there are at least two models, $\mathfrak{A}_1$, $\mathfrak{A}_2$, of T such that $\mathfrak{A}_1$ and $\mathfrak{A}_2$ are non-equivalent/nonisomorphic/isomorphic/automorphic, and the ontology of T in $\mathfrak{A}_1$ is different from its ontology in $\mathfrak{A}_2$.
(D) *The theory T is NE-/NI-/I-/A-indeterminate* iff either some term of L is *NE-/NI-/I-/A*-indeterminate in T or the ontology of T is *NE-/NI-/I-/A*-indeterminate.

Definition (D) is reducible to.

(D′) The theory T is *NE-/NI-/I-/A-indeterminate* iff some term of L is *NE-/NI-/I-/A*-indeterminate in T.

Centering our attention on *theory* indeterminacy, we note the following theses:

(I) *Thesis of NE-indeterminacy:* A theory is *NE*-indeterminate iff it is incomplete.
(II) *Thesis of NI-indeterminacy:* Every theory with an infinite ontology is *NI*-indeterminate in the standard first-order framework.

(III) *Thesis of I-indeterminacy:* Every consistent theory is I-indeterminate.

(IV) *Thesis of A-indeterminacy:* A theory T is A-indeterminate iff there is at least one model $\mathfrak{A}$ of T and at least one term ξ of L such that the referent of ξ in $\mathfrak{A}$ is not closed under all permutations of the universe of $\mathfrak{A}$. (Keenan forthcoming.)[13]

It follows from these theses that every consistent theory formulated within the framework of standard first-order logic (with identity) suffers loss of information: every incomplete theory suffers loss of information about truth; every theory admitting an infinite ontology suffers loss of information about quantity; every (consistent) theory whatsoever suffers loss of information about the identity of its ontology; and every theory with at least one name and an ontology of cardinality larger than 1 suffers loss of information about "who is who" within the said ontology. Thus, take for example, first-order Peano arithmetic: Peano arithmetic fails to determine (i) the truth of some arithmetic statements, (ii) the size of the class of natural numbers, (iii) the identity of the natural numbers [what kind of objects the natural numbers are: von Neumann sets, Zermelo sets, some other kind of object (Benacerraf 1965)], and (iv) what element plays the role of what natural number in what ontology. The loss of information involved in formulating arithmetic within the framework of standard first-order logic is large indeed.

Indeterminacy, however, is loss of information in a more intricate way than suggested so far. Suppose we formulate Peano arithmetic in a first-order language with the logical quantifier 'there are exactly $\aleph_0\, x$ such that....'[14] Within that framework, Peano arithmetic can easily be expanded to a theory that gives us precise information on the quantity of natural numbers: If we add to Peano arithmetic the axiom '$(\aleph_0\, x)x \approx x$', all models of the expanded theory will contain exactly $\aleph_0$ elements. Using the new framework will not free Peano arithmetic from all its indeterminacies, but its *NI*-indeterminacy will be significantly reduced. In particular, the "new" arithmetic will have no models of cardinality larger than $\aleph_0$.

Indeterminacy, then, comes in degrees: the stronger (the more expressive) a given logical framework is, the weaker the indeterminacy it generates. But indeterminacy is not restricted to logical frameworks. Logical frameworks are not the only kind of framework, and relativity to different kinds of framework gives rise to different kinds of indeterminacy, hence to loss of different kinds of information. *Logical* frameworks are marked by loss of *extralogical* information, *experiential* frameworks (frameworks fixing the experiential content of various terms), by loss of *theoretical*, including *logical*, information. Much of the philosophical literature on indeterminacy is concerned with relativity to *logical* frameworks. In Putnam's parable, for example, God mixes up men's and women's understanding of *extralogical terms*, not their understanding of *logical terms*. The *logical* content of human discourse does not vary from gender to gender, only its *extralogical* content does. Likewise, in logical frameworks, *extralogical* content varies from model to model; *logical* content is uniquely determined by the logical framework.

Logical frameworks, however, are (as we have noted) not the only background frameworks. An example of discourse conducted within a different kind of framework – an *experiential* framework – is found in Quine's (imaginary) case study of the field linguist.[15] Spotting a rabbit, the natives (consistently and repeatedly) utter the word (sentence?) 'Gavagai'. The linguist, who shares their observational standpoint, jots 'rabbit?' in his manual. The question mark conveys his conundrum regarding the individuation criterion associated with 'Gavagai': Is 'Gavagai' a word for discrete rabbits? Rabbit stuff (undetached rabbit parts)? As long as the linguist's framework is purely experiential, the individuative status of 'Gavagai' is indeterminate. His triangular experience of observing a rabbity thing,[16] observing the natives observing a rabbity thing, and observing the natives uttering 'Gavagai', does not suffice to individuate the referent of 'Gavagai'. Individuation, Quine notes, is a matter of *logical* parameters: identity, quantification, Boolean operations. But no amount of *experiential evidence* will determine the logical parameters of the native's utterances. Suppose the linguist queries: 'Is this Gavagai the same as that?' A positive answer to this question will not adjudicate the matter. The import of such an answer depends on the native's interpretation of the linguist's question, and as far as the purely experiential data available to the linguist go, the native may be interpreting the question as: 'Does this Gavagai belong with that?' ('Is this Gavagai part of the same stuff as that?') (Quine 1969a, p. 33) The problem is not evidential; the problem is factual. There is simply *no experiential fact of the matter* as to whether a stimulus generated by a rabbit sighting is a stimulus of a discrete rabbit or a stimulus of rabbit stuff. Introducing the idea of *experiential models* (models preserving the experiential features of a given discourse), we can say that *stimulus meaning* (meaning as determined by experiential stimuli) is fixed throughout experiential models, but logical parameters – hence individuation – are not. In all experiential models, 'Gavagai' denotes a rabbity thing, but in some experiential models, it denotes discrete rabbits whereas in others, undetached rabbit parts; in some experiential models 'is the same as' functions as identity; in others, as the relation of belonging with. The *logic* of purely experiential discourse is underdetermined.

Indeterminacy is loss of information: loss of information about who is who, who possesses what properties, who stands in what relations to whom, and so on. Indeterminacy as loss of information is a universal predicament of *partial* frameworks: frameworks that "fix" the logical parameters of discourse but not its observational parameters, frameworks that fix its observational parameters but not its logical parameters, frameworks that fix its logical and observational parameters but not its theoretical parameters,[17] and so on. To the extent that human discourse is commonly conducted within partial frameworks, indeterminacy is a universal predicament of human discourse.

Indeterminacy, we have seen, is a barrier to knowledge, and our account explains why this is so: knowledge means an increase in information, but indeterminacy means loss, or dilution, of information. Indeterminacy, however, is not just a negative element in the generation of knowledge. Indeterminacy, like other phenomena of human cognition, plays a positive as well as a negative role in producing knowledge and information.

IV. Indeterminacy as Specificity of Information

To see what positive role indeterminacy plays in the production of information, let us turn once again to Quine:

> Imagine a fragment of economic theory. Suppose its universe comprises persons, but its predicates are incapable of distinguishing between persons whose incomes are equal. The interpersonal relation of equality of income enjoys, within the theory, the substitutivity property of the identity relation itself; the two relations are indistinguishable. (1969a, p. 55)

Quine's economic theory in effect generates a framework in which individuals with equal economic attributes are not differentiated. But the absence of a more discriminating apparatus of individuation does not detract from the efficiency of the theory. On the contrary: to discriminate individuals according to economically irrelevant features – for example, hair color, hour of birth, favorite movie star – would only introduce clutter into the theory, obscure its content, and decrease its efficiency. Given its goal, the theory's nonstandard method of individuation is well motivated: its identity principle is not undiscriminating; it is tailored to the needs of a highly specialized theory. Indeterminacy, as far as our economic theory is concerned, is *specificity* of information rather than *loss* of information.

The shift from loss of information to specificity of information can be explained as follows: Indeterminacy, in its most general form, is *partiality* of information, but partiality of information is both the presence and the absence of information. Indeterminacy occurs as an element in a pair: *indeterminacy* as the loss or absence of information, complemented by *determinacy*, the presence and specificity of information. Quine's economic theory does not distinguish between individuals with different hair colors (indeterminacy), but does distinguish individuals with different income levels (determinacy); the native's utterance fails to inform us of the identity of that which has passed by an instant ago (indeterminacy), yet by uttering 'Gavagai' the native has succeeded in conveying to us a very specific bit of information, namely, that that which has just passed by is a *rabbity-thing* (rather than an *elephanty-thing* or a *snaky-thing* or . . .) (determinacy). The meaning of 'Gavagai' is partially determinate, partially indeterminate, and it is only by sorting out its determinate elements from its indeterminate ones that its net informative value can be calculated.

Relative indeterminacy is doubly relative: relative to an external standard of meaning, and relative to a task at hand. Relative to one task (e.g., the task of generating a Native-English manual of translation), indeterminacy of individuation is loss of information; relative to another (the task, say, of signaling the presence of edible things), it is not. Relative to some goal, failure to distinguish individuals with different hair color is loss of information; relative to another – specificity of information. To view indeterminacy as specificity of information is to view it as a tool for the generation of determinacies. We can illustrate the positive role that indeterminacy plays in knowledge by a new parable of "God and His Pursuit of Knowledge":

The New Parable of Indeterminacy. When God was young, His interests were all-encompassing. God was interested in every detail of every happening in every corner of the world: whose cat was black, whose dog was white, which pebble lay on what riverbed, and so forth. As God grew older, His interests became refined. Today, God's interests are restricted to the universal laws of nature. And God knows that someday His interests will reach the pinnacle of purity. Someday God will arrive at the age of wisdom, and from that day on His interests will focus on the logical structure of the world. In preparation for this day and the obvious limitations associated with old age, God is spinning a clever plot. We can sum it up by one maxim: Let each human speak his/her own language, but let all use the same logic. God reasons as follows: If all humans assign the same reference (interpretation) to the logical particles of their language but differ in the assignment of reference to the nonlogical particles, then, by sifting through the common elements of their discourse, one could find what logical features they attribute to the world. So God decrees: Let all humans use the same syntax and let them all use the same semantic rules for the logical constants of their language. But let each human use his/her own ontology and his/her own scheme of nonlogical reference. Listening to humans talk, God will hear a human say: ‡Snow is white *or* snow is *not* white‡, and the reverberations of his/her utterance: ‡Grass is green *or* grass is *not* green‡, ‡Sand is blue *or* sand is *not* blue‡, ... And God will know that, given a universe X, an object y *in* X, and a subset Z of X, y is in the union of Z and its complement in X. In symbols: $y \in ZU(X - Z)$. God will have learnt, or will have relearnt, the ontological version of the law of excluded middle. This is God's plan for obtaining logical information. Had God been interested in obtaining physical information, He would have "fixed" the physical constants of human language. Since His plan is to obtain logical information, he is "fixing" its logical constants.[18]

Extralogical indeterminacy is a means of extracting logical information. Logical frameworks are designed to convey a special type of information, namely, logical information; therefore the logical structure of statements and theories formulated within them are fully determinate (relative to their intended meaning), but their extralogical content is not. Extralogical variation is a tool for identifying logical regularities: logical truths (laws), logical consequences, logical consistencies, logical equivalences, and so on. Whether a given indeterminacy is loss of information or specificity of information is thus relative to context:

relative to God's youthful aims, indeterminacy of extralogical vocabulary is loss of information; relative to his golden-age aims – specificity of information.

Our claim that logical frameworks are intrinsically limited to the transmission of logical information challenges a widely held view on the utility of logical frameworks.

V. Does Logic Provide a General Framework for the Construction of Theories?

What is the role of logical frameworks in the development of theories (the transmission of information[19])? Since the birth of modern logic over a century ago, two competing views have emerged. According to one view (the "generalist" view), logic provides a *general, all-purpose* framework for the construction of theories, designed to improve their overall conceptual clarity, enhance their descriptive as well as explanatory capabilities, and increase their predictive power (where applicable). According to the second view (the "specialist" view), logic provides a *special* framework for the formulation of theories, designed to facilitate the discovery of their (distinctively) *logical* consequences, their *logical* consistency or inconsistency, the *logical* dependence or independence of their axioms, and so on. An early locus of both approaches is Frege's (1879) *Begriffsschrift*, where the idea of an artificial symbolic notation is justified on the basis of two kinds of considerations: general "Leibnizian" considerations stressing the need for a universal tool for the precise expression of ideas, and special logical (or, rather, meta-logical) considerations calling for the construction of a specialized device for carrying out logical proofs.

The general benefits of a symbolic notation were emphasized by Frege in the preface to his monograph. Frege opened *Begriffsschrift* with general methodological considerations:

> In apprehending a scientific truth we pass, as a rule, through various degrees of certitude. Perhaps first conjectured on the basis of an insufficient number of particular cases, a general proposition comes to be more and more securely established by being connected with other truths through chains of inferences. . . . Hence we can inquire . . . how we can finally provide it with the most secure foundation. (Ibid., p. 5)

A secure foundation for knowledge cannot be provided within natural language, however, due to its *imprecision* of expression. A more precise notational system is required, one related to natural language as a scientific instrument is to the "naked" eye.

> I believe that I can best make the relation of my ideography to ordinary language clear if I compare it to that which the microscope has to the eye. . . . as soon as scientific goals demand great *sharpness* of resolution, the eye proves to be insufficient. The microscope, on the other hand, is perfectly suited to precisely such goals [Ibid., p. 6].[20]

Frege likened his idea of a symbolic ideography to Leibniz's idea of a universal symbolism: "Leibniz, too, recognized . . . the advantages of an adequate system of notation. His idea of a universal characteristic, . . . a *calculus philosophicus* or *ratiocinator*" was a "worthy goal" which, if realized, would lead to an "immense increase in the intellectual power of mankind." (Ibid.)[21] Even the name that Frege gave to his notation suggests a general goal: 'Begriffsschrift', "a notation of ideas," rather than 'Logikbegriffsschrift', "a notation of *logical* ideas."

Another line of thought developed in the preface, however, points to a narrower conception of Frege's symbolism. Among the statements emphasizing the *logical* nature of Frege's ideography are the following:

> [W]e divide all truths that require justification into two kinds, those for which the proof can be carried out purely by means of *logic* and those for which it must be supported by facts of experience.
>
> The most reliable way of carrying out a proof, obviously, is to follow pure *logic*.
>
> [The] first purpose [of the proposed ideography] is to provide us with the most reliable test of the validity of a [*logical*] *chain of inferences*.
>
> [We exclude from this ideography] anything that is without significance for the [*logical*] *inference sequence*. (Ibid., pp. 5–6, Frege's emphasis)

It is quite clear from these statements that Frege conceived his *Begriffsschrift* as a specifically *logical*, rather than "universal," language. Moreover, it follows from Frege's Logicist Project – the project of reducing mathematical knowledge to *purely logical* knowledge – that the language in which mathematical theories are to be reconstructed is a *purely logical* language. Finally, Frege himself drew a sharp distinction between a "purely logical system" and a "universal system of notation." The two systems, according to Frege, differ in their treatment of objects: "*pure logic . . . disregard[s] the particular characteristics of objects*," while *a Leibnizian "system of notation [is] directly appropriate to objects themselves*" (Ibid., pp. 5–6, my italization). Frege explained the difference between a purely logical language and a general Leibnizian language by differences in their goals: logical languages are designed to express a special kind of laws whereas Leibnizian languages are designed to express laws of objects in general. More specifically, logical languages are intended to express those "laws upon which *all* knowledge rests" and which therefore "transcend all particulars" (Ibid., p. 5); Leibnizian languages are intended to express the whole gamut of laws constituting our knowledge, including laws applicable to objects directly and in their particularity.

The Leibnizian, or generalist, approach to logical frameworks is exemplified in Skolem's explanation of the utility of logic. Starting with intuitive and vague mathematical (or scientific) ideas, we use the resources of modern logic to generate sharply delineated "images" (representations, counterparts) of these ideas. For example, by axiomatizing set theory within the framework of standard first-order logic, we are able to define the intuitive set-theoretical notions of 'function', 'ordinal number', 'cardinal number', 'finitely many', 'uncountably many', and so on in a sharp and precise manner, reducing them to the membership relation whose properties are specified by the axiomatized theory.

The relative indeterminacy of extralogical notions within logical frameworks, however, challenges Skolem's approach. Skolem regarded indeterminacy as the price we pay for increasing the precision of our (extralogical) concepts, but the price we pay is, in effect, a *decrease* in their precision. How can the axiomatization of set theory within the framework of standard first-order logic be said to yield a precise notion of uncountability if any consistent statement of the form 'S is uncountable' is satisfied by a countable model? And how can a first-order formulation of number theory be said to yield a precise notion of natural number, if the quantity (not to say the identity) of objects falling under it is highly indeterminate?

We can explain the failure of logical frameworks to transmit accurate non-logical information by reference to "Frege's principle": it is because logical frameworks do not distinguish 'the particular characteristics of objects' (see above) that notions based on such characteristics cannot be accurately formulated within these frameworks. A contemporary version of Frege's principle is the invariance principle for logical constants:

(*LI*) *Logical Invariance:* Logical constants are invariant under isomorphic argument-structures. (Mostowski 1957, Lindström 1966, Tarski 1966, Sher 1991, and others)

This principle says that if C^n is a logical constant (a logical predicate, quantifier, or function) of a language L of a logical framework $\mathscr{L}$, and $\langle A, \beta_1, \ldots, \beta_n \rangle$ is an argument-structure for C^n – namely, a structure consisting of a universe A followed by n elements of types corresponding to those of the arguments of C^n – then,

If $\mathfrak{A}$ and $\mathfrak{A}'$ are models for L (in $\mathscr{L}$) with universes A and A', respectively, and the argument-structures $\langle A, \beta_1, \ldots, \beta_n \rangle$ and $\langle A', \beta'_1, \ldots, \beta'_n \rangle$ are isomorphic, then $\langle \beta_1, \ldots, \beta_n \rangle$ satisfies C^n in $\mathfrak{A}$ iff $\langle \beta'_1, \ldots, \beta'_n \rangle$ satisfies C^n in $\mathfrak{A}'$.[22]

The intuitive meaning of (*LI*) is that logical constants do not distinguish formally identical objects, or, logical constants discern only formal patterns of objects possessing properties and standing in relations (not their "material" features). Since the expressive power of logical frameworks is largely determined by their logical constants, it follows from (*LI*) that the expressive capability of logical frameworks is restricted to the formal.[23] This restriction is captured by the *thesis of logical I-indeterminacy*:

Extend the notion of term of L to include logical constants of L and extend the notion of 'reference in $\mathfrak{A}$' accordingly.[24]

The Thesis of Logical Indeterminacy. Let $\mathscr{L}$ be a framework satisfying (*LI*), L a language in $\mathscr{L}$, T a theory in L, and ξ a term of L. Then:

(i) ξ is I-indeterminate in L;
(ii) ξ is I-indeterminate in T;
(iii) the ontology of T is I-indeterminate;
(iv) T is I-indeterminate.

Although logical frameworks do not allow the precise expression of non-formal notions, formal notions are, in principle, accurately expressible in such frameworks. This fact is (partly) reflected in the absolute A-determinacy of such notions.

Define 'logical notion' as:

ξ is a *logical notion* of a language L in a logical framework $\mathscr{L}$ iff ξ is either a logical constant of L or an open formula of L with no nonlogical constants.

The Principle of Logical Plenum. If ξ is a logical notion of a language L in a logical framework $\mathscr{L}$, then ξ has exactly the same reference in all automorphic models for L, i.e., ξ is A-determinate.

This principle says that the extension of logical notions in any given model is formally "full" in the sense of:

Closure under Permutations. A notion (term) ξ of L is A-determinate in *L/T* iff for any model $\mathfrak{A}$ for *L*/ of T with universe A, the reference of ξ in $\mathfrak{A}$ is closed under all permutations of A. (Lindenbaum and Tarski 1934–5, Sher 1991, Keenan forthcoming, and others. See Thesis IV, above.)[25]

It is characteristic of formal, or mathematical, notions in general that they can be so formulated as to satisfy the principle of logical plenum and, more generally, the principle of logical invariance (*LI*). For that reason logical frameworks are naturally suited for the expression of formal as well as meta-formal ideas, that is, the ideas of formal law (truth), formal consequence, and so on.[26]

Our analysis of logical frameworks is in the spirit, if not in the letter, of Frege's narrower conception. A logical framework is an instrument designed for a particular purpose. Its primary task is to identify the logical properties and relations of theories, and to this end it is tuned to those features of theories (their referents, ontology) that are relevant to the logical task but not to others. The extralogical information transmitted by a logical framework is largely indeterminate, but the logical skeleton of that information is highly determinate. The logical skeleton of a piece of information is, however, itself a piece of information; therefore, a logical framework can be viewed as a tool for the transmission of *logical information*.

Turning back to Skolem and the axiomatization of set theory, I would say that what this axiomatization achieves is not a general sharpening of the set-theoretical notions (the indeterminacy of '∈' is hardly a sign of sharpness), but rather a sharpening of the logical kernel of these notions. Whether a given indeterminacy means loss or specificity of information (or neither) is largely a matter of what the framework is designed to accomplish. Relative to the "standard" conception of logical consequence (a conception according to which standard first-order logic fully captures the intended notion of logical consequence) the indeterminacy of 'uncountably many' is *not* loss of information (since it does not impede the derivation of any logical consequence), relative to other conceptions [e.g., Sher (1991, 1996a)] it is.

A logical framework is, in general, not an all-purpose framework for the construction of theories, yet sometimes a logical framework is so aligned with a given (preformalized) theory that it is possible to fully express the theory's content by purely logical means. When such an alignment occurs, we say that, for this theory and that logical framework, the *logicist project* is realized. The Logicist Thesis constitutes a bridge between the logical and the Leibnizian projects. Since Frege's goal was to capture the content of mathematical (or, more narrowly, arithmetical) concepts by purely logical means, his language was designed to be at once a logical language and a general language for the expression of mathematical ideas. But even so, Frege's *Begriffsschrift* is inherently logical: it only due to the logical nature of arithmetical notions (according to Frege's position) that *Begriffsschrift* can serve as a general framework for the construction (or reconstruction) of arithmetic.

VI. Full Determinacy as the Absence of Knowledge

Is it possible to construct an altogether general framework for the formulation of theories, a framework in which their logical, experiential, and theoretical constituents are all uniquely determined? Contemplating the possibility of realizing Leibniz's ideal, Frege says:

> The enthusiasm that seized [Leibniz] when he contemplated the ... system of notation [he envisaged] led him to underestimate the difficulties that stand in the way of such an enterprise. But, even if this worthy goal cannot be reached in one leap, we need not despair of a slow, step-by-step approximation.... It is possible to view the signs of arithmetic, geometry, and chemistry as realizations, for specific fields, of Leibniz's idea. The ideography proposed here adds a new one to these fields, indeed the central one, which borders on all the others. If we take our departure from there, we can with the greatest expectation of success proceed to fill the gaps in the existing formula languages, connect their hitherto separated fields into a single domain, and extend this domain to include fields that up to now have lacked such a language.
>
> I am confident that my ideography can be successfully used ... when the foundations of the differential and integral calculus are established.

It seems to me to be easier still to extend the domain of this formula language to include geometry. We would only have to add a few signs for the intuitive relations that occur there. In this way we would obtain a kind of *analysis situs.*

The transition to the pure theory of motion and then to mechanics and physics [where "besides rational necessity empirical necessity asserts itself"] could follow at this point. (Frege 1879, pp. 6–7)

Considering Frege's program in the present context, we can distinguish four ways of transforming a given logical framework into a general conceptual framework. The first two methods have to do with axiomatization of theories within the framework, the last two with adding new "distinguished" constants to the framework (i.e., new constants whose "intended" interpretation is "hardwired" into the framework). The four methods are: (i) axiomatizing theories within the framework, (ii) specifying an intended model (or models) of axiomatized theories, (iii) adding new logical constants to the framework, (iv) adding new extralogical distinguished constants (and making appropriate adjustments in the apparatus of models).[27] Each of these methods has its uses, but each also has its limitations. We have already noted the limitations of the first method. The second method renders the axiomatic method (as a method for capturing the exact content of theories) redundant: if it is possible to single out a model (which, from the point of view of the axiomatization, is indistinguishable from a host of other models), as capturing the precise content of a given theory, the axiomatization itself is superfluous. The third method does lead to a considerable gain in the expressive capabilities of the framework, but this gain is, as we noted earlier, limited to formal notions. Cardinality statements can be expressed with full precision and determinacy, but physical statements cannot.

The fourth method amounts to adding a new layer to the initial logical framework, that is, combining the logical framework with one or more other frameworks, for example, a theoretical physical framework, an experiential framework, or an everyday objectual framework. (A physical framework has physical distinguished constants satisfying a principle of physical invariance[28] and an apparatus of models representing all physically possible structures of objects relative to a given language.) The "layering" method is familiar from other contexts. To design an artifact, for example, an airplane, we integrate a number of scientific theories into a single application guide. Likewise, to arrive at a unique interpretation (unique model, unique reference, etc.) of a real-life discourse or a real scientific theory, we integrate various conceptual frameworks into a single whole. The new conceptual framework treats all constants (or rather, all undefined constants) as distinguished, eliminating relative indeterminacy and zeroing in on a "standard" model. Here, singling out a model is not an act of "deus ex machina"; rather, the selection of models is based on a set of background guidelines brought together deliberately by the combination method.

The combined framework, however, is parasitic upon the constituent frameworks. Just as an applicational system in science validates, rather than cancels, the independent existence of the constituent theories, each accounting for some specific aspect of nature and overlooking all others, so an applicational framework in semantics mandates the existence of the partial constituent frameworks, each designed for the determinate expression of some notions, some elements of theories, but not others. A single model, a single referent, means absolute particularity, but knowledge requires some degree of generality, hence some degree of indeterminacy. Consider, once again, logic. Not only does metalogical knowledge (e.g., knowledge of what follows logically from what) require the existence of a broad array of models and a broad array of referents of extralogical notions (i.e., a high degree of extralogical indeterminacy, including relative indeterminacy), but the logical notions themselves obtain their meaning through the abundance of models and referents (i.e., through the indeterminacy of their extralogical counterparts). Take primitive logical notions, that is, logical constants, first. The standard logical constants are relatively determinate within the standard logical framework, but their relative determinacy involves absolute indeterminacy: it follows from (*LI*) that the standard logical constants are at least I-indeterminate, and in fact, the standard logical constants are also *NI*-indeterminate. Thus, take the extension (reference) of '$\exists$' in two nonisomorphic models, $\mathfrak{A}_1$ and $\mathfrak{A}_2$, whose universes are $\{a\}$ and $\{a, b\}$, respectively, $a \neq b$. The extension of '$\exists$' in $\mathfrak{A}_1$ is $\{\{a\}\}$, and its extension in $\mathfrak{A}_2$ is $\{\{a\}, \{a, b\}\}$; obviously the two extensions are not equal. In a similar way we can show that '$\exists$' is *NE*-indeterminate. It is only in terms of A-determinacy (the Plenum principle) that the standard logical constants are absolutely determinate. The absolute indeterminacy of the logical constants extends to logical notions in general. 'Exactly one' is I-, *NI*-, and *NE*-indeterminate in the absolute sense, just like '$\exists$' ('at least one') and '$\approx$' ('is identical to').

The absolute indeterminacy of the logical notions, however, does not involve the loss of logical information. On the contrary: the pattern of indeterminacy of a given logical notion constitutes its meaning. The meaning of identity is a pattern across models (the pattern '$\langle\langle a_1, a_1\rangle, \langle a_2, a_2\rangle, \ldots, \langle a_\alpha, a_\alpha\rangle, \ldots\rangle$', where '$a_1$', '$a_2$', ..., '$a_\alpha$', ... represent members of the universe of an arbitrary model); the meaning of the existential quantifier is another pattern across models; the meaning of 'exactly one' is a third pattern, and so on and so forth. The logical laws delineate another constant pattern across models. The pattern displayed by the law of the excluded middle – $\forall x(\Phi x V \neg\Phi x)$ – consists, as we have seen, in the universality of a union: the union of any subset of a given ontology with its complement in that ontology. The pattern displayed by the law of noncontradiction – $\neg(\exists x)(\Phi x \mathbin{\&} \neg\Phi x)$ – consists of the emptiness of an intersection: the intersection of any subset of a given ontology and its complement in that ontology. We can characterize a logical law as determinacy bounded by indeterminacy: the determinacy of the pattern represented by $\neg\exists x(- \& -)$ against the indeterminacy of the pattern represented by Φx.

A logical law is a path across a field of indeterminacy. A physical law is a different kind of path, across a different field of indeterminacy. Full indeterminacy is the absence of knowledge, but so too is full determinacy. Knowledge is a network of determinacies against a background of indeterminacies. To generate a concept is to abstract from something (to overlook something). To draw a pattern is to relegate some details to the background. We can see a shooting star in the darkness of night, but not in the brightness of daylight. . . .

NOTES

The impetus for this paper came from Parsons' comments (in conversation) on the interest of Quine's indeterminacy thesis and his numerous observations on the interrelations between logic, ontology, and language. See, for example, Parsons (1965, 1971, 1982, 1983a,b). An earlier version of this paper was read to the Workshop in the Philosophy of Logic and Mathematics at the University of California at Irvine. I am thankful to the participants for insightful comments. I also thank Peter Sher for comments and advice.

1. More literally, the theorem says: Given a formula ("equation") Φ of the first-order calculus of relations ("relatives") with identity such that Φ is finitely valid but not valid (Φ is a "fleeing equation"), then: if D is a domain of objects of *any* infinite cardinality (D is "*at least* denumerably infinite"), Φ is not valid in D ["it is no longer the case" that Φ is "satisfied for arbitrary values of its relative coefficients" (its relational symbols) in D]. That is, if $\neg\Phi$ is satisfiable in *any* infinite domain, it is satisfiable in *every* infinite domain.
2. Free translation:

   "The distinction between the simple or absolute notion of set and the notion obtained by a determinate method for making the former notion precise is essential. The second notion is more precise but, at the same time, is relative to the manner in which it is delimited."

3. Free translation:

   "The fact that axiomatization leads to relativism is sometimes considered the weak point of the axiomatic method. But without reason. Analysis of mathematical thought, determination of fundamental hypotheses, and modes of reasoning, are nothing but an advantage for the science in question. It is not a weakness of a scientific method that it cannot do the impossible."

4. Free translation:

   My point of view is, then, that we ought to use formal systems for the development of mathematical ideas. In this way we will be able to render the mathematical notions and methods precise. . . . If, then, we wish to have a general theory of sets, this theory should also be developed as a formal system. . . . I do not understand why most mathematicians and logicians seem to be unsatisfied with such a notion of set defined by a formal system, but on the contrary speak about the insufficiency of the axiomatic method. Naturally this notion of set has a relative character: it depends on the chosen formal system.

5. Free translation:

   The Löwenheim – Skolem theorem itself is true only within a certain interpretation of the symbols. In particular, if we interpret the symbol $\in$ of a formalized set theory as a two-place predicate analogous to any other [nonlogical] predicate, the Löwenheim–Skolem theorem applies and there exists a denumerable model. But on the other hand, if we treat $\in$ like a logical symbol (quantificational etc.) interpreted as signifying membership, we will, in general, not have a denumerable model.

6. For a semantic account of what it means to treat a constant as logical, see Sher (1991, 1996a, 1996b). Tarski himself (in his 1966 lecture) regarded the higher-order, but not the first-order, membership relation as an admissible logical constant.
7. Text enclosed in "‡" represents God's point of view. The relevant instances of Putnam's star (*) mapping should be obvious from the context.
8. Of course language still suffers from well-known problems of ambiguity: homonymy, amphibology, and so on, but these do not concern us here.
9. (a) Here and later, my general statements apply to reasonably rich languages. Thus, suppose Quine's linguist seeks to translate the natives' "Gavagai" to a language with no logical (hence no individuative) terms. In a translation to such a language, indeterminacy may not arise. (b) The present chain of reasoning is challenged in Sher [1998/99] on the basis of considerations developed in Sections IV–VI. (c) By "standard first-order logic," I mean a system of logic similar to those presented in most textbooks of mathematical logic [e.g., Enderton (1972)]. The adjective "standard" is intended to connote, among other things, the traditional choice of logical constants in such systems.
10. Absolute indeterminacy is, of course, relative to choice of framework.
11. For generalized first-order languages, see Mostowski (1957), Lindström (1966), Barwise and Feferman (1985), and Sher (1991).
12. Models $\mathfrak{A}$, $\mathfrak{B}$ for L with universes A, B, respectively, are non-equivalent iff for at least one sentence σ of L, the truth value of σ in $\mathfrak{A}$ is different from its truth value in $\mathfrak{B}$. $\mathfrak{A}$, $\mathfrak{B}$ are isomorphic iff there is at least one 1-1 function from A onto B that preserves functions and relations in $\mathfrak{A}$. $\mathfrak{A}$, $\mathfrak{B}$ are automorphic iff $\mathfrak{A}$ and $\mathfrak{B}$ are isomorphic and $A = B$. *Note:* The thesis and characterizations based on the notion of *NE*-indeterminacy are less pertinent for my discussion than are those based on *NI*-, *I*-, and *A*-indeterminacy. [The idea of treating *NE*-indeterminacy as a special case of indeterminacy appears in Hansen (1987).]
13. When the extension of ξ in $\mathfrak{A}$ is a subset of A^n or a function from A^n to A, we mean by '$R_{L,\mathfrak{A}}(\xi)$ is closed under a permutation p of A' that $R_{L,\mathfrak{A}}(\xi)$ is closed under the permutation p^* of $\mathcal{P}(A^n)$ – the power set of A^n – or $\mathcal{P}(A^n \times A)$, induced by p'.
14. See references in note 11.
15. (a) There are two ways of approaching Quine's case study as an example of indeterminacy: (i) the linguist himself detects the indeterminacy of the natives' discourse; (ii) it is we, the observes, who detect the indeterminacy of the linguist's understanding of the natives' discourse; the linguist is part of the observed situation. In the present construal, I adopt (i), but this choice is not essential for my point.
    (b) The present construal of the 'Gavagai' indeterminacy as representing loss of information is, of course, offered as a *new* interpretation of Quine's "case study" rather than as a *neutral* report of it.
16. I use 'a rabbity thing' as an individuation-wise neutral expression, that is, an expression that does not distinguish between discrete rabbits and rabbit stuff.

17. We may view some of Putnam's discussions of indeterminacy as relating to background frameworks of this kind.
18. This parable should to be taken with a grain of salt (i.e., as a parable rather than as a foolproof method for determining the logical structure of the world). For example, we did not take into account human fallibility, we assumed human language is rich enough and the number of people large enough to cover all formally possible unions of sets and their complements, we assumed God is not subject to the limitations of ω- (and higher) incompleteness, and so on.
19. In this paper, I treat knowledge essentially as information.
20. (a) Here and in later citations, the emphasis is mine (unless otherwise indicated). (b) The microscope analogy can be interpreted either as supporting a Leibnizian conception of logical languages or as supporting a specialist conception of such languages. In the first case, we view the microscope as an instrument for observing small things in general; in the second – as an instrument for observing things of a special kind.
21. The belief that a universal symbolic language would lead to an "immense increase in the intellectual power of mankind" is attributed by Frege to Leibniz. But Frege himself appears to endorse this belief.
22. To apply (*LI*) to connectives as well as to functions and predicates, we can either add a special entry saying that logical connectives are invariant under identical truth-structures or we can construe the connectives as designating set-theoretical operators: '$\neg$', complement; '&', a family of Cartesian product operators, including intersection (since $A \cap B = \{a : \langle a, a \rangle \in A \times B\}$). [In this connection, see Lindström (1966) and Sher (1991, 1996a,b).]
23. By 'formal' in this paper, I mean 'formal in a semantic rather a syntactic sense'. See Sher (1996a).
24. Given a model $\mathfrak{A}$ with a universe A, the referent of '$\approx$' in $\mathfrak{A}$ is $\{\langle a, a \rangle : a \in A\}$, the referent of '$\exists$' in $\mathfrak{A}$ is $\{B \subseteq A : |B| > 0\}$, and the referents of the truth-functional connectives are based on their analysis either as truth-functional operators (in which case models will be assigned two distinguished elements, T and F) or as set-theoretical operators. (See note 22.) The referent of an open formula containing no nonlogical constants is its extension in $\mathfrak{A}$ based on the Tarskian definition of satisfaction in a model.
25. More precisely, the condition is that the reference of ξ in $\mathfrak{A}$ – $R_{L,\mathfrak{A}}(\xi)$ – is closed under all automorphisms of $\mathfrak{A}$. For example, consider the logical notion 'exactly one', construed either as a primitive or as a defined logical notion in a language L in a logical framework $\mathcal{L}$. Let $\mathfrak{A}$ be a model for L with a universe $A = \{a_1, a_2, a_3, a_4\}$. Then, $R_{L,\mathfrak{A}}$('exactly one') $= \{\{a_1\}, \{a_2\}, \{a_3\}, \{a_4\}\}$. $R_{L,\mathfrak{A}}$ constitutes a *plenum* in A in the sense that, for any permutation p of A and any $X \in R_{L,\mathfrak{A}}$('exactly one'), the image of X under p is already in $R_{L,\mathfrak{A}}$('exactly one').
26. A more detailed account of the "anatomy" of indeterminacy in logical frameworks (and other frameworks of analogous structure) will center on three principles: (i) the principle of distinguished constants, that is, constants whose interpretation is "hard-wired" into the framework, versus nondistinguished constants; (ii) the invariance principle characterizing the distinguished constants (a principle that determines the kind of distinctions that distinguished constants are capable of making), and (iii) the principle of variability of models [a principle that says what structures of objects (relative to a given language) are represented by models (for the language)]. I have discussed these principles at length elsewhere. [See Sher (1991, 1996a,b).]

27. For example, by adding to our logical framework 'metal' and 'conducts electricity' as extralogical distinguished constants, we rule out the existence of models in which the extension of 'metal' is not included in the extension of 'conducts electricity'.
28. A physical invariance principle will essentially say that distinguished physical constants are invariant under physically equivalent conditions.

REFERENCES

Barwise, J., and Feferman, S. (eds.) (1985). *Model–Theoretic Logics* (New York: Springer–Verlag).

Benacerraf, P. (1965). "What Numbers Could Not Be," *Philosophical Review*, 74: 47–73.

Enderton, H. B. (1972). *A Mathematical Introduction to Logic* (New York: Academic Press).

Field, H. (1974). "Quine and the Correspondence Theory," *Philosophical Review*, 83: 200–28.

Frege, G. (1879). *Begriffsschrift: A Formula Language, Modeled upon That of Arithmetic, for Pure Thought*, reprinted in van Heijenoort (1967), pp. 1–82.

Hansen, C. (1987). "Putnam's Indeterminacy Argument: The Skolemization of Absolutely Everything," *Philosophical Studies*, 51: 77–99.

Keenan, E. (forthcoming). "Logical Objects," in *Alonzo Church Memorial Volume*, M. Zeleny and C. A. Anderson (eds.) (Dordrecht: Kluwer).

Keisler, H. J. (1970). "Logic with the Quantifier 'There Exists Uncountably Many'," *Annals of Mathematical Logic*, 1: 1–93.

Lindenbaum, A., and Tarski, A. (1934–5). "On the Limitations of the Means of Expression of Deductive Theories," reprinted in A. Tarski, *Logic, Semantics, Metamathematics*, 2nd Ed., (Indianapolis: Hackett, 1983), pp. 384–92.

Lindström. P. (1966). "First Order Predicate Logic with Generalized Quantifiers," *Theoria*, 32: 186–95.

Löwenheim, L. (1915). "On Possibilities in the Calculus of Relatives," reprinted in van Heijenoort (1967), pp. 228–51.

Mostowski, A. (1957). "On a Generalization of Quantifiers," *Fundamenta Mathematicae*, 44: 12–36.

Parsons, C. (1965). "Frege's Theory of Number," reprinted in Parsons (1983a), pp. 150–75.

Parsons, C. (1971). "Ontology and Mathematics," reprinted in Parsons (1983a), pp. 37–62.

Parsons, C. (1982). "Objects and Logic," *The Monist* 65: 491–516.

Parsons, C. (1983a). *Mathematics in Philosophy: Selected Essays* (Ithaca, NY: Cornell University Press).

Parsons, C. (1983b). "Quine on the Philosophy of Mathematics," in Parsons (1983a), pp. 176–205.

Putnam, H. (1977). "Models and Reality," reprinted in H. Putnam (1983b), pp. 1–25.

Putnam, H. (1978). *Meaning and the Moral Sciences* (London: Routledge and Kegan Paul).

Putnam, H. (1981). *Reason, Truth and History* (Cambridge, UK: Cambridge University Press).

Putnam, H. (1983a). "Introduction: An Overview of the Problem," in Putnam (1983b), pp. vii–xviii.

Putnam, H. (1983b). *Realism and Reason: Philosophical Papers*, Vol. 3 (Cambridge, UK: Cambridge University Press).

Quine, W. V. O. (1958). "Speaking of Objects," reprinted in Quine (1969b), pp. 1–25.

Quine, W. V. O. (1960). *Word & Object* (Cambridge, MA: MIT Press).

Quine, W. V. O. (1969a). "Ontological Relativity," reprinted in Quine, (1969b), pp. 26–68.

Quine, W. V. O. (1969b). *Ontological Relativity & Other Essays* (New York: Columbia University Press).

Sher, G. (1991). *The Bounds of Logic: A Generalized Viewpoint* (Cambridge, MA: MIT Press).

Sher, G. (1996a). "Did Tarski Commit 'Tarski's Fallacy'?" *Journal of Symbolic Logic* 61: 653–86.

Sher, G. (1996b). "Semantics and Logic," in S. Lappin (ed.), *The Handbook of Contemporary Semantic Theory*, (Oxford: Blackwell), pp. 511–37.

Sher, G. (1998/99). "On the Possibility of a Substantive Theory of Truth," *Synthese*, 117: 133–72.

Skolem, T. (1920). "Logico–Combinatorial Investigations in the Satisfiability or Provability of Mathematical Propositions: A Simplified Proof of a Theorem by L. Löwenheim and Generalizations of the Theorem," reprinted in van Heijenoort (1967), pp. 252–63.

Skolem, T. (1922). "Some Remarks on Axiomatized Set Theory," reprinted in van Heijenoort (1967), pp. 290–301.

Skolem, T. (1929). "Über einige Grundlagenfragen der Mathematik," reprinted in Skolem (1970), pp. 227–73.

Skolem, T. (1941). "Sur la Porté du Théorème de Löwenheim-Skolem," reprinted in Skolem (1970), pp. 455–82.

Skolem, T. (1958). "Une Relativisation des Notions Mathématiques Fondamentales," reprinted in Skolem (1970), pp. 633–8.

Skolem, T. (1970). *Selected Works in Logic*, Ed. J. E. Fenstad (Oslo: Universitetsforlaget).

Tarski, A. (1966). "What Are Logical Notions?" In *History and Philosophy of Logic*, Vol. 7 (1986): 143–54.

van Heijenoort, J. (ed.) (1967). *From Frege to Gödel: A Source Book in Mathematical Logic, 1879–1931* (Cambridge, MA: Harvard University Press).

Wang, H. (1970). "A Survey of Skolem's Work in Logic," in Skolem (1970), pp. 2–52.

The Logic of Full Belief

ISAAC LEVI

I

That logic has a prescriptive use as a system of standards regulating the way we ought to think is fairly noncontroversial. Frege (1967, pp. 12–13), the resolute antipsychologist, did not deny it. But Frege also thought that logic was a system of truths. The prescriptive force of logic derived from this together with a fundamental value commitment to believe what is true. According to this view, we ought also to believe the true laws of physics and to reason in conformity with them. To be sure, logic is distinguished from physics by virtue of the special status of its truths as "the most general laws, which prescribe universally the way in which one ought to think if one is to think at all." (Frege 1967, p. 12.) Still, the prescriptive use of logic is not what distinguishes it from physics. Both logic and physics as human activities aim to ascertain truths regarding certain subject matters. Logic differs from other natural sciences only insofar as its subject matter is objective truth itself. The imperative to seek truth and reason in conformity with it ought to be obeyed regardless of whether the true laws invoked are laws of objective truth as Frege took them to be or laws of physics.

Ramsey (1990, p. 80) rightly pointed out that undertaking to fulfill a value commitment to believe what is true is jousting at windmills:

> We may agree that in some sense it is the business of logic to tell us what we ought to think; but the interpretation of this statement raises considerable difficulties. It may be said that we ought to think what is true, but in that sense we are told what to think by the whole of science and not merely by logic. Nor, in this sense, can any justification be found for partial belief; the ideally best thing is that we should have beliefs of degree 1 in all true propositions and beliefs of 0 degree in all false propositions. But this is too high a standard to expect of mortal men, and we must agree that some degree of doubt or even of error may be humanly speaking justified.

Even if we grant that in some sense or other the laws of logic are conceptual necessities, they are not conceptual necessities in a sense that guarantees that we are capable of assenting to them, upon demand, either in language or behavior. In this respect too, they are like the true laws of physics.

In one respect, however, they may be understood to be different. To the extent to which we fail to fulfill the commitment to believe logical truths, the remedy for the failure is to be found in more training (in logic and mathematics and their use in applications), in the use of prosthetic devices (such as paper and pencil or tables and computer technology), and in various forms of psychotherapy. When we fail to fulfil the commitment to believe extralogical truths, we must supplement the use of these therapeutic devices with *inquiry*. We consult the testimony of the senses and reliable witnesses and authority and we draw ampliative inferences as in statistical and inductive inference or in choosing which of rival theories to come to believe.

The laws of deductive logic are, I am suggesting, different from the laws of physics not only because they are true but because rational agents are committed by the standards of rational health to full belief that they are true. We are not committed at any given time or period of time t to fully believe (or to be certain of) all truths at that time t, but we are committed at that time t to have full beliefs that are logically consistent and to fully believe all the logical consequences of what we fully believe at that time. We no doubt fail to fulfill this commitment; indeed, we cannot do so in forming our doxastic dispositions and manifesting them linguistically and in our other behavior. We are, nonetheless, committed to doing so in the sense that we are obliged to fulfill the commitment insofar as we are able to do so when the demand arises and, in addition, have an obligation to improve our capacities by training, therapy, or the use of prosthetic devices, provided that the opportunity is available and the costs are not prohibitive. (Levi 1991, Ch. 2.)

Thus, we are committed to fully believing at every time t the truths of logic because we are committed as rational agents to having full beliefs at t that are at once consistent and closed under deductive consequence. Insofar as we fail to fulfill this commitment, the defect is in us and we must seek remedies for our deficiencies. Failure to believe the true laws of physics reflects no such defect in our doxastic condition. Indeed, it is often a mark of a healthy mind to be able to acknowledge our ignorance. Removing such ignorance calls for a *change* in doxastic commitment – that is, a change from a commitment to one system of consistent and closed full beliefs to another such commitment. This view stands in opposition to Frege's (and to Russell's) view according to which our commitment to believing the truths of logic derives from our commitment to believing truth, including in particular the true laws of the science of truth. According to this view, all doxastic changes alter the extent to which we fulfil our single commitment to believe what is true. The commitment never changes. Only our attempts to fulfill it do.

The opposition might not seem as sharp as I am advertising it to be, given a certain explication of the special status of logical truth. According to the standard of doxastic health, rational agents are committed to fully believing all logical theses and, hence, to judging (i.e., to being certain) that theses of deductive logic are true. In this sense, a logical truth is any candidate belief that all rational agents are committed to fully believing according to the standards of conceptual health.

The opposition may be mitigated but it is not eliminated in this way. As I understand Frege's view, the status of logical truths as truths does not derive from standards of rational doxastic health and does not characterize our doxastic commitments. We may be committed to believing logical truths, but our commitment is the same with respect to the laws of physics. Frege does not allow for a distinction between failures of commitment that are removed by therapy, training, and the use of prosthetic devices on the one hand and failures that are removed by inquiry on the other.

The contrast I charge Frege with neglecting or denying can be seen as a contrast between two senses of change of belief. Some changes of belief are best seen as attempts to fulfil unfulfilled commitments. These are of the sort that Peirce described as "explicative." Deductive inference is the classic example. According to Ramsey, who endorsed Peirce's distinction, deduction "is merely a method of arranging our knowledge and eliminating inconsistencies or contradictions" (1990, p. 82.). I prefer to understand such changes over time as changes in belief that incur no change in doxastic commitment during the process but only a change in fulfilling such commitments. Ampliative reasoning, by way of contrast, seeks to justify changing from a relatively uninformative commitment to full belief to a stronger one. Ramsey, following Peirce, saw ampliative reasoning as a way of acquiring new knowledge. For Ramsey, the standard ways of doing so are through observation, memory, and induction. I would delete memory from this list or at least accord it an ambiguous status since the use of memory is often in the service of fulfilling commitments already undertaken. But setting aside the ambiguous status of memory, my proposed gloss on the Peirce–Ramsey contrast seeks to draw the distinction between explicative and ampliative changes in belief in terms of changes that are improvements in the fulfillment of commitments already undertaken or are the product of improvements in capacity to fulfill such commitments and changes in commitments that are the product of inquiry.

Elaboration of this view of the matter calls for three main lines of investigation. First, one must provide some sort of account of what a state of doxastic commitment (commitment to full belief) is. In terms of such an account, one can then discuss failures to fulfill doxastic commitments and changes in belief that represent improvements in fulfilling commitments. One also can provide an account of changes in doxastic commitment. Second, one must provide some sort of account of the techniques involved in improving fulfillment of commitment. Armchair philosophical reflection will not be sufficient here. Investigation of the psychology of learning, the development of appropriate psychotherapies, the identification of relevant technologies, and the study of logic and mathematics are all important. Finally, one needs an account of the intelligent conduct of inquiry. In this connection, philosophical reflection on the invariant features of intelligently conducted inquiry can play some role, provided it is recognized that the proposed features of inquiry should be made to square somehow with what is noncontroversially recognized as the intelligent conduct of inquiry.

Ramsey proposed to distinguish between a lesser logic or the logic of consistency and a larger logic or the logic of discovery or induction (Ramsey, p. 82.). As I am glossing him, the logic of consistency for full beliefs is logic in the sense in which logic characterizes the doxastic commitments undertaken in a given state of full belief. Given any agent X at any given time t, the logic of consistency specifies conditions that X's full beliefs at t should meet to satisfy the commitments that X has undertaken at t. This logic of consistency is, therefore, addressed to partially answering the first of the three questions just itemized: What are the commitments generated by a state of full belief?

Ramsey conceded that insofar as we are in a position to characterize the set of logical truths independently of their role in characterizing doxastic commitments, we might proceed as Frege had done by first specifying the set of logical truths as "objective" truths and then prescribing a commitment on the part of rational agents to those truths and thereby obtaining the logic of consistency. Ramsey registered an unexplained reservation with this concession endorsed by him for the sake of the argument. One possible reason is that insofar as logical theses can be sensibly explicated without appeal to their role in characterizing doxastic commitments, logical truths are characterized without any indication of their intended application. They constitute a formal structure that may and does have many applications including the prescriptive one under consideration here.[1] As a constituent of a formal structure, a logical thesis is not a truth at all. Frege's claim may then be understood as claiming that one application of a pure formal logic is to the systematic characterization of the objective logical truths. Skeptics might reasonably wonder whether Frege succeeded in furnishing an application of pure logic at all. Waiving that objection (which may not have been Ramsey's worry), when it comes to deductive logic, there is a prima facie case for saying that the "logic of consistency" coincides with the "logic of truth" – that is, a science of logical truths.

Ramsey's chief concern was not with the details of probability logic – that is, a logic regulating degrees of belief in the sense of degrees of subjective or credal probability. He argued that there is a logic of consistency for judgments of credal probability just as there is for full beliefs. If an agent judges hypothesis h to be probable to degree $1/3$, the agent is, thereby, committed to judging $\sim h$ to be probable to degree $2/3$ just as the agent who fully believes that h is committed to full belief that $h \vee f$. But the commitment to credal probability judgment involves no commitment to judgments that are true or false except those required by the logic of full belief. (1990, p. 93.) Probability logic requires the agent to fully believe and, hence, assign credal probability 1 to all truths of deductive logic. If the agent fully believes some extralogical propositions, the agent is committed to judging them probable to degree 1 and to fully believing and assigning probability 1 to their logical consequences. These are the only truths to which probability logic requires a commitment. For this reason, Ramsey denied that the logic of consistency for probability coincided with a logic of truth for probability. Ramsey explored a logic of truth for probability appealing to reasoning about frequencies. But, as Ramsey appreciated, the relation of frequencies to subjective probability provides no basis for equating the logic of consistency and of truth for probabilities.

I wish to press Ramsey's point still further and to argue that his concession to Frege's view in the case of full belief needs to be modified. The logic of consistency for full belief, so I will argue, does not coincide with the logic of truth.

II

Why should we insist on requiring deductive closure and consistency as part of the standard of doxastic health? What is the point of insisting that rational agents are committed to meeting such a standard?

Decision theorists, economists, statisticians, and philosophers interested in rational deliberation and inquiry commonly presuppose that agents who assign positive credal probability to a hypothesis judge that that hypothesis is possibly true. Presumably, the linguistic and behavioral manifestations of the judgment of possibility, like such manifestations of judgments of credal probability, reveal propositional attitudes that are subjective in the banal sense that they are propositional attitudes. I contend that commitment to judgments of subjective possibility and impossibility is tantamount to a commitment to a set of full beliefs. Because of the central relevance of judgments of serious possibility to deliberation and inquiry, I call such judgments, judgments of *serious possibility*. The *standard for serious possibility* (SSP) *condition* formally characterizes the bridge between commitments to full belief at a time and judgments of serious possibility at that time.

(SSP) For every X and *t*, X is committed at *t* to fully believe that *h* if and only if X is committed at *t* to judge $\sim h$ impossible. For every X and *t*, X is committed not to believe that *h* at *t* if and only if X is committed at *t* to judge that $\sim h$ is possible.

According to the SSP principle, an agent at a time may have commitments to fully believe and commitments not to fully believe. SSP by itself does not rule out cases in which X lacks both a commitment at *t* to believe that *h* and a commitment at *t* not to believe that *h*. If such a lack of commitment were to arise, X would have no commitment to judge *h* possible or to judge *h* not possible. X's state of full belief would fail to serve the function of a standard for serious possibility with respect to *h*.

It would, of course, be undesirable to rule out cases in which the standard for serious possibility does not assess serious possibility with respect to h. SSP states only that if assessing h with respect to serious possibility is off the charts, so is commitment to believe that h and commitment not to believe that h.

However, when attention is restricted to the domain of objects that are candidates for doxastic commitment and modal judgments, SSP is too weak and needs supplementation by the following *principle of disbelief* (~B):

(~B) X is not committed at t to full belief that h if and only if X is committed at t not to believe that h.

SSP and ~B alone do not prevent X at t from being committed to judge that both h and $\sim h$ are impossible. Supplementing SSP with the principle of deductive consistency does. But consistency and SSP do not prohibit X at t from being committed to judging $\sim h$ and judging $h \wedge \sim f$ both impossible while judging both f and $\sim f$ possible. The addition of ~B and intuitionistic deductive closure prevents this. But unless we insist on closure under classical deductive consequence, there is no way that X could be committed to judging h possible or impossible. X could not be committed to suspend judgment (be in doubt) concerning h and $\sim h$; for this requires judging both h and $\sim h$ to be possibilities. If X is committed to fully believing (not to fully believing) that $\sim h$, X would be committed to judging $\sim\sim h$ impossible (possible) without commitment to judging h impossible (possible). We could add a clause to SSP providing for judging h impossible (possible) if and only if $\sim h$ is (not) fully believed. But adding this clause would ensure closure under classical logical consequence. The principles of classical deductive consistency and closure are needed as supplements to SSP and ~B to ensure that judgments of serious possibility can provide a basis for making coherent credal probability judgments in a self-critical way.[2] As I shall argue shortly, another principle is needed as well. Before turning to that matter, I shall elaborate some more on the difference between commitment to full belief and fulfilling such commitments.[3]

III

The previous discussion suggests that the sentence "X fully believes at t that h" may be understood in three ways: It may represent a feature of X's doxastic commitment at t, it may represent a set of dispositions to assent and to act that generate such a doxastic commitment or fulfillment of the doxastic commitment already undertaken, or it may represent manifestations of such dispositions. I shall follow the policy of expressing the first sense by "X is committed at t to full belief that h," the second sense by "X fully believes at t that h," and the third sense by "X manifests at t full belief that h."[4]

The sentence h that appears in the that clause is to be understood here as part of a linguistic apparatus used to represent X's state of full belief or doxastic commitment. It need not be ingredient in the language used by X in manifesting X's full beliefs. It does not matter whether h is a sentence in some natural language or in some suitably regimented language L. The sentences in L can be used to represent X's doxastic commitments, the extent of X's fulfillment of these commitments by having the requisite full beliefs, and, indeed, the manifestations of such commitments. The sentences and sets of sentences are not used to manifest beliefs as one does in sincerely making a statement. They are used instead to represent doxastic commitments just as a geometric structure is used as a phase space to represent mechanical states.

Thus, I propose to represent X's state of full belief (insofar as it is representable in L) by a set $K_{X,t}$ of sentences in L that I shall call X's *corpus* at t. X is committed at t to full belief that h if and only if h is a member of $K_{X,t}$. I employ a regimented language L that comes with a consequence relation $A \vdash h$ between a sets of sentences A in L and a sentence h in L for a sentential or first order deductive logic. The logical theses of L are consequences in L of any set of sentences in L. I shall use sets of sentences closed under deductive consequence and consistent sets of sentences. These products of the structure L are introduced without any consideration of truth or falsity.

X believes at t that h if and only if X at t has dispositions directly generating the commitment to full belief that h. Two features of this characterization need to be explained at least briefly.

First, having the dispositions is not sufficient for full belief. Just as, on some occasions, a person may sign on the dotted line without contracting to do something, so too X might have and manifest dispositions to assent and behave that in some contexts directly generate commitment to full belief that h but fail to do so in the given situation. Whether having and manifesting certain dispositions fulfill or fail to fulfill the conditions for directly generating commitment to full belief that h cannot be settled without appealing to normative considerations. Settling the issue is, however, to decide whether the dispositions in question are interpretable as believing at t that h. The task of interpretation is to identify the obligations that the agent has incurred in acquiring or having those dispositions just as it is when signing on the dotted line is interpreted as making a promise.

The second feature of my understanding of "X believes at t that h" is that it claims that the X's dispositions *directly* generate doxastic commitments. The reason for insisting on the qualification that the commitments be incurred directly is that X might be committed at t to full belief that $h \vee f$ by virtue of fully believing that h without fulfilling that commitment by fully believing that $h \vee f$. X has dispositions at t that incur a commitment at that time to believe that h and, as a consequence, a commitment to believe that $h \vee f$. He may not then have the dispositions to assent and act that would have generated the commitment to full belief that $h \vee f$ whether or not he fully believed that h (or that f) or was committed to doing so. Thus, the dispositions that directly generate the commitment to full belief that $h \vee f$ partially fulfill the commitment undertaken by X to believe that h without generating the latter commitment either directly or indirectly. The very same dispositions directly generating commitment to full belief that $h \vee f$ simultaneously partially fulfill that commitment. Consequently, the notion of direct generation of a doxastic commitment may be replaced in the formula characterizing full belief as follows:

X fully believes at t that h if and only if X has dispositions to assent and behave at time t that not only commit X at time t to fully believe that h but partially fulfill the commitment.

I take for granted here that commitment to full belief that h is also commitment to full belief that $h \vee f$. In so doing, I invoke some normative principles of the logic of consistency for full belief. Without a logic of consistency for full belief (and judgments of serious possibility), there is no system of normative principles characterizing commitment to full belief or the notion of full belief as a set of dispositions to assent and act, directly generating such commitments. We return, therefore, to the discussion of this logic of consistency.

IV

X's state of doxastic commitment (representable in L) at t shall be represented by the corpus $K_{X,t}$ of sentences in L; h in L belongs to $K_{X,t}$ if and only X is committed at t to full belief that h (whether X has dispositions directly generating such commitment or not). By the ~B principle, sentence g in L but not in $K_{X,t}$ represents X's commitment at t not to fully believe that g. SSP commits X at t to judge h seriously possible if and only if $\sim h$ is not in $K_{X,t}$. The principle of deductive consistency and closure requires $K_{X,t}$ to be deductively consistent and closed. X's full beliefs at t are "fully coherent" just in case X's set of full beliefs at t are representable by the same set of sentences that represents his doxastic commitments at t – that is, X's corpus at t. The logic of consistency for full belief is the set of normative principles spelling out those conditions that a set of sentences in L should satisfy to represent X's state of doxastic commitment – that is, X's state of full belief. In that sense, they characterize potential states of full belief in terms of potential corpora.

The principle of deductive consistency and closure commits every rational agent X at all times (1) to full belief that T, where T is a logical thesis; (2) to full belief that h if X is committed to full belief that x for every x in A such that $A \vdash h$. By principle ~B, X is committed at all times (3) not to believe any sentence F that is the negation of a logical thesis. These requirements together with SSP commit X at t (4) to judge seriously possible all and only those sentences in L that are logically consistent with $K_{X,t}$.

Notice that all rational agents at all times are committed to full beliefs representable by logical theses in L. In this sense, logical theses are universally valid. Are they true? More important, are the full beliefs they represent true? It is noncontroversial that if X at t fully believes that h, X fully believes that h is true. That is, believing that h is judging that h is true. So, every agent is committed to judging that logical theses are true. So, rational agents are committed to a consensus that theses of deductive logic are true. To the extent that an answer to the question manifests a judgment (full belief) by a rational agent fulfilling his or her doxastic commitment, the answer by universal commitment is: Yes! Logical theses in L and the beliefs they represent are true.

Similarly, by universal commitment, logical consequence is truth preserving, and logical contradictions are false.

Thus, the theses of deductive logic in L represent beliefs to which every agent at all times is committed if their beliefs satisfy the requirements of coherence specified by the principles of deductive consistency and closure. They also characterize a logic of truth by unanimous commitment. If the principles of deductive consistency and closure were to constitute a complete logic of consistency for full belief, there is an important sense in which the logic of consistency for full belief would coincide with a logic of truth. But a logic of truth so conceived is a logic of truth by unanimous commitment. The logic of truth by unanimous commitment presupposes the normative principles of the logic of consistency just laid down. Whether this characterization of a logic of truth for full belief would have been sufficiently nonpsychologistic to satisfy Frege is a matter I leave to Frege scholars. It is for me. Even so, I do not think the logic of consistency for full belief does coincide with a logic of truth for full belief; for the principle of deductive consistency and closure does not exhaust the logic of consistency for full belief; and the additional principles do not impose commitment to full beliefs representable by additional sentences that are universally valid and, hence, true by universal commitment.

Given a universally mandatory commitment to a deductively consistent and closed system of full beliefs and to ~B, it follows that rational agents are committed to satisfying the following *Opinionation Condition* (OC):

For every X at every time and for any h, either X is committed at t to full belief that h, to full belief that $\sim h$, or to suspension of judgment between h and $\sim h$.

For every h in L, either X is committed to full belief that h or X is not so committed. By ~B, it follows that either X is committed to full belief that h or committed not to believe that h. Consistency rules out being committed to full belief that h and at the same time to $\sim h$. Commitment to full belief that h generates a commitment not to believe that $\sim h$. But being committed not to believe that h and also not to believe that $\sim h$ is not ruled out. Even so, deductive closure requires commitment to full belief that $h \vee \sim h$. OC thus follows from ~B: deductive consistency and closure when commitment to suspension of judgment between h and $\sim h$ is understood to be a commitment to full belief that $h \vee \sim h$ without commitment to full belief that h and without commitment to full belief that $\sim h$. Commitment to suspension of judgment between h and $\sim h$ becomes, via the SSP and ~B principles, equivalent to a commitment to judge both h and $\sim h$ seriously possible.

I take for granted that X's belief at t that h and, indeed, X's commitment to belief at t that h may coherently be judged (by X and Y at some time or other) true or false. I also take it for granted that the full belief of someone or other at sometime or other that X at t judges it possible that h may be judged by someone at sometime to be true or false. But it would be incoherent of anyone at anytime to judge X's judgment of serious possibility that h at t to be true or false. Full beliefs may be judged true or false but not judgments of serious possibility. In judging both h and $\sim h$ to be serious possibilities at t, X is no doubt committed to full belief that $h \vee \sim h$ and by deductive consistency and closure is committed to fully believing it. So this full belief is truth valued and, indeed, by unanimous commitment, is judged true. But X is also committed not to believe that h and not to believe that $\sim h$. Commitments not to believe are not themselves truth valued. Consequently, insofar as X's state of doxastic commitment at t serves as X's standard for serious possibility via SSP and ~B, the logic of consistency of full belief is simultaneously a logic of consistency for judgments of serious possibility. Possibility judgments cannot carry truth values without implying that probability judgments do so as well. Ramsey rightly concluded that the logic of consistency for credal probability judgment cannot coincide with whatever may pass as a logic of truth for credal probability. A parallel observation is appropriate regarding the logic of consistency for judgments of serious possibility and a logic of truth for such judgments. But the logic of consistency for judgments of serious possibility is automatically a logic of consistency for full belief. The logic of truth for full belief (if there be such) cannot coincide with the logic of consistency.

It seems to me, however, that this is not the end of the story. Granted that the logic of consistency for full belief combined with SSP and ~B cannot be a logic of truth for judgments of serious possibility, the logic of consistency for full belief may still remain a logic of truth for full belief. This claim needs some examination.

V

Rational agents are not merely agents who have doxastic commitments that they fulfill to some partial extent and fail to fulfill otherwise. Nor are the prescriptions characterizing doxastic commitments designed merely to be used by other agents to determine the extent to which the agent under study is or is not fulfilling his or her commitments. Rational agents are supposed to use the logic of consistency to criticize their own performance. They are not merely rational automata who may or may not be well designed to conform to the demands of the logic of consistency but are self-critical agents who can apply the logic of consistency to their own point of view. Self-reflection, so I argue, calls for a logic of consistency that includes deductive logic but calls for a larger network of commitments to full belief than those demanded by deductive logic. It is in this setting that it becomes apparent that the logic of consistency for full beliefs cannot coincide with a logic of truth (if there be such) for full belief.

Self-reflecting agents are committed to deductively closed and consistent sets of full beliefs but are also committed to identifying what those states of commitment are. In particular, all agents at all times are committed to beliefs in conformity with the following principle:

(BB) If agent X at t is committed to fully believing that h, X at t is committed to fully believing that X at t fully believes that h.

Since fully believing that h presupposes a commitment to fully believe that h, fulfilling the commitments generated by this BB-principle requires that an agent who fully believes that h fully believes that he or she fully believes that h. If X fully believes that h but fails to fully believe that X fully believes that h has not completely fulfilled the commitments undertaken in fully believing that h, just as X has not succeeded in fulfilling such commitments in failing to fully believe that $h \vee g$. That is, X has dispositions to assent interpreted as directly generating X's commitment to full belief that h but does not have such dispositions interpretable as directly generating commitment to full belief that $h \vee g$.[5]

An automaton might be designed to have dispositions to assent and behave that we interpret as simulating manifestations of a doxastic state that is consistent and deductively closed. But we should not understand the automaton as having commitments to full belief that it either fully or partially fulfills (as having full belief in the strict or weaker sense) unless we also are prepared to interpret the automaton as being self-critical – perhaps, by virtue of its exhibiting manifestations of dispositions to assent and behave in ways that we can interpret as detecting and removing inconsistencies, failures of deductive closure, and elimination of self-deception. If an automaton is interpretable as a self-critical rational agent, then we interpret the automaton as interpreting its own behaviors as generating obligations and partially fulfilling them or not doing so as the case might be.

Whether automata are constructible or exist in nature and may be accorded the status of rational agents rather than that of rational automata that conform to a logic of consistency by design or by natural law is an issue into which I shall not enter. I do take for granted that at least human beings are, with relatively rare exceptions, rational agents for much of their careers and that some social institutions sometimes are. Rovane (1994) has pointed out the interesting possibility that, on some occasions, so-called multiple personalities may be interpretable as several rational agents housed in a single human body. The point I mean to belabor here is that rational agents are understood as self-critical agents who, among other things, are held up to the standard of rational health requiring consistency and logical closure of full belief and satisfaction of the BB principle.

The self-critical dimension of rational agency demands that rational agents have additional obligations. To identify the state of doxastic commitment, the agent is also committed to fully believing that *h* is not part of the doxastic commitment, if it is not.

(B~B) If X at *t* is not committed to full belief that *h*, X at *t* is committed to fully believing that X does not fully believe at *t* that *h*.

As in the case of the BB principle, deliberating agents do fail to fulfill this B~B principle fully. An agent may fail to believe that he or she does not believe that *h* even though he or she does not have the commitment to believe that *h*. Lack of a commitment to fully believe that *h* can arise because the agent is committed to fully believing that ~*h* or because the agent is committed to being in suspense between *h* and ~*h* – that is, to judging both *h* and ~*h* to be serious possibilities. If, in such situation, the agent fully believes that he or she fully believes that *h*, the B~B principle is violated. Such violation may be due to lack of the requisite computational capacity, or it may be attributable to some psychological disorder. These forms of self-deception are just as much failures to satisfy the demands of the logic of consistency for full belief as is failure to satisfy the BB principle or the requirements of deductive consistency and closure. To understand X's full belief that X fully believes that *h* to be a failure to fulfill a commitment is to claim that the dispositions that would normally be interpreted to generate a commitment to believe that X believes that *h* cannot be so interpreted on this occasion but are better construed as a failed attempt to do so. As before, the remedy for this is to be sought through training, therapy, and the use of prosthetic devices.

Imposing the B~B principle as a feature of the logic of consistency for full belief has seemed more controversial than imposing the BB principle. However, it must be imposed if we are to supplement the logic of consistency for full belief with a logic of consistency for probability judgment. A self-reflective agent will be required in deliberation to identify his or her credal probability judgments, including credal probability judgments assigning positive probability to hypotheses and to their negations. That is to say, the deliberating agent will be committed to full belief that his or her credal probability that *h* is positive if it is positive. But assigning positive probability that *h* commits the agent to judging

h to be seriously possible – that is, to not fully believing that $\sim h$. The agent who is committed to full belief that his or her credal probability that h is positive and, therefore, to not fully believing that $\sim h$ should also be committed to fully believing that he or she does not fully believe that $\sim h$. Otherwise, the agent cannot be using his or her state of full belief as a standard for serious possibility defining the space of possibilities over which credal probability judgments are made. The B$\sim$B principle seems to be required by appeal to the demands of self-criticism extended to cover credal probability judgment.

The appeal to the requirement that rational agents are committed to the exercise of critical control over their own doxastic performance to determine how well they fulfill their doxastic commitments thus justifies supplementing the constraints on doxastic commitment generated by deductive logic with two additional constraints: the BB principle according to which an agent committed to fully believing that h is committed to fully believing that he or she fully believes that h, and the B$\sim$B principle according to which an agent who is not committed to fully believing that h is committed to fully believing that he or she does not fully believe that h.

Commitment to conformity to the BB and B$\sim$B principles does not, however, generate a commitment to believe a certain set of extended logical truths additional to the truths of deductive logic in the way that the principles of logical consistency and closure do. To come to grips with this point, we need to examine more closely what a logically true or valid proposition should be taken to be in developing a logic of consistency for full belief.

VI

According to the account sketched initially, the standard of rational health for full belief spelled out by the requirements of logical closure and consistency requires that all rational agents be committed to full beliefs that are closed under logical consequence and are consistent. This requirement demands that rational agents be committed to full belief in all logical truths no matter what other full beliefs they may have. This condition may be spelled out slightly more formally as follows:

The logical theses expressible in L are *positively valid for agent* X *at t* in the sense that no matter what potential corpus K in L is X's corpus at t, all of the theses of deductive logic expressible in L are in K. Deductive closure ensures this to be so.

The logical theses expressible in L are *negatively valid for agent* X *at t* in the sense that the negation of a logical thesis is never a member of X's corpus at t no matter what that potential corpus might be. Consistency and positive validity requires this.

The logical theses expressible in L are *omnitemporally valid for agent* X in the sense that, for given X, they are positively (and, hence, negatively) valid for X at every time t.

The logical theses expressible in L are *universally valid* in the sense that they are omnitemporally valid for every agent X. The universally valid theses in L are those that are judged true by universal commitment. Here "universal" means no matter who the agent is, no matter the time the beliefs are held, and no matter the potential belief state that the agent is in at the time – that is, no matter what the agent's creed is at the time. So, no one should resist acknowledging that they are true and, indeed, not only true but valid in the sense that they would be true no matter what might be the case.[6]

Extralogical theses expressible in L are neither universally valid, omnitemporally valid, positively valid, nor negatively valid.

Consider now a language ML that contains L but also contains sentences of the form X believes at t that h (where h is a sentence in ML together with all the truth functional and, perhaps, quantifications over the variables X and t). A potential corpus in MK is a deductively consistent and closed set of sentences in ML that for each X and t satisfies the requirements of BB and B~B. X's corpus $K_{X,t}$ at t in L is to be understood as the intersection of X's corpus $MK_{X,t}$ at t in ML and L. As before, a sentence g in ML is in $MK_{X,t}$ if and only if X is committed at t to fully believe that g. Thus, if X is committed at t to fully believe that X fully believes at t that h, $MK_{X,t}$ contains "X fully believes at t that X fully believes at t that h." The distinction between negatively valid theses for X at t, positively valid theses for X at t, omnitemporally valid theses for X, and universally valid theses carries over to sentences in ML.

If h (in L) is in $K_{X,t}$, the sentence "X believes at t that h" is in $MK_{X,t}$ by virtue of the BB principle. So is h. Hence, sentence BB(X, t, h) in ML that asserts that $h \supset$ X fully believes at t that h is in $MK_{X,t}$. If $\sim h$ is in K, the sentence BB(X, t, h) is still in $MK_{X,t}$. But BB(X, t, h) is not positively valid for X at t. When neither h nor $\sim h$ is in K, X at t is not committed to full belief that BB(X, t, h). Hence, BB(X, t, h) is not positively valid for X at t. BB(X, t, h) is negatively valid for X at t. Its negation is not in $MK_{X,t}$. That is, it would violate the standards of doxastic consistency for X at t to fully believe that ~BB(X, t, h). Clearly BB(X, t, h) is neither omnitemporally valid for X nor universally valid. BB(X, t, h) does not qualify as a logical truth in any respect that the theses of deductive logic do except with respect to negative validity.

Consider, however, the sentence BB(X, t, X fully believes at t that h) that asserts the following in ML: X fully believes at t that $h \supset$ X fully believes at t that X fully believes at t that h. This is a sentence in $MK_{X,t}$ whether X is committed at t to full belief that h, to full belief that $\sim h$, or to being in suspense. If X is committed at t to full belief that h, X is committed at t to full belief that

X believes at t that h. That is required by the BB principle. "X fully believes at t that X fully believes at t that h" is in $MK_{X,t}$ if and only if X is committed at t to full belief that h. So, "X fully believes at t that X fully believes at t that h" is in $MK_{X,t}$. Deductive closure then requires that BB(X, t, X fully believes at t that h) should also be in MK. In case X at t is committed to full belief that $\sim h$ or to suspense between h and $\sim h$, X is committed not to believe that h. So, h is not in MK. By deductive closure, BB(X, t, X fully believes at t that h) is in MK. It is positively valid for X at t, but it is not omnitemporally valid. Let MK' be X's corpus at t'. BB(X, t, X fully believes at t that h) need not be in MK' as representing X's part of X's commitment to full belief at t' concerning X's full beliefs at t. X at t' might be committed to doubt or disbelief as to whether X at t fully believes that X fully believes at t that h even though X at t' is committed to full belief that X at t fully believes that h. It is no part of the doxastic commitments of X at t' that he fully believe that X at t' fulfills X's doxastic commitments at t. So, BB(X, t, X fully believes at t that h) is not positively valid for X at t' and, hence, is not omnitemporally valid. A fortiori, BB(X, t, X fully believes at t that h) is not universally valid. BB(X, t, X fully believes at t that h) lacks the universal validity of a logical thesis even though it is positively valid for X at t.

To obtain a form of universality, we need to consider the BB principle itself. It characterizes constraints on the doxastic commitments of every rational agent at all times. But the BB principle is a *prescription* specifying the doxastic obligations of rational agents. All agents are committed to have full beliefs conforming to it, but commitment to full belief is here distinguished from the full beliefs that fulfill the commitment. The BB principle is a principle of normative doxastic logic but is not a universally valid proposition. Like the requirement that commitments to judgments of credal probability conform to the requirements of the calculus of probabilities, it is not a universally valid proposition even if it is a universally obligatory prescription. The requirements of deductive consistency and closure are also universally mandatory principles of the logic of consistency for full belief. They too are not universally valid propositions but, unlike the BB principle, they secure the universal validity of the theses of deductive logic and their truth by universal commitment.

Turn now to the B$\sim$B principle. If h is not in X's corpus K at t, it is not in $MK_{X,t}$. However, X is committed at t to believe that he does not believe at t that h. So "X does not believe at t that h" is in $MK_{X,t}$. So is the sentence "X does not fully believe at t that $h \supset$ X fully believes at t that X does not fully believe at t that h." Call this B$\sim$B(X, t, X does not believe at t that h). This sentence is positively valid for X at t, but is neither omnitemporally nor universally valid. Yet, B$\sim$B is a principle of normative doxastic logic – that is, of the logic of consistency for full belief.

The upshot is that the BB and B$\sim$B principles do not generate commitments to full belief additional to the theses of deductive logic that, like logical theses, are universally valid in *ML*. Every rational agent is committed at every time to believe that logical theses are true. Every rational agent is committed at every time to avoid full beliefs inconsistent with such logical theses. But there are no additional propositions possessing truth by universal commitment even though the logic of consistency imposes doxastic obligations additional to the commitment to deductive closure and consistency. In this respect, therefore, the logic of consistency for full belief no more coincides with a logic of truth than the logic of consistency for credal probability judgment does *even if we overlook judgments of serious possibility and restrict ourselves to truth value bearing full beliefs*.

We can, to be sure, collect all the propositions that are positively valid for X at *t* and claim them to be logical truths for X at *t*. But logical truths for X at *t* may be falsehoods for X at *t'* or for Y at *t*. Talk of logical truths for an agent at a time grates. It is not surprising that this should be so. Logical truth is truth by universal commitment. If *h* is positively valid for X at *t*, it is true according to X's commitment at *t* no matter what potential corpus *MK* in *ML* satisfying the requirements of the logic of consistency is identical to $MK_{X,t}$. But it does depend both on X and the time. Truth according to X's creed at *t*, no matter what it is, remains a far cry from truth by universal commitment.

Still, theses positively valid for X at *t* reflect doxastic obligations that X has at *t* no matter what X's creed at *t* might be. And these doxastic obligations are universalizable in the sense that Y at *t'* has doxastic commitments structurally similar to X's at *t* even if they are not identical. We might be prepared to accept some notion of truth according to X's creed at *t*, no matter what it is, as capturing logical theses if positive validity for X at *t* coincided with negative validity for X at *t*. If validity in the logic of truth is true no matter what might be the case, *h* is valid if and only ~*h* could not be the case. Positive and negative validity coincide in the logic of truth.[7] They do not in the logic of consistency for full belief.

VII

Consider Moore's paradox of saying and disbelieving. When X assents to the sentence "*h* but *I*, X, do not believe that *h*," X is a manifesting dispositions to assent that might, prima facie, be interpreted as directly committing X to full belief that ~BB(X, *t*, *h*). But that violates the conditions for commitment. The behavior so interpreted is incoherent. BB(X, *t*, *h*) is negatively valid. X is prohibited by the logic of full belief from having a commitment to full belief at *t* that the negation is true. But X is not, thereby, committed to full belief that BB(X, *t*, *h*) by the logic of consistency for full belief. BB(X, *t*, *h*) is not positively valid. X is not committed at *t* to full belief that it is true. Moreover, it would be incoherent for him to be so committed no matter what *h* in *ML* might be involved.

The valid sentences of a logic of truth are supposed to be judged true by universal commitment. Part of the paradoxical seeming character of Moore's example derives from the fact that it is incoherent for X at t to be committed to full belief that ~BB(X, t, h) while there is no difficulty with Y's believing it at t or t'. But there is worse to come: I conjecture that much of the puzzlement concerning Moore's paradox derives from the failure of the positive and negative theses for X at t to coincide in the case of claims like BB(X, t, h) as they are expected to do if a logic of truth for full belief coincides with the logic of consistency for full belief.

Neither the relativity of validity to X at t nor the failure of positive and negative validity to coincide is paradoxical or even paradoxical seeming when the logic of consistency is not conflated with the logic of truth. For those imbued with the traditions of Frege and Russell, Moore's paradox looks more troublesome simply because they have endorsed the conflation of the two. The obvious conclusion is: So much the worse for Frege's objectivism.

VIII

$MK_{X,t}$ at t uniquely determines X's $K_{X,t}$ at t. If h in L is in $MK_{X,t}$ so that X is committed at t to fully believe that h, h is also in $K_{X,t}$. This claim clearly does not generate any new commitments positively valid for X at t. It commits X to $h \supset h$, which is a universally valid thesis in deductive logic to which closure already commits X.

Does $K_{X,t}$ determine $MK_{X,t}$? I think it does.

Suppose that X is committed at t to fully believing that X fully believes at t that h. Given such commitment, it is generally conceded that X at t is committed to fully believing that h. We do not, however, need to assume this as an additional principle of the normative logic of consistency for full belief. We already have the resources to defend it. If X is not committed to believing that h, the OC principle (which is itself derivable from the requirements of logical consistency and closure, SSP and ~B) demands that either X is committed to believing that ~h or to suspending judgment. In either case, X is committed at t not to fully believe that h. The B~B principle then requires X to be committed to full belief that X does not believe that h. Since X is, by hypothesis, committed to full belief that X does fully believe that h, we have the conclusion that X's doxastic commitments violate the requirement of deductive consistency. $MK_{X,t}$ contains both the claim that X does fully believe at t that h and its negation. Hence, the principles of consistency and closure together with B~B require that X's doxastic commitments at t satisfy the following BT condition:

If X at t is committed to full belief that X at t believes that h, then X is committed at t to full belief that h.

The BT principle requires that if "X believes at t that h" is in $MK_{X,t}$ at t, h is $K_{X,t}$ and, hence, in $MK_{X,t}$. Consequently, BT(X, t, h) (= "X believes at t that $h \supset h$") is in $MK_{X,t}$. When "X does not believe at t that h" is in MK, BT(X, t, h) will still be present.

Hence, BT(X, t, h) is positively valid for X at t. It is not, however, omnitemporally valid or universally valid for reasons paralleling those offered for the positively valid propositions supported by BB and B~B. That is, if we consider the corpus MK^* representing X's corpus at t' or Y's corpus at t, then the BT(X, t, h) need not be present in MK^*.

Thus, the BT principle does not add any further universally valid logical theses. It does yield new positively valid theses to X's corpus at t. And it does something more. It allows us to say that X's corpus $K_{X,t}$ at t expressible in L uniquely determines that portion of his corpus in $MK_{X,t}$ that represents X's doxastic commitments as to what he believes – insofar as this does not pertain to other agents' doxastic commitments. When it comes to the commitments of other agents (say, agent Y's commitments at t' to full belief), the set of constraints introduced allows unique determination from X's commitments at t to what the contents of Y's corpus $MK_{Y,t'}$ in ML that are not expressed with an initial "Y believes at t that" operator. This includes Y's commitments to full belief at t' as to what X's full beliefs at t or some other time are as well as Y's commitments to full belief representable by sentences in L.

Last but not least, the requirements of consistency and closure on potential corpora in ML together with the BB, B~B, SSP, and ~B principles guarantee that every consistent potential corpus in ML contains as positively valid formulas for X at t theses of an S5 modal logic using "X fully believes at t that" as the necessity operator. In this sense, the logic of consistency for full belief yields an S5 doxastic logic for the full beliefs of X at t. It yields a different S5 doxastic logic for the full beliefs of X at t' and for Y at both t and t'. In all cases, the structure of the doxastic logic is the same, but different theses qualify as theses of the S5 doxastic logic. Except for the theses of deductive logic, there are no universally valid theses. Except for the theses of deductive logic, there are no truths by universal commitment of S5 doxastic logic. To have a coherent or consistent system of beliefs, X's beliefs should conform to S5 requirements, but consistency of full belief is a far cry from truth of full belief.

According to the account offered earlier, h is in $MK_{X,t}$ if and only if X is committed to full belief at t that h; h is not in $MK_{X,t}$ if and only if X is committed at t not to believe that h. The presence of h in $MK_{X,t}$ does not imply that h is true (although it does imply that X is committed to full belief at t that h is true). The absence of h in $MK_{X,t}$ does not imply anything about the truth or falsity of h either. It implies only that X at t is committed not to believe that h. The only issue of fact is whether X at t has dispositions requisite to generate commitments to full belief at t in the truth of the items listed in

$MK_{X,t}$ and has commitment not to believe items that do not belong to that set. The logic of consistency characterizes conditions every such corpus for every X at every time should satisfy. None of these conditions stipulates that any theses positively valid for an agent X at t ML are true except, perhaps, for the truths by universal commitment or theses of deductive logic. In particular, BT(X, t, h) is not dictated by the logic of consistency for full belief to be true. X and X alone at t is rationally obligated to fully believe that BT(X, t, h). No one, not even X, is committed to do so at any other time. Still, philosophers, computer scientists, game theorists, and other would-be users of doxastic logic deny that the logic of full belief is S5.

Now it is often an affront to the sensibilities of others to declare all of one's own beliefs to be true. Political correctness in philosophical discourse leads to confusion just as directly as it does in other contexts. I do not think that proper speech is the only concern here – or, at least, I would like to think that something philosophically more important is at stake. My conjecture is that resistance to claiming that the logic of consistency for full belief for an agent at a time yields an S5 structure for beliefs positively valid for the agent at that time is a concern on the part of those who think we should have the logic of consistency for full belief coincide with the logic of truth for full belief. I believe I have said enough already to establish the futility of the effort, but I have done so without mentioning Kripke semantics for doxastic or epistemic logic where, so it might be hoped, salvation of the Frege–Russell vision of logic might be sought when the logic of belief is considered a logic.

IX

According to a Kripke-style model theoretic semantics for the sentences of ML, X's state of full belief (and disbelief) is specified by a set $T(MK_{X,t})$ of sentences, all of which are prefixed either by "X at t fully believes that" or "X at t does not fully believe that" and it is then understood that X undertakes to have beliefs that secure the truth of all of these statements.

$T(MK_{X,t})$ may be obtained from $MK_{X,t}$ as follows: (a) Convert every sentence h in $MK_{X,t}$ that is not prefixed by "X fully believes at t that" or "X does not fully believe at t that" to "X fully believes at t that h," (b) convert every sentence f in ML that lacks a prefix of either kind and is not in $MK_{X,t}$ to "X does not fully believe at t that h," and (c) close under deductive consequence.

The result will be a deductively closed theory in ML/L explicitly stating what the agent believes and does not believe to be true at t. It is a representation of the state of full belief equivalent to the representation I have been using; for an inverse transformation can be uniquely defined by deleting initial prefixes of the form "X fully believes at t" and removing sentences lacking such prefixes. $T(MK_{X,t})$ will satisfy the requirements of S5 among sentences in ML/L as does $MK_{X,t}$ in ML. Moreover, for every g in L, either "X believes at t that g" or "X does not believe at t that g" is in both $T(MK_{X,t})$ and $MK_{X,t}$.

Consider then adding to $T(MK_{X,t})$ all the sentences of $K_{X,t}$. Doing so is not intended to add redundant information. The sentences in $T(MK_{X,t})$ are assertions identifying X's full belief state at t. "X believes at t that Albany is the capital of New York" does not mean that X believes at t that X believes at t that Albany is the capital of New York. It means that it is true that X believes at t that Albany is the capital of New York. We should understand forming the union of this set with $K_{X,t}$ in the same way. So, if "Albany is the capital of New York" is in $K_{X,t}$, it no longer means "X believes that Albany is the capital of New York" but "It is true that Albany is the capital of New York."

On the assumption that X at t is perfectly fulfilling X's doxastic commitments at t no matter what potential belief state X might be in at that moment and as long as the set of sentences in L that are added to $T(MK_{X,t})$ are the sentences of $K_{X,t}$ all now declared to be true, then no matter what the original $MK_{X,t}$ might have been, the S5 conditions will have been satisfied. The result of this exercise will be the same set of sentences in ML, $MK_{X,t}$, as before. But this set of sentences no longer represents X's commitments at t to full belief but in addition asserts that all of these commitments are true, so that it is also claimed that X at t actually has the beliefs fulfilling the commitments and that all sentences in $K_{X,t}$ are true. When so construed, I shall characterize the set as $S(T(MK_{X,t}), K_{X,t})$.

Take any potential corpus in $MK_{X,t}$ in ML and introduce $S[T(MK_{X,t})K_{X,t}]$. For any such $MK_{X,t}$, consider all maximally consistent extensions in L of $K_{X,t}$. $S[T(MK_{X,t})w]$ for any such extension w represents a "possible world" in which X's state of full belief is specified by $MK_{X,t}$ and all of X's full beliefs are true. All the sentences that are instances of theses of S5 modal logic are true under the assumptions we are making no matter what the contents of the original $MK_{X,t}$ happen to be. That is to say, the theses of an S5 modal logic in the full ML can be shown to be universally valid or true in all $S[T(MK_{X,t})w]$ or possible world constructed in this manner. Each possible world specifies a state of doxastic commitment for X at t. The alternatives to each such possible world z are precisely those possible worlds in which that state of doxastic commitment by X at t that holds at z obtains. We have a species of semantic models structured along lines pioneered by Kanger (1957), Hintikka (1969, Sec. III), and Kripke (1963) in variant ways. The theses of S5 are now valid (i.e., true in all possible worlds) as a logic of truth might require when alternativeness is an equivalence relation. Equating the logic of truth with the logic of consistency would require that every rational agent Y at every t' is committed to believing all of these theses true.

If the S5 doxastic logic were interpretable as both a logic of consistent or coherent full belief and a logic of truth, some vision such as this would be required. The vision is, of course, a sheer fantasy. It requires every rational agent to fully believe that what anyone at any time fully believes is true and this is noncontroversially madness.

The sensible response to this circumstance would be to abandon the identification of the logic of truth with the logic of consistency for full belief and retain S5 logic of consistency for full belief for reasons such as those I have listed previously as conditions on $MK_{X,t}$ or the corresponding restrictions on $T(MK_{X,t})$ but allowing all maximally consistent sets, w, of sentences in L to be arguments in $S[T(MK_{X,t}), w]$. The possible worlds or maximally consistent sets in ML (represented by the logical consequences of the union of $T(MK_{X,t})$ and w for every S5 satisfying $MK_{X,t}$ in ML and every maximally consistent w in L) would exhibit a common structure securing truth in all possible worlds for sentences that amount to the S5 structure without validity for sentences BT(X, t, h) unless h is a sentence possessing a prefix of the form "X at t believes that" or "X at t does not believe that." The resulting logic of truth for full belief is S5-manqué or KD45.

Of course, this "logic of truth" is so-called only with some charity, because the only "possible worlds" contemplated are ones where the agent X actually fulfills the requirements of the S5 logic of consistency for full beliefs – a condition that it is incredible to suppose any agent X satisfies. At best, we have a model for ideally rational agents and not a logic of truth. Even if such "worlds" are logically possible, there are worlds where the requirements specified fail. This observation need not, however, deter us too much if we claim that the logic of truth is counterfactual, focusing on what might be true in any situation in which all the agents concerned were ideally rational.[8]

The main point is that even with this charitable construal, the logic of consistency specifying what a rational agent ought to fully believe to be ideally consistent remains S5 and, hence, different from the S5-manqué logic of truth. Not only in the case of probability judgment but also in the logic of full belief, there is no coincidence between the logic of truth and the logic of consistency.

Weakening the logic of consistency to S5-manqué while keeping the S5-manqué logic of truth intact is not workable. BT(X, t, h) will no longer be positively valid for X at t. Hence, BT(X, t, X believes at t that h) will not be in $T(MK_{X,t})$ in some possible cases even though it remains in $MK_{X,t}$ and is positively valid for X at t. Hence, some sentences positively valid for X at t will not be valid in the sense of being true in all possible worlds. We can no longer assume following the approach adopted above that BT(X, t, X believes at t that h) will be in $S(MK_{X,t}, w)$ for every possible belief state for X at t and every w. Both the logic of consistency and the logic of truth for full belief will be weakened in a way that prevents coincidence of the two.

Moreover, weakening the S5 logic of consistency to S5-manqué or something else entails giving up either SSP, ~B, or B~B in a way that abandons the idea that the state of full belief defines the space of serious possibility over which credal probability judgments are made. Or at least this cannot be done without abandoning the use of standard requirements that, for probability judgment to be coherent, the requirements of a finitely additive probability measure be satisfied by numerically precise probability judgments.

For X at t to fully believe that what X at t fully believes is true smacks of doxastic immodesty. If so, it is immodesty mandated by doxastic coherence. I have been arguing that one cannot give up this coherence lightly. In any case, the demand for such coherence does *not* require that X be committed to full belief that X has fulfilled the demands imposed by X's commitments to full belief. Such arrogance would be monstrous. Moreover, X's rationally mandatory doxastic immodesty need not be accompanied by Y's credulity. Y is not obligated as a rational agent to full belief that whatever X fully believes at some specific time is true.

X

If, as I have been arguing, the logic of full belief characterizing an agent's commitments to full belief at t is S5, and insofar as we may ignore X's commitments to full belief regarding other agents, $MK_{X,t}$ is uniquely determined by $K_{X,t}$. Consequently, changes in X's doxastic commitments can be represented as changes in X's corpus K in L. This point is important to keep in mind when exploring the ways in which a doxastic commitment may be modified. In particular, if neither h nor $\sim h$ in L is in $K_{X,t}$, $K_{X,t}$ can be changed by adding h and forming the deductive closure. Expansion of the initial corpus by adding h (perhaps through observation, receiving information from others, or by ampliative inference) cannot be represented formally as an expansion if h is added to $MK_{X,t}$ and taking the deductive closure. The transformed corpus in ML will not represent the doxastic commitments generated by the corresponding corpus in L. Indeed, there is no way, satisfying the requirements of doxastic consistency, that $MK_{X,t}$ can be expanded with doxastic consistency. By virtue of the S5 structure of the logic of consistency for full belief, we may happily focus on expansion of corpora in L.

Similar observations apply mutatis mutandis to suppositional reasoning, the understanding of conditional judgments of serious possibility and the logic of conditionals that regulates the coherence of such judgments. (Levi 1996).

XI

Shifting gears somewhat, what may we say about epistemic logic or the logic of knowledge? Much depends upon how one understands knowledge. According to the pragmatist view that I favor, what X knows is, according to what X fully believes, what X fully believes; for what X fully believes, X is committed to judging to be true. And it is common ground among the classical pragmatists that, from X's point of view, what X fully believes (does not seriously doubt) stands in no need of justification. Justification is required for changing one's full beliefs and not for the full beliefs one has. Since the logic of consistency of full belief for agent X at *t* requires X to equate what X is committed at *t* to know with what X is committed at *t* to fully believe, there can be no logic of consistency for knowledge distinct from the logic of consistency for full belief. The positively valid theses for X at *t* of the logic of consistency for knowledge are the same S5 theses as apply to the logic of consistency for full belief.

Even if Y at *t′* fully believes that X at *t* completely satisfies the requirements for doxastic and, hence, epistemic coherence, however, Y's views regarding X's full belief need not equate them with X's knowledge. Y will be committed to the supposition that X's knowledge satisfies a KK principle corresponding to BB and a KT principle but not with a K∼K condition.[9] Suppose that Y fully and correctly believes that X's full beliefs at *t* satisfy the S5 requirements for doxastic consistency. If Y fully believes that X at *t* fully believes that *h* and also that *h* is false, Y should fully believe that X fully believes at *t* that X fully believes at *t* that *h* and, indeed, fully believes at *t* that X knows at *t* that *h*. But Y believes that X does not know at *t* that *h* is false because Y judges *h* to be false. Y also believes that X does not know at *t* that X does not know that *h*. So the K∼K condition fails. That is, K∼K fails as a model of Y's beliefs concerning X's knowledge at *t*. But, insofar as epistemic logic, like doxastic logic, is a logic of consistent evaluation by an agent X concerning what X knows at *t* or evaluation by agent Y of X's consistency in making such evaluations, the logic of full belief coincides with the logic of knowledge and both are S5.

Of course, the claim that X knows that *h* at *t* may mean even for a pragmatist that X not only truly believes that *h* at *t* but does so authoritatively in the sense that, in some way or other, he can justify to others the truth of what he fully believes. But whether X can do so or not even by X's own lights depends upon what by X's lights the others already fully believe. We should be skeptical of the availability of a logic of consistency for knowledge as distinct from the logic of consistency for full belief and even more skeptical of a useful epistemic logic of truth.

XII

Suppose that X and Y at *t* share a common corpus *K* in *L* and both fully believe that this is so. Finally, they both fully believe that they are both doxastically coherent in the S5 sense I have been advocating. For each *h* in *K*, X is committed by S5 to "X believes at *t* that *h*" in $MK_{X,t}$ as well by the suppositions made above to "Y believes at *t* that *h*." Y is committed by parallel reasoning to the membership of these sentences in $MK_{Y,t}$. By S5, X is also committed to "X believes at *t* that X believes at *t* that *h*" and to "X believes at *t* that Y believes at *t* that *h*." By X's full belief at *t* that Y is doxastically consistent, X is also committed to full belief represented by "Y believes at *t* that X believes at *t* that *h*" and "Y believes at *t* that Y believes at *t* that *h*." And by parallel reasoning, Y is committed to these at *t* as well. It is easy now to see that iteration of this process can be carried on indefinitely. Both X and Y have a commitment to common belief that *h* at *t* and, indeed, to all elements in *K*. Moreover, their common belief corpus has an S5 structure where what is positively valid for X is positively valid for Y.

In this case, we may, indeed, say that BB(X, *t*, X believes at *t* that *h*), B∼B(X, *t*, X does not believe at *t* that *h*), and BT(X, *t*, *h*) are positively valid at *t* for X and Y. From their common perspective, their common full belief is common knowledge.

With this understanding, if everyone at all times regarded everyone else at all times to be doxastically coherent, and believed that everyone else always regarded them to be doxastically coherent, they all would be committed to common knowledge of the universally valid theses of deductive logic.

Of course, if Z outside the consensus fully believed that *h* is false, Z would deny that X and Y have common knowledge that *h* even though Z believed that they have common belief and that the common belief is doxastically consistent in the S5 sense even though it is false. It is an old commonplace that consistency does not imply truth. The main burden of this paper has been to suggest that those who do not think that the logic of consistency for full belief is S5 are so wedded to the coincidence of the logic of consistency with the logic of truth that they have forgotten this commonplace.

NOTES

For my good friend and admired colleague, Charles Parsons. Thanks are due to Horacio Arló Costa for good advice.

1. Logic as understood by Koslow (1992) studies diverse "implication structures" and their properties. Deductive logics (either propositional or first-order quantificational logic) are implication structures of certain kinds. I am considering cases in which the implication structures are atomless Boolean algebras of potential states of belief closed under meets and joins of sets of states of any cardinality (Levi 1991, Ch. 2). Here the "implication relation" is the binary relation partially ordering the potential states in the algebra with respect to how well potential states relieve doubt. Using techniques described by Koslow (1992, Ch. 20), the given implication structure for potential belief states can be extended so that quantification can be introduced while continuing to avoid attributing syntactical structure to the potential belief states. The additional elements belonging to the extended implication structures are neither additional states of belief nor constituents of states of belief.

   Belief states belonging to subalgebras of the set of the algebra of potential states of full belief may be represented by sentences in a regimented language *L* in which the logical consequence relation between sentences or sets of sentences in L may be understood as generating an implication structure that represents the subalgebra of potential states of full belief by preserving the consequence relation between belief states. Hence, the implication structures at the focus of attention when we think of logic of full belief are algebras of potential belief states and secondarily the structures determined by the relation of logical consequence for the language L used to characterize such structures. Frege apparently was interested in implication structures in which the domains of entities are thoughts distinct both from potential states of full belief and from linguistic entities. I have no understanding of the applications intended when thoughts are introduced.

Nothing in this paper is intended to question the importance of the study of implication structures in general. I am focusing on structures relevant to the understanding of logic as characterizing standards of doxastic health – the logic of consistency for full belief – and its relation to the study of implication structures characterizing logics of truth.

Hintikka (1969, p. 5) writes: "A branch of logic, say epistemic logic, is best viewed as an *explanatory model* in terms of which certain aspects of the workings of our ordinary language can be understood." A logic so conceived need not even be an implication structure and rightly so when we consider logics of probability judgment, preference, and so on. But even in the case of logics of knowledge and belief where Hintikka and I might agree that the study of implication structures is central, we seem to differ regarding the role of logic in explaining "the workings of our ordinary language." For me, unlike Hintikka, the primary focus is on the critical control of the attitudes in general and of belief in particular. I use language in offering a systematic account of the standards for doxastic health. I favor and acknowledge, as one must, that linguistic means are used to communicate beliefs and other attitudes. I do not deny that the regimented languages I use for systematic discussion are implication structures of a certain kind. But their logics are used for the purpose of examining the doxastic commitments of rational agents. Explanation of "the workings of language" are of, at best, marginal interest and relevance to this project.

2. There may be contexts in which intuitionist logics (implication structures) gain useful application; but not as the logic of consistency for full belief. Closure under a classical deductive consequence relation should be required when states of commitment to full belief are understood to be states of commitment to standards for judging serious possibility. Another version of the case against an intuitionistic understanding of deductive closure is based on the arguments advanced by Levi (1991, Ch. 2) for requiring that the set of potential or candidate belief states have a structure that is at a minimum that of a Boolean algebra rather than of a pseudo-Boolean algebra. My argument in brief is that if x and y are potential belief states, there ought to be a potential belief state that is a state of doubt or suspense with respect to x and y – that is the join of x and y. Those who insist that the structure is that of a pseudo-Boolean algebra say that a potential belief state x may have a "pseudocomplement" $\neg x$ such that join of x and $\neg x$ is strictly stronger than the maximally skeptical belief state of being committed to judging only logical truths to be true. Yet, there is no potential belief state relative to which one could judge the join of x and $\neg x$ impossible. One could not fully believe it to be false. Prohibiting such potential belief states constitutes a dogmatic foreclosing of inquiry and ought to be rejected. The set of potential belief states ought to constitute at a minimum a Boolean algebra reflecting the demands of a classical two-valued logic.

3. One could argue that satisfying the strict standard of commitment characterized by deductive consistency and closure, SSP, and ~B is really unnecessary. The extent of logical omniscience demanded of rational agents need only suffice to meet the demands of the problem being addressed. One can get by with something substantially short of full omniscience and still come out with the same conclusions one would have reached were one fully omniscient. That is often true. Still there is no upper bound on the complexity of the problems we may face. The commitments enjoined by the strict standard are an acknowledgment of that point.
4. In Levi (1991, 1996, I reserved "X fully believes at that h" for full belief in the commitment sense and "X fully recognizes at t that h" for full belief in the dispositional (and sometimes manifestation) sense. For the purposes of this discussion, I wish to focus attention on the consistency or coherence of beliefs understood as attempts at partial fulfillments of doxastic commitments. Hence, the change in terminology.
5. Recall that, for the purposes of this discussion, "X fully believes at t that h" is to be understood to assert that X at t has dispositions to assent and behave that directly generate a commitment at t to full belief that h. X may be committed at t to full belief that h without fully believing at t that h – that is, having the requisite direct commitment generating dispositions to assent and behave. X may, indeed, have the dispositions that directly generate a commitment to full belief that $\sim h$ or, indeed, that directly generate a commitment to suspense between h and $\sim h$. In such cases, X is failing to live up to X's commitments to full belief. X's full beliefs are in this sense inconsistent. As a consequence, at least one of the system of normally direct commitment-generating conditions must be considered to have failed to be commitment generating at all. For, the direct commitment-generating conditions for commitment to full belief that h and the corresponding conditions for commitment to full belief that $\sim h$ cannot both be direct commitment-generating conditions when the two sets of conditions are jointly satisfied. It becomes a problem for interpretation to determine which of the two sets of conditions, if either, is to be interpreted as commitment generating in the given context. Even so, we may claim that X fully believes that h and also fully believes that $\sim h$ in the sense that, at the same time, X satisfies both the conditions that *would be* directly commitment generating for full belief that h if no failure to have commitments required by the logic of consistency were thereby mandated and the corresponding conditions for full belief that $\sim h$.

   By the same token, X may have the dispositions at t directly generating a commitment to full belief that h, but may lack at t the dispositions directly generating a commitment to full belief that X fully believes that h. In this case, too, a failure to fulfil commitments arises that warrants judging X's full beliefs at t as inconsistent. We are not to conclude that his doxastic commitments violate the logic of consistency but only that *prima facie* direct commitment-generating conditions fail to be commitment generating at all. A problem for interpreting the agent's dispositions to assent and to behave arises.

   The dispositions to assent and to other behavior exhibited by X at t may be sufficiently confused at a given time that neither X nor anyone else can identify a doxastic commitment that X has undertaken. X cannot be understood to believe that h, to believe that $\sim h$, or to be in suspense regarding h and $\sim h$. Here, too, X has failed to fulfill commitments and X is doxastically inconsistent. The OC has been violated. Given any proposition h, X is committed either to full belief that h, to full belief that $\sim h$, or to suspension of judgment between h and $\sim h$. No fourth commitment alternative is available. Thus, X is committed to fully believe that 9 is the integer in the billionth place of the decimal expansion of π or to fully believe that it is not, or to be in suspense on this issue. But only those who have obtained the printout of a program for generating this decimal expansion are capable of satisfying this commitment. Assuming that X is committed to full belief in theses of arithmetic, any behavior interpretable as suspense will violate X's commitments; for either it is an arithmetic thesis that 9 is the integer or it is an arithmetic thesis that it is not. X might recognize this and refuse the posture of being in suspense without being able to make up his or her mind on whether 9 is or is not the right integer. Nonetheless, all rational agents are committed to full belief, full disbelief, or suspense on this issue at all times.

6. I contend (Levi 1996) that suppositional reasoning of the sort expressed in belief-contravening ("counterfactual") conditionals is to be represented by a transformation of a deductively closed and consistent corpus K in L to another such corpus K_h^{*r}. This transformation (Ramsey revision) is a variation of the more familiar AGM revision transformation (Alchourrón, Gärdenfors, and Makinson 1985). Supposing for the sake of the argument or fantasizing, no matter how fantastical it might be, cannot be coherent unless the transformation is from the current belief state meeting the requirements of the logic of consistency for full belief to another potential belief state meeting the same standards. Thus, any thesis positively valid for X at t should be in every belief state that is representable by a corpus obtainable via Ramsey revision of X's corpus $K_{X,t}$ by adding any consistent sentence h in L. A universally valid thesis has this property for every X and t. In this sense, a thesis judged true by universal commitment may be considered valid in the sense of being true in all possible worlds or under all interpretations or models whether they embrace Frege's objectivist view of logic or the more moderate view endorsed here.
7. As in note 2, I am assuming a classical nonintuitionistic logic of consistency for full belief. I also take for granted, therefore, that the logic of truth is a classic two-valued logic as should be appropriate, in my judgment, to truth value as applied to belief. The case for proceeding in this way, like the argument of note 2, is based on the views that I advanced in Levi (1991, Ch. 2). Shifting from the current belief state to one that is weaker incurs no risk of error. Every other shift does (1991, p. 12). In particular, if an agent were in the weakest potential belief state, shifting to the join h and $\neg h$ should incur a risk of error if negation is intuitionist; for such suspense is stronger than maximal ignorance. But there is no consistent belief state that falsifies the join of the states represented by these propositions. So, the notion of risk of error makes no sense. The set of potential belief states ought to constitute at a minimum a Boolean algebra reflecting the demands of a classic two-valued logic. In spite of this rejection of intuitionism, positive and negative validity for an agent at a time do not agree in the logic of consistency for full belief.
8. Even if one has qualms about regarding this structure as a logic of truth, it is a logic in the structural sense of Koslow (1992), where we have a classical implication structure and its dual with negation, disjunction, and conjunction, and where "X fully believes at t that" and "X does not fully believe that ~" are modal operators with respect to the implication structure and its dual that are Y modals (Sec. 35.4), K_4 modals (Sec. 35.3), and S_5 modals (Sec. 35.9). The assumptions we have specified to hold ensure that the conditions for such an implication structure are satisfied so that we have a logic in Koslow's structural sense. What one may well doubt is that the implication structure so generated is suited for characterizing a logic of truth.
9. See Hintikka (1962) for a development of an epistemic logic without K~K and Lamarre and Shoham (1994) for a counterinstance.

REFERENCES

Alchourrón, C., Gärdenfors, P., and Makinson, D. (1985). "On the Logic of Theory Change, Partial Meet Functions for Contraction and Revision," *Journal of Symbolic Logic*, 50: 510–30.

Frege, G. (1967). *The Basic Laws of Arithmetic*, translated and edited by M. Furth (Berkeley: University of California Press).

Hintikka, J. (1962). *Knowledge and Belief* (Ithaca, NY: Cornell University Press).

Hintikka, J. (1969). *Models for Modalities* (Dordrecht, The Netherlands: Reidel).

Kanger, S. (1957). *Provability in Logic* (Stockholm: Almqvist and Wiksell).

Koslow, A. (1992). *A Structural Theory of Logic* (Cambridge, UK: Cambridge University Press).

Kripke, S. (1963). "Semantical Considerations on Modal Logics," *Acta Philosophica Fennica*, 16, pp. 83–94.

Lamarre, P., and Shoham, P. (1994). "Knowledge, Certainty, Belief and Conditionalization," Unpublished. Manuscript,

Levi, I. (1991). *The Fixation of Belief and Its Undoing* (Cambridge, UK: Cambridge University Press).

Levi, I. (1996). *For the Sake of the Argument* (Cambridge, UK: Cambridge University Press).

Ramsey, F. (1990). "Truth and Probability," *Philosophical Papers*, ed. D. H. Mellor (Cambridge, UK: Cambridge University Press).

Rovane, C. (1994). "The Personal Stance," *Philosophical Topics*, 22: 351–96.

II. INTUITION

Immediacy and the Birth of Reference in Kant: The Case for Space

CARL J. POSY

FOR CHARLES PARSONS

> All cognition, that is, all presentations consciously referred to an object, are either *intuitions* or *concepts*. Intuition is a singular presentation (*repraesentatio singularis*), the concept is a *general* (*repraesentatio per notas communes*) or *reflected* presentation (*repraesentatio discursiva*). (Jäsche Logic, §1)

> In whatever manner and by whatever means a cognition may relate to objects, *intuition* is that through which it is in immediate relation to them, and to which all thought as a means is directed. But intuition takes place only in so far as the object is given to us. (*Critique of Pure Reason*, A19/B33)[1]

> Space is not a discursive or, as we say, general concept of relations of things in general, but a pure intuition. (A25/B39)

Charles Parsons has taught us that the Kantian conception of intuition is a multifaceted notion and that this complexity affects Kant's philosophy of mathematics. In this essay, I focus on these two lessons, but also broaden them a bit. Specifically, I have three goals:

(1) Parsons has taught us that the notion of immediacy – which he interprets phenomenologically[2] – must be separately added to the traditional criterion of singularity that has been stressed by all commentators on Kant's definition of intuition. In this essay, however, I shall point out that Kant offers not two but *three* marks of human intuition: There is singularity; there is, as Parsons insists, immediacy; and there is also something I shall call "reference." Kant calls it the "object givingness" of intuition. It is there quite clearly at A19.

(2) Parsons uses his account of Kantian intuition to explain Kant's conception of arithmetic. I, however, shall apply my expanded notion of intuition to explain some of Kant's views about the mathematics of space, notably his notorious claim that the representation of mathematical space is a pure (or a priori) intuition in its own right.

(3) As for my third goal, I would also like to show you that each of the three components of Kant's notion of human intuition is derived from a

corresponding Leibnizian notion. In particular, Leibniz has a picture of God's intuitive grasp of a complete concept, a picture in which singularity, reference, and immediacy are seamlessly united. The Kantian account of intuition, I shall argue, is the result of replacing Leibniz's picture of divine intuition with the notion of human perception and then making the adjustments that are demanded by this replacement. I shall claim, indeed, that we can understand Kant's views about the pure intuition of space only in the light of this Leibnizian background.

This is not an exercise simply in reparsing the components of an isolated Kantian notion, or even merely an excursion into his philosophy of mathematics. These issues run deeply throughout Kant's "critical philosophy." Thus Kant's versions of the three criteria of intuition will go to the heart of the *Critique of Pure Reason*:

- Singularity will take us into Kant's transcendental psychology, into his doctrines about the "categories of the understanding," and into his metaphysical views about the empirical world.
- Reference is at the core of his full conception of truth, a conception, we shall see, that is the birth of a modern referential semantics.
- And we shall find that Kantian immediacy fuels his famous distinction between intuitions and concepts.

But more than that, these issues about intuition will take us into the third *Critique* as well. In particular, I shall show you that Kant's notion of immediacy is not only separate from but is apparently incompatible with the first *Critique* notions of singularity and reference. It is only in the third *Critique*, I shall argue, that Kant demonstrates how all of these aspects can coexist in a human intuition.

I shall occasionally illustrate my points with allusions to twentieth-century phenomenology and logic. I think these allusions appropriate in this *Festschrift* for Charles Parsons not merely because the topics of logic and phenomenology are central themes throughout his work on Kant and much of the rest of his work, but also because perhaps his greatest contribution has been to illustrate the fruitful interplay between modern philosophical tools and their historical origins.

The paper has five parts.

- In the first, I shall sketch in very broad terms the Leibnizian picture of God's grasp of a complete concept, the picture that Kant is attempting to adapt.
- In the second, I will show how Kant's move from God's intuition to human perception affects the notion of singularity.
- In the third, I will do the same for reference.

- In the fourth part, I will do the same for immediacy and will also show why we must look to the *Critique of Judgment* to relieve the tensions apparent in Kant's first *Critique* notion of immediacy.
- Finally, in the fifth part, I will show how these aspects of intuition in general come together to explain Kant's claim that the representation of space is a pure intuition.

I. Leibniz

A. Complete Concepts

Leibniz had a central image about the arrangement of concepts. This image is not the whole story, and I shall not tell you all the details of Leibniz's philosophy. But I do want to show enough so that you can see the ways in which it inspires much of what Leibniz has to say about ontology, epistemology, and semantics.

Leibniz pictures a vast latticework of concepts, the top level of which is populated by a set of maximally general, primitive predicates and in which one moves downward, so to speak, by combining these basic predicates and their negations into increasingly specific compound concepts.[3] The higher, more general, components are "marks" (*Merkmale*) of the more specific compound concept and are said to be "contained" in it.

At the very bottom, says Leibniz, are the maximally specific combinations, concepts that are fully determined with respect to each elementary predicate. Serially order these combinations and you get a sequence of "stages," each complete in itself. That serial arrangement is a "complete concept," a vast conjunction containing each predicate or its denial for every stage along the way.

To use a modern simile, a Leibnizian complete concept is like a cable of mental optic fibers. There is one fiber for each elementary predicate, and each fiber contains information about the states at which its simple predicate is affirmed or denied. Figure 1 captures this Leibnizian picture.

B. Singularity

Now a complete concept is an *infimum species*: nothing further can fall under it. Speaking quite formally, such a concept satisfies Aristotle's definition of substance. So, we can (and Leibniz does) speak of it as an "individual concept" and describe the object it represents as an individual substance. God's grasp of such an individual concept is clearly, then, a "singular" representation. But this notion of singularity is not merely part of God's "psychology," a description of a particular kind of divine mental state. It is, rather, an ontological, epistemic, and semantic notion as well.

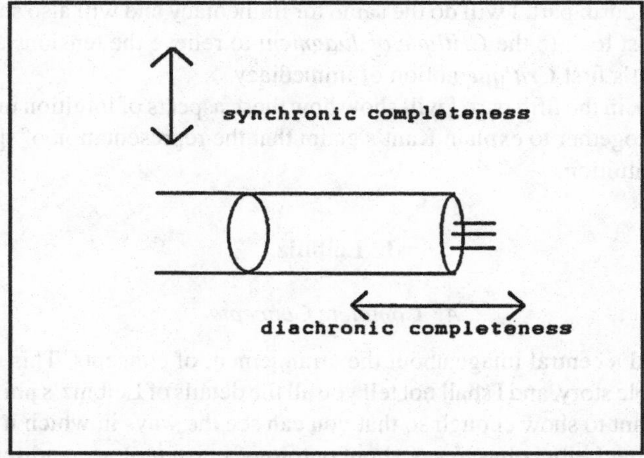

Figure 1

Ontology. The only true individuals (monads) are those that are described by complete concepts dictating all the properties of the individual at each stage of its development.[4] That conceptual completeness is the "unity" that, for Leibniz, marks a true object.

It is a synchronic unity: Each stage contains information about the status of every possible elementary predicate. And it is a diachronic unity: To use Leibniz's famous example, everything about Julius Caesar (from his birth, to crossing the Rubicon, to his assassination on the Ides of March) is coded in his individual concept.[5]

Any concept less unified than such an individual concept – anything that lacks this completeness in one or both dimensions (i.e., any concept that leaves some elementary predicate undetermined at some stage) – at best describes a mere compound or "aggregate." The same holds a fortiori for relational concepts. That is why, for Leibniz, space itself and spatially extended things are never true objects and always aggregates.

To be sure, some aggregates may have a sort of unity (indeed, a "sortal" unity), for they may contain individuals all of a single type, or individuals that stand in a well-defined relation. But that is an ontologically inferior unity. Indeed, it is an externally imposed, "phenomenal," unity, says Leibniz. And such aggregates are merely phenomenal entities.[6]

Epistemology. God knows the truth of a singular judgment solely by virtue of grasping the complete concept that serves as that judgment's subject. It is, for instance, this grasp of Caesar's individual concept that underwrites God's knowledge that Caesar will cross the Rubicon. All the necessary evidence is in the concept, and the concept alone thus provides "sufficient reason" for the truth of the judgment.

Semantics. For Leibniz, a simple predicative judgment is true only when the analysis of its subject concept produces the predicate concept. That is, as Leibniz would say, the judgment is true when the subject concept contains the predicate concept.

"All humans are rational" is true because the concept *human* contains the concept *rational*. And "Caesar crossed the Rubicon" is similarly true because unpacking Caesar's full individual concept reveals that to be the case. This famous containment theory of predication is the heart of the Leibnizian notion of truth.[7]

Leibniz himself described his containment theory as an "intensional" theory of predication. For, the set of general concepts that are contained in a given concept is called the "intension" of that concept. But since, as we have seen, this conceptual containment is also an epistemic notion, it follows that Leibniz had an epistemic theory of truth for these singular and general categorical judgments. His theory of predication is thus what we would today call "assertabilist." A singular judgment is true if and only if there are sufficient grounds to assert it.[8]

C. Reference

Nevertheless Leibniz also says – equally clearly – that the truth of a judgment stems from a "relationship amongst the objects of the ideas" composing that judgment. (*New Essays*, Book IV, Chapter ii) "Let us content ourselves," he says, "with seeking truth in the correspondence of the proposition in the mind with the things in question." (Ibid. §3) This is what we today call a "reference-based" theory of truth: a semantic theory that rests the truth of a judgment on objects and their interrelations rather than on evidence.

And that is a problem. Referentialism and assertabilism as we now understand them are incompatible semantic theories. In particular, the referentialist account of simple predicative truth contains two components that cut against the assertabilist grain:

(1) There is *reference* itself, a semantic relation between the parts of a judgment on the one hand and objects and properties on the other hand.[9]

(2) There is *predication*, the actual connection, whatever that might be, between the object referred to by the subject term and the property denoted by the predicate term. This is an ontological interaction or relation between objects and their properties.

According to the referential theory of truth, the truth of a simple predicative judgment rests on these two relations. And though assertabilism need not deny that either of these relations actually holds, it does leave them out of the semantic equation. Objects and their relations do not enter into the pure assertabilist theory of truth.

But for Leibniz, reference and assertability actually coincide quite naturally. This is because Leibniz simply equates predication (the "ontological" component of truth) with conceptual containment. Thus, for instance, in the continuation of the first *New Essays* passage I cited above, Leibniz tells us that this ontological relation that grounds the truth of a judgment is just the relationship "by virtue of which one idea is or is not included within the other." (Ibid.) To put it in other words, one ontological property of a monad is that its concept is analyzable in the way that it is. So, in this respect, assertabilism and referentialism coincide.

As for what I called above, the "semantic" reference relation, the relation between the components of a judgment and their referents, for Leibniz this is straightforward:

- When the component in question is a general concept, then the referential relation is, in fact, simple identity. For, general concepts are simultaneously components of judgments and predicates of objects.
- If the component in question is an individual concept, then the referential relation consists in the fact that this concept completely describes the corresponding individual substance. God's grasp of such a complete concept is the vehicle of reference to an individual substance, precisely because that concept "agrees" in every possible detail with that substance.

To be sure, we must not mistake this isomorphism for a sort of idealist constructivism. Isomorphism is not identity. Monads are not themselves complete concepts, nor do they have concepts as components.[10] It is just that there is no hidden aspect of a monad's individual nature (what Leibniz calls its "*haecceity*") than what God grasps in grasping that monad's individual concept.[11]

D. Immediacy

In his essay on "Knowledge, Truth, and Ideas" Leibniz remarks that as concepts become more and more complex, it becomes increasingly difficult (for us humans at least) to grasp the arrangement of all their elementary components in one attentive act.[12] For that reason we tend to introduce "symbolic" abbreviations for compound arrangements of more elementary concepts. These can, of course, be several layers deep, and the process of layer-by-layer analysis is our means of unpacking those abbreviations to recover the more elementary components of a given concept.[13]

God, by contrast, has no limit on the number of elementary concepts or the amount of complexity that He can comprehend at once. So He can – without inference, and without mediation by these symbolic abbreviations – grasp the infinite complexity of an individual concept in a single cognitive act. That inferentially immediate aspect of divine cognition – God's direct awareness of all the components of an infinite individual concept – is, according to Leibniz, the paradigm of intuition.[14]

From Leibniz's perspective, we humans, by contrast, have almost no intuitive knowledge of true individuals.[15] We depend ultimately on sensory perception, which, according to Leibniz, is merely confused thought. Individual perceptions, he held, are concepts, but deeply inferior ones. And perceiving is certainly not intuiting. For, an individual perception is always perspectival: it never presents its object all at once, and it is never complete either synchronically or diachronically.

Indeed, our human perceptions present only composite (and therefore phenomenal) entities. And empirical concepts, which we create from repeated perceptions, inherit this inferiority. They cannot have more in them than experience has provided. Analyzing these concepts, says Leibniz, can never provide "sufficient reason" for any judgment. And we can have "immediate" grasp of only the most elementary and most general items, never of a true object.

Thus, to continue Leibniz's example of Caesar, suppose that Caesar were to see a tree bough floating on the Rubicon. This perception would not reveal to him whether that bough was hollow or how it might appear from the other side. Indeed, even if we were to sum up into a single prolonged stare all that Caesar and everybody in the vicinity perceived about that truncated limb, it would still be possible to extend that attenuated perception in at least two incompatible ways: In one, the bough snags on a rock and stops; in the other, it continues to coast. In fact, these two extensions detail the properties of different tree boughs. That is the technical meaning of saying that perception – which cannot distinguish these entities – is a "confused" concept. And no combination of perception-based concepts will explain with "sufficient reason" why the tree bough should stop or turn about in a moment. Indeed, even Caesar's present perception of the bough and its motion is not legitimately veridical; for, strictly speaking, there is no reality corresponding to it.

II. Singularity in Kant

A. *Kant the Leibnizian*

It is important to see how much of this Leibnizian picture Kant keeps in his mature philosophy, and the challenges that this poses for him:

- Kant agrees, for instance, that spatially extended objects like that tree bough have a sortal unity. That is why, quite like Leibniz, he describes them as "phenomena." That indeed is the ontological thrust of Kant's "transcendental idealism."[16]

- He even agrees that true objects are predicatively complete. This is the point of his claim (at A571-2/B599-600) that

> every *thing*, as regards its possibility, is likewise subject to the principle of *complete* determination, according to which if *all the possible* predicates of *things* are taken together with their contradictory opposites, then one of each pair of contradictory opposites must belong to it.[17]

- He accepts the definition of "intuition" as the mind's presentation of such an individual object, and as evidence for true synthetic predications about that object.
- And Kant will also agree that Caesar's glance is perspectival and shows nothing about the future (even near future) positions of the tree bough. How could he deny that?

But despite all of this, Kant claims that extended objects are legitimate objects, and – in a deep Leibnizian blasphemy – that human perceptions are always legitimate intuitions. They are not confused concepts; they are not concepts at all. Indeed, whereas Leibniz allows general and singular concepts, for Kant our human concepts are always and only general. They present only species. Moreover, for us humans there are no infinitely complex concepts, and thus no *infima species*.[18]

Indeed, as a consequence, there are no judgments that are automatically singular. Even Caesar's judgment "That floating log is brown" is not singular on its face. Only when he accompanies it with a verifying perception does Caesar have a "singular use" of that judgment.[19] But when he does, then Caesar will take that perception to be the presentation of a legitimate object, and the evidence for a true judgment. Leibniz says that Caesar is wrong on both counts. Perceptions do not present legitimate objects and do not ground true judgments. Kant will say that Caesar is right in both claims.

That is the challenge of singularity: How can Caesar's fleeting, and incomplete, perceptions count as the sole presentations of full-blooded, complete objects? And how can these perspectival bits of sensory experience serve to ground true objective judgments?

In the *Critique of Pure Reason*, Kant gives a two-tiered answer. First, there is a fine-grained analysis of the "phenomenal" unity granted by human perceptions. This is Kant's theory of empirical synthesis. It belongs to "transcendental psychology." And then there is a semantic account of how human intuitions serve as the grounds of true predications. This belongs to "Transcendental Logic." I will describe each briefly.

B. Empirical Synthesis

The question in transcendental psychology is how can Caesar distinguish objective spatial groupings such as the arrangement of leaf and branch on that floating log from what Kant calls "accidental collocations" (A121) or mere "subjective unities" (B142), that is, simple reports of the internal arrangements of Caesar's momentary state of mind. And the answer is that Caesar's perceptual "state of mind" (his *Gemütszustand*) contains implicit projections about further glimpses of that tree limb: projections that these glances (by himself and others) will confirm the currently perceived arrangement of branch and leaf, and will confirm the order of the limb's relative positions on the river. These projections – which Kant says must be imagined – are part and parcel of the perception itself.[20]

Here is a place where modern phenomenology can help. For, we can say that this is akin to what Husserl calls a "horizon," a barely felt set of projections that accompany a conscious act and that depict the possible ways that the act can be continued.[21]

This horizon of expectations, I suggest, is the main tool in Kant's theory of empirical synthesis, the connection of bits of sensation into the representation of an object.[22] For the fact that Caesar noticed the branch before the leaf is a subjective fact about Caesar. And he will project other experiences (by himself and by other observers) in which these parts are observed in different orders. But in all of these imagined experiences, the bough will maintain its treelike shape. That is what makes the shape "objective"; and that, indeed, is how the intuition – though subjective and perspectival – contains the marks of an aperspectival object. Quite similarly, the fact that the bough was first upstream and then downstream is an *objective* claim. So, in projecting further reports about the bough's position, Caesar will keep that relative order of observation intact. Moreover, Caesar's horizon will contain expectations about the bough's further career down the river, and even an array of possibilities concerning its past history as well.[23]

Figure 2 schematically pictures Caesar's state of mind with some of the imagined projections made explicit. In Figure 2, only the head of the stick figure represents Caesar's actual sensory state. It is like an incomplete cross section of the Leibnizian concept depicted in Figure 1. The upward graph represents his imagined projections, which propose more knowledge about the bough's present condition. The horizontal graph represents his projections about that limb's future states. The nodes in these graphs depict imagined and not actual perceptions. The graphs extend indefinitely, for there is always more to experience.[24] *With* this horizon, we have an intuition (the grasp of an object); *without* it, we have just a subjective jumble of sensations.

So, though Caesar's momentary glance may be short, perspectival, and incomplete, *qua* "intuition" it is embedded in a network of imagined possible

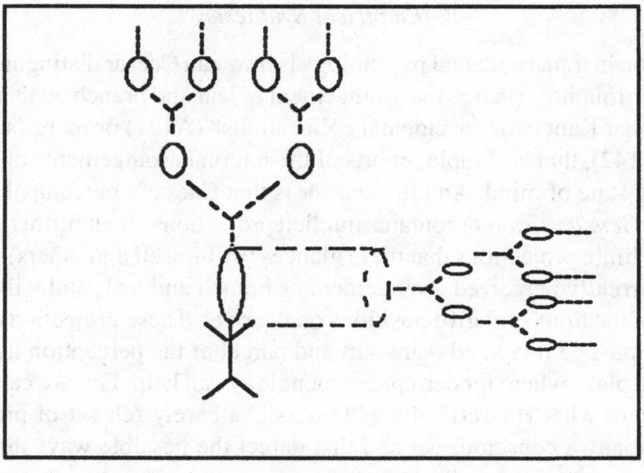

Figure 2

future glances that fill in the other perspectives and that complete the missing information. That is how Kant adapts the Leibnizian claim that an intuition presents a legitimate (and thus predicatively complete) object.

C. *Empirical Truth*

For Leibniz, a divine intuition also provides sufficient reason for the *truth* of singular judgments about its presented object. Kant needs to capture that semantic aspect of intuition as well, but all this analysis – the whole business of projections, horizons, imagination, and the rest – belongs solely to transcendental psychology. It has no semantic content, no connection, for instance, to the actual truth of the judgment that the tree bough is brown. And without that connection, Caesar's mental state does not fulfill the Leibnizian semantic task.

To give human intuitions a semantic role, Kant does two things: First, he translates Leibnizian assertabilism into his own "human" version. *Human* perception and the evidence based upon it are now the criterion of truth, rather than God's conceptual grasp. Second, he makes what I call a Peircean move.[25] Once again, I shall describe each briefly.

Humanizing Assertability. Kant considers Caesar's *Gemütszustand* not merely as a mental state but as an "epistemic situation." Caesar's horizon of projections, the nodes of Figure 2, are not just expectations but predictions of future evidence about that floating bough. And warranting these predictions is a necessary condition for the truth of his judgment that the bough is brown.

Going from the psychology of expectation to the semantics of predictions is the move that introduces the categories. For, having made this identification of evidence with truth, we must ask what justifies the predictions associated with Caesar's empirical syntheses? What justifies his prediction, for instance, that the bough will continue to appear brown, and treelike?[26]

One thing is certain. Under Kant's transcendental idealism, Caesar cannot appeal to objects outside the circle of his own representations in order to justify his claims. He cannot, that is, appeal to what Kant calls a "transcendental object ($= x$)" to explain and necessitate his view of the future evidence about the bough's shape or motion. Unknown x's do not empirically ground descriptive claims about the future.

The only thing that can justify a claim about further evidence is information forthcoming from the evidential state itself. When Caesar's empirical claim concerns time order – the claim, for instance, that further evidence will confirm the observation that the bough was first upstream and then downstream – then the only thing that would justify this claim would be Caesar's discovery within his experience of an empirically satisfactory explanation of the change in position. And, according to Kant, the heart of any such explanation will be a causal physical law that identifies the log's latter state as a necessary effect of its prior condition. Only if he is certain that he can discover some such explanation – only, that is, if he presupposes the truth of the general principle of causality – can he be certain that future experience will corroborate the time order: log upstream, then downstream.

And the same holds for the projections associated with his judgment about the bough's shape: Caesar implicitly predicts that further evidence will uphold his belief that he is seeing a truncated tree-shaped object. These predictions are well grounded only by virtue of the truth of the principle of extensive magnitude, the principle that says that all extended objects will have measurable (and thus relatively constant) spacial dimensions.

So, the point is that, without such categorial principles, our mental states cannot serve their semantic role as intuitions. The projections of objectivity and completeness cannot, by themselves, guarantee that the mental states of which they are part truly do ground singular judgments.

The Peircean Move. But even if we take categorially secured evidence as truth and even if we add all Caesar's accompanying projections, the fact remains that his perceptual state will always leave some unanswered questions about that traveling bough. What color is its underside? Will it be snagged or continue in its present path? And, indeed, where had it been before it arrived at this point in the river? Without determinations of these and other open predications, Caesar's representation of the bough remains unfinished and therefore still falls short of the Leibnizian requirement that it completely establish the truth or falsity of each elementary predicative claim.

Now, to be sure, when we move from Leibniz's God-style intuitions to Kant's emphasis on human perception, then we must assess the assertability of a judgment at the time that the judgment is made or at least within a narrow specious present. But that does not mean that truth and falsity are established solely by evidence present at that time. For Kant, a predicative statement is true, not merely by virtue of its being assertible right now, but rather by being assertible in the fullness of time.[27] Truth is thus a tenseless notion. If Caesar should discover tomorrow that the bough is hollow, then it is simply true, tenselessly, that the bough is hollow.

Adding this "Peircean" element is the second semantic move. Caesar's present glance says nothing about whether the bough will stay on trajectory or will veer off path or even will be snagged and stop. But our expanded notion of truth allows us to say that, nevertheless, one of these will be assertible, and thus is true *simpliciter*. Semantically, then, if not psychologically, our predicative grasp of that bough is complete.

Thus it is Kant's humanized assertabilism – bolstered by the categories and extended in a Peircean way – that allows human intuition, though fleeting, incomplete, and perspectival, to serve as the semantic grounds for objective predications about complete and aperspectival objects.

D. *Assertability and Physical Space*

So, again, given Kant's Peirce-like semantic theory, we can see why the bough is a complete object. Indeed, so too would be the tree from which it came, the whole forest for that matter, and even the entire planet. All of these can be intuited, both from a transcendental psychological and a logical point of view.

But not, Kant will say, the whole spatial universe! In the case of the totality of physically occupied space, says Kant, our grasp is conceptual, not intuitive and not singular.[28] The representation of that totality captures an aggregate, not an object. That is the point of the "First Antinomy."

Kant's reason is not merely that we cannot grasp the totality all at once,[29] for there are very large masses that do count as graspable objects. In denying that the representation of the world-whole is an intuition, Kant is once again resting on his humanized assertabilism.

If a cognitive grasp of this world-whole were an intuition, then that grasp (together with the categories) would have to provide a guarantee that we could answer any elementary question about the world and its parts. It would guarantee, in particular, that we could determine whether any given part is the furthest removed from us, or whether the series of occupied spacial regions extends infinitely into the distance. In the first case, the world would have a finite size.

That is the thesis claim in the "First Antinomy." In the second, the world would be infinite. That is the antithesis claim.

But Kant argues in the "First Antinomy" that our finite human grasp cannot possibly determine either of these things. And so, true to his assertabilism, he says that both claims are false. The world thus has no size at all, he says. And – this time true to his Leibnizian metaphysics – he claims that the world-whole is an aggregate and not an object.

Though we may believe that judgments about the world are singular judgments – or, to employ Kant's terminology, though we might want to use these judgments in a singular way – the fact is that there is no singularity here. There is only a pair of false general judgments.

Here is another place where twentieth-century techniques can help us. For the modern intuitionistic logic that attends L. E. J. Brouwer's twentieth-century assertabilism explains most neatly how Kant's idealist can simultaneously assert that the world has no utmost part and that it is still not infinitely expandable.

The claim that the physical world has a finitely distant ultimate part is

$$(\exists y)(\forall x)[(x \neq y) \rightarrow (y > x)] \tag{1}$$

(where the variables range over distances of occupied spatial regions from some fixed point), while the infinity of the physical world is

$$(\forall x)(\exists y)(y > x). \tag{2}$$

And the disjunction of these claims is classically valid but not intuitionistically valid.

Indeed, the modern assertabilist reading of the quantifiers gives a precise diagnosis of what fails in each disjunct:

Formula (2) requires a general method that, for any observed region, produces the actual observation of a yet more distant region.[30] That is the "constructive reading" of the universal quantifier. And, Kant argues, no such method can be given. The best we can do for any given spatial region x is to put ourselves in a position to perceive some further region y. But that alone does not guarantee the perception of y. Indeed, we can well imagine what it would be like to have such a perception. Imagining, however, is not perceiving; and when it comes to warranting existential claims, imagining is not enough.

Formula (1), in its turn, requires the observation of an occupied spatial region together with a proof that no observation of further such regions is possible. And once again, Kant argues that no observation, survey, or reasoning can satisfy these demands.[31]

Finally, Kant, just like Brouwer, ties the applicability of intuitionistic logic to the existence of unsolvable problems. For within assertabilism itself, the assumption that all relevant questions eventually can be answered, together with the Peircean semantics, yields a classical and not an intuitionistic logic.

Brouwer knew this. That is why he repeatedly rejected Hilbert's claim that all well-formed mathematical problems are solvable.[32] And Kant applies the same insight – in his case to empirical knowledge and truth – when he states that "[i]n natural science ... there is endless conjecture, and certainty is not to be counted upon." (A481/B509)

III. The Birth of Modern Reference

A. *Kant the Referentialist*

So, as I said, Kant's adaptation of Leibniz's assertabilism accounts for why we can make singular judgments about the bough but not about space. It is the assertability theory of truth that establishes the falsity of both the thesis and the antithesis. And it is the connection between assertabilism and unanswerable questions that provides the logical structure for his reasoning in the "Antinomy." But the fact is that, when Kant does speak explicitly about truth, he defines it not as assertability but as "agreement with the object."[33] The purpose of intuition, he says, is to "give" that object. An intuition does not merely represent an object, it *presents* its object. Moreover, through that presentation, it grounds the truth of our judgments about that object. This is the language of the referential theory of truth; and that is why I spoke of this "object giving" aspect of intuition as the criterion of "reference."

Though Kant says his notion of correspondence is merely a "nominal definition" of truth (A58/B82), I must say that his referentialism is not simply verbal obeisance to an old tradition. It is a pervasive systematic position about truth. Here are four places where that is clear:

(1) On the famous issue of excluded middle, Kant seems wedded to the referential camp. One of the hallmarks of Brouwer's intuitionism is its rejection of excluded middle, the principle that, for any sentence p, $(p \vee \sim p)$ is a logical law. Yet, we have already seen that, at least for elementary predications, Kant quite explicitly embraces that principle. Or, to be more precise, he does so in those cases in which the object is present. The logical schema that underlies his remarks is actually

$$(\exists x)(x = t) \rightarrow (Pt \vee \sim Pt), \qquad (3)$$

where P may be any elementary predicate, even one that is not yet decided for t. So, once again, it is the connection to an existent object that underlies the truth of an as-yet undecided disjunction.[34]

(2) Kant's epistemic pessimism in empirical science – the excuse for adopting an intuitionistic logic for empirical discourse – that position itself stems from the nature of our contact with empirical objects.

> In natural science . . . there is endless conjecture, and certainty is not to be counted upon. For the natural appearance are objects which are given to us independently of our concepts, and the key to them lies not in us and our pure thinking, but outside us; and therefore in many cases, since the key is not to be found, an assured solution is not to be expected. (A480/B508)

Empirical truths elude us not because (or not merely because) our cognitive epistemic abilities are finite, but because the objects may not be found. This bald referentialism is quite different from the considerations of complexity and infinite information that often underlie our modern epistemic pessimism.

(3) In the theory of synthesis where I have spoken about the objectivity of connecting the leaf and the branch in Caesar's evidential state, Kant himself speaks of their "combination in the object" (B142). That is what underlies the truth of Caesar's judgment about the shape of the floating bough. Indeed, Kant describes the success of the categories in bolstering that truth as their "objective validity" – their application to objects.

(4) Perhaps most prescient of all is Kant's view of natural-kind concepts:

> And indeed what useful purpose could be served by defining an empirical concept, such, for instance, as that of water? When we speak of water and its properties, we do not stop short as what is thought in the word, water, but proceed to experiments. The word, with the few characteristics which we attach to it, is more properly to be regarded as merely a designation than as a concept of the thing, the so-called definition is nothing more than a determining of the word. (A728/B756)

This Kantian account of our ability to refer to natural kinds is "anti-descriptivist." Kant's point is that we may learn new facts and revise our catalogue of properties while referring to the same object or kind. We might even make mistaken claims about objects and natural kinds. The object or kind – and thus what is true about them – cannot be defined. Leibniz's God, with his grasp of complete concepts, might be able to *define* the objects of His judgments, but we humans can only *refer*. And so, if our human judgments are to be legitimate truth bearers, they must derive their truth referentially – from the objects, so to speak.

Thus Gordon Brittan is right when he says that "[o]nly reference gives us objects in the appropriately full-blooded, yet transcendentally ideal, sense necessary to Kant's various purposes."[35]

B. The Birth of Modern Reference

Now, all this object talk should be quite foreign to the modern assertabilist, for whom truth is strictly a matter of one's (or at least of the scientific community's) epistemic state. But, once again, Kant is simply adopting the Leibnizian stance.

Like Leibniz's, Kant's notion of truth is simply simultaneously assertabilistic (though in this case humanly assertabilistic) and referential.

When you stop to think of it, this is perfectly proper. For, it is not the God-centered containment theory that allows Leibniz to merge evidential and referential accounts of truth. Leibniz's semantic merger works simply because he believes that there is nothing more to the "*haecceity*" of the object than what is contained in the corresponding evidential state. And Kant preserves that "onto-epistemic" isomorphism. To be sure, it is now human perceptual intuition, rather than God's grasp of a complete concept, that counts as the appropriate evidential state. But still, to return to my example, Kant allows no secret properties conferred by the tree limb's actuality beyond those betokened by Caesar's intuition of it.[36]

Indeed, the cases of referentialism in Kant that I just gave are instances of cooperation rather than conflict between reference and assertability in the truth conditions for Caesar's judgment that the tree bough is brown. Thus, for instance, it is reference – the presence of the tree bough – that guarantees that the material will be available for all those projected future glances. That is the real force of principle (3). When we have an empirical intuition – and thus can refer to an existent empirical object – then and only then can we be certain that even currently unrevealed aspects can become known to us.

And we can say, just as we did for Kant's assertabilism, that the categories serve to bolster the referential role of our perceptions. If principles such as causality and extensive magnitude were not valid, then Caesar's intuitive state of mind – with all its projective halo – would be a mere pretense of objectivity. It could not count as the presentation of an object. That is why Kant describes the semantic success of the categories as their "objective validity."

So, when Kant says that the correspondence theory is a "nominal" account of truth, he is saying that it is a theory that operates in both the Leibnizian semantic scheme and his own. But that is only a formal or "nominal" similarity. For the Leibnizian notion is a thin, epistemically transparent notion of referential agreement. In the very act of referring to Caesar, God already determines all that is or can be true of Caesar. Kant's notion of "empirical" reference, by contrast, is epistemically opaque, or "thick." It allows conjecture, uncertainty, and even revision. That, recall, is the point of Kant's remarks about natural-kind concepts.

Moreover, when we speak of empirical objects, then reference is even more clearly a "thick" notion: reference to an object brings with it the existence of the object; and that existence entails quite directly that there will be more to the object than any cognitive act can grasp. This, indeed, follows from the recognized limitations of empirical synthesis.[37]

Finally, this thickness – this outstripping of our cognitive grasp – is the heart of our modern referentialist truth theories. These are theories that harp upon the fact that success in referring to an object, and ultimately the truth or falsity of our judgments about that object, is largely independent of our cognitive hold upon that object.[38] That is why Kant's robust referential theory represents the birth of a modern conception of reference.

IV. Immediacy

A. Kantian Immediacy

Recall that Leibniz equates intuitivity with an inferential immediacy: God's grasp of Julius Caesar's complete concept is called "intuitive" because – no matter how complex that concept might be – He can directly apply to Caesar each of the concept's component predicates, without intervention by symbolic abbreviations and without inference.

To see how Kant adapts this Leibnizian notion, we need to notice two additional things about God's grasp of that complete concept.

(1) Insofar as the object of this grasp is a concept, then the grasp is "immediate" in the sense that there are no intervening mental representations. The full individual concept is directly present to God's mind.[39]
(2) But precisely because the object of this grasp is a concept this is still at best an indirect grasp of the Julius Caesar himself.

Now Kant's notion of intuition accepts the directness of the first point and most definitely rejects the indirectness of the second. Caesar's intuition of that floating bough is a direct awareness of the bough. The bough itself (and not some conceptual intermediary) is the object of Caesar's mental state. For Kant, intuitions are *immediate* presentations of their objects simply because they are nonconceptual. Kant will say that a representation is "immediate" if it presents its object without mediation by *Merkmale*, which might equally well apply to other objects, that is, without mediation by any concept, be it simple or compound. Kant defines immediacy by this nonconceptuality rather than by counting the number of steps in an inferential process.

This is not an artificial maneuver of Kant's, made in order formally to adapt just one more component of Leibnizian intuition. It is, rather, a direct consequence of Kant's central move. Concepts are always general; only intuitions can present true individuals. Nonconceptuality is precisely what defines intuitions for Kant.

Moreover it is immediacy that guarantees Kant a "full-blooded" notion of reference. For, if we had only conceptual means to pick out the objects to which we refer, then we would effectively be defining those objects in the act of referring to them. We would be capable only of descriptive – and therefore "thin" – and not of referential use of intuition.[40]

So, Kant's definition of immediacy – immediacy as nonconceptuality – is crucial to the role of human intuition both as singular representation and as the vehicle of reference.

B. Is Immediacy Possible?

The problem is simply this: If you look at how intuitions come into being – that is, how we go from a subjective jumble to the representation and presentation of an object – then you will see that the process is inescapably conceptual.

For, to have a legitimate intuition, Caesar must have a horizon of projections. We may interpret these projections phenomenologically, epistemically, and semantically, but in all of this the fact remains that this horizon is shaped by a concept. The completeness that makes the bough, the tree, and the planet into legitimate objects is governed by the sortal concepts *bough, tree*, and *planet*. The concept *bough*, for instance, tells Caesar how to imagine expected future perceptions. It tells him what sorts of things may vary and what sorts of things he must imagine as constant. Indeed, that is how it becomes a bough perception rather than a branch or leaf perception. So too, in the intuition of the bough's motion from upstream to down, it is the concept of motion (positional change) that shapes the objectifying horizon.

It would seem, then, that intuitions without concepts are not merely "blind" (B49), they are impossible. There is no theoretical opening to achieve that concept-independent immediacy. The very notion of an intuition is soaked to the core with those allegedly banished mediating *Merkmale*. Or, to put the matter in other words, though I have employed Husserlian phenomenology to characterize Kant's notion of intuition, there seems no room for what Parsons calls the "phenomenological presence" of the object.[41]

C. Immediacy and Reflective Judgment

We can find our answer in the aesthetic theory that Kant presents in the *Critique of Judgment*: His notion of beauty and his description of the process of reflection, which underwrites judgments of beauty, provide examples of perceptual experiences that go beyond their conceptual components.

According to Kant, a judgment of beauty is based on the presentation of an individual object in a special way. The experience underlying such a judgment is tied to an act of reflection, which itself is supposed to be generalizable but without concepts, and subjective but without interests or desires.

Now when Caesar judges that the bough is brown, or that it is moving, or even that this motion has a cause, in each of these cases he is subsuming a particular object under a more general concept. He starts with the concept and uses it to organize the sensory manifold by guiding the imaginative projections. This sort of synthesis underlies what Kant calls "determining" judgments. "Reflective" judgment, by contrast, works in the opposite direction: In this case, "only the particular is given and the universal has to be found for it." (AK, V 179) Kant's favorite instances of this sort of judgment come from his analysis of scientific discovery and of the "logical origin of concepts," activities that he believes have a special attendant delight.

Kant describes these reflective processes similarly: You start with a group of distinct data and then pass to a single principle that "unifies" the data, and explains some things about them. In science the data are specific scientific laws and the unifying principle is a general scientific theory that organizes and links them. In concept formation the data are individual objective perceptions, and the unifying principle is an empirical concept, under which they all fall.

In both science and concept formation, a unifying principle once found is a "hardened" sort of thing: it can apply to new data, to as yet unexperienced objects and situations.

As I said, there is supposed to be a certain delight attendant to this sort of discovery of a systematic unity. It is a delight that comes from the search itself for a unifying principle and from its successful completion.

In the ordinary course of things, this delight is a fairly rare commodity. Large-scale scientific innovation is not an everyday affair, and, for the most part, we are out of the business of generating new empirical concepts. So, we do not often have the delight of a discovered unity.

But, says Kant, some experiences do excite in us this delight. This is perhaps because they are experiences of objects that have an aspect of style or unity that is not exhausted by any hardened concept. That happens when there is more in the horizon than is given by a describing concept. Kant's examples include flowers and freely drawn patterns. He also speaks of appreciating music without invoking the underlying technical concepts. Or, says Kant, the delight may be excited for quite the opposite reason. It might occur because the experience itself is so intense and uncluttered that "its uniformity is not disturbed or broken by any foreign sensation." (AK, V. 224)

The point in both the complex and the intensely simple cases is that the perceiver feels (again, perhaps implicitly) a sense of order that guides his or her imaginative projections. It is a feeling that some parts of the manifold simply belong together – indeed, in a horizon-forming, rulelike way. But unlike ordinary perception, the order and the guidance are not given by a "hardened" concept, which may well apply to other objects as well. It is peculiar to this particular experience, and the feeling is akin to the delight of discovering a new scientific uniformity. Objects that occasion this feeling of discovered unity are the ones that Kant calls beautiful.

To be sure, I might apply a concept while enjoying a beautiful object. In music and in poetry, for instance, the medium itself sets up a field of expectations about how a line or a phrase might be continued. That actually enhances the appreciation: Having a sense of what choices the artist faced highlights our awareness of the absolute "rightness" of the work's construction. And, without some sense of the possibilities, we might be unable to appreciate a thing of beauty. But we must keep in mind that in beautiful music and a beautiful poem the artist makes a choice that is right in a way more specific than anything the medium or any rules of craft could predict. In these cases, Kant will insist that the unifying horizon is not exhausted by any concept with general application. The feeling of unity in hearing a great symphony, for instance, is not resolvable to the score or to any other general description.

This is how Kant's theory of beauty provides the possibility of intuitions without concepts. Just as conceptual description may well be part of aesthetic experience, so too, in ordinary perception, conceptual shaping may be essential to objectivity. But we see from the aesthetic theory that this conceptual component leaves room for individuality. The third *Critique* theory of beauty, gives meat to the notion of immediacy: an intuitive grasp with a full horizon that outstrips any descriptive concept. And Kant's treatment of beauty shows cases in which that nonconceptual intuition actually plays an important role in Kantian philosophy.

This is the opening for the revisability that characterizes our empirical concepts. For, though not all objects are beautiful, and though not all perceptual experiences are experiences of beauty, nevertheless, Kant's aesthetic theory does show that intuitive unity can outstrip conceptual unity, and thus that there is an aspect to an intuitively presented object that can survive redescription and revision.

V. Space and Intuition

Now, having explored the criteria of intuition in general, let us proceed, as I promised, to the original representation of space, a representation that is supposed to be a "pure," or a priori, intuition. Here we must address four issues: There is first the question of what Kant means when he calls this a "pure" representation, and then there are the issues of singularity, reference, and immediacy. I will briefly address each of these in order.

A. *Figurative Synthesis and Pure Intuition*

Purity. Consider again the complex mental state underlying Caesar's perception of that floating tree bough, a state that contains a horizon of imagined continued perceptions of that same floating limb. This, I have argued, is the core of Caesar's *empirical* synthesis. Caesar's *figurative* synthesis differs from his empirical synthesis in two specific ways.

First, Caesar will abstract away all the details specific to this particular limb. Indeed, he must ignore its special treelike features, its color, texture, and weight.

Instead, he will concentrate only on those features that this particular object shares with all similarly shaped physical objects. He will do this, in particular, if he is interested in exploring the geometry of cylindrical objects. In this case he will also abstract away the log's motion in the river. Under other circumstances – specifically if he is concerned with properties of continuity – he would keep the motion in mind, and ignore the size and shape. And in yet another case, he will abstract from all of these and concentrate only on the iterative process involved in measuring the length of a side. Common to all of these is Caesar's abstraction from the particulars; and thus the conclusions warranted by that abstraction will be conclusions about the concept of cylinder general, or of linear extension, or of size itself.[42]

The second difference is that, if Caesar were acting as a mathematician, he need not begin with an actually perceived bough. An imagined limb, or for that matter, any imagined cylinder will do just as well. To be mathematically cogent, the projective horizon need not be anchored in an actually present physical object.[43] The mathematician could do just as well with an imagined line and an imagined motion as well. This is the sense in which the work of the mathematician is supposed to be "pure."

A Priority. But this purity alone – or at least the purity that comes from unanchored imagination – is not yet *a priority* in the deep Kantian sense. For, "a priority" connotes that the findings derived from this figurative synthesis are unavoidable. Thus, if we gather up all this mathematical work and bind it together into a characterization of space,[44] we must be justified in predicting that, come what may, the same findings will be unrevisable and will be preserved in all perceptual circumstances.

Were we to view these figurative syntheses by which we discover and abstract mathematical properties in the same way as we view the reflection that engenders empirical concepts, then we could never justify that prediction of unrevisability. Indeed, we have just seen that one of the powers of that reflection is precisely that it is thus revisable.[45]

Instead, Kant must provide an argument that geometric theorems will not suffer revision or be overturned as we encounter new circumstances beyond those in which we reflectively discovered the properties of space. His argument is that the representation of space is itself presupposed by the very possibility of representing any outer object at all. And thus it could not itself depend on a process of abstraction from the experience of outer objects. This is the point made (perhaps successfully, perhaps not) by the first pair of arguments in the "Metaphysical Exposition of the Concept of Space" (A23-24/B37-39). It is not a psychological argument. It is an argument, ultimately, about epistemic presupposition.

B. The Singularity of Space

Here we must compare the representation of space (considered, for instance, in the "Aesthetic" of the *Critique of Pure Reason*) with the "Antinomy" notion of the spatial profile of the world-whole.

Space itself – the subject matter of geometry and the topic of the "Aesthetic," – is a collection of spatial regions. In this respect, it is like the empirical world-whole. But we have seen that the empirical world's spatial profile is only an aggregate, not a unified object. And we have learned that this is so because our representation of this empirical profile lacks the epistemic and semantic completeness that Leibniz and Kant demand in an intuition. Physical space will always be incomplete, because there will always be unanswered questions about it.

Unlike physical space, however, Kant says that the pure representation of *mathematical* space does have the epistemic completeness that he denies to our concept of the world. In mathematics in general, he says, we can "demand and expect none but assured answers to all the questions within its domain." (A480/B508) And the reason that we have this epistemic completeness is precisely because the abstraction that characterizes figurative synthesis removes the element of "receptivity" that had blocked our access to full empirical knowledge. The "key," says Kant, is "within us." (Ibid.) With that epistemic completeness comes semantic completeness: the logic of mathematics is classical, as assertability theory demands.[46]

And with semantic completeness comes a certain ontological singularity as well. The individual regions of mathematical space, unlike the regions of the empirical world, *are* part of an encompassing whole. They are individuated by their positions in the whole. Indeed, the grand synthesis of all these parts into a single whole – a synthesis shown to be impossible for physical space – is perfectly possible for mathematical space. Thus, in the mathematical case, the relation between the representation of space and that of individual spaces is not the relation of predication; it is, rather, the relation that obtains between a unified whole and each of its parts.[47]

C. Immediacy

One consequence of this singularity is that Kant is now free to say that space itself has a size; indeed, that it extends (as Euclidean geometry demands) infinitely in each dimension. And he does say exactly that at A25/B39: "Space is represented as an infinite, *given* magnitude." Moreover, when we speak of mathematical space (stripped of any empirical content), then we can actually assert formula (2) of §I.D above, the infinity claim:

$$(\forall x)(\exists y)(y > x).$$

This fact, says Kant in the continuation of that famous passage, is a proof that the original representation of space is an intuition, and not a concept. And his point is really about the criterion of immediacy.

Here is the full passage:

> Space is represented as an infinite *given* magnitude. Now every concept must be thought as a representation which is contained in an infinite number of different possible representations (as their common character), and which therefore contains these *under* itself; but no concept, as such, can be thought as containing an infinite number of representations *within* itself. It is in this latter way, however, that space is thought; for all the parts of space coexist *ad infinitum*. Consequently, the original representation of space is an *a priori* intuition, not a concept. (A25/B39-40)

Kant's reasoning in this passage consists of four points:

(1) The claim that space has infinitely many parts is an a priori claim. As such it must derive from the "original representation" of space.
(2) If the original representation of space were a concept, then that claim about infinite size would have to be analytic.
(3) For the claim to be analytic, it would require that the concept have an infinite intension. (That is what it would mean for a concept to contain infinitely many representations *within* itself.)
(4) And that last is impossible. The central tenet of Kant's revolt from Leibniz is that, in the realm of human knowledge and truth, there are no infinitely complex concepts. (That, recall, is why there are no *infima species*.)

The infinity claim of formula (2) is established rather by a process of figurative synthesis. For each region x that has been measured (actually or in the imagination), all it takes to establish the existence of a further region y is to imagine it and imagine measuring its distance from us. The empirical constraint of actual perception has been lifted.

So, in the end, we must appeal to the constructive reading of the quantifiers – together with the abstraction of figurative synthesis – rather than to any conceptual analysis in order to establish the infinity of space.

This does not preclude our subsequently axiomatizing geometry, and including formula (2) [or some other formula from which (2) is derivable] among the axioms. That would be conceptual description. But, Kant insists, the *original* grounds for (2) constitute an aspect of the unity of space that outstrips any conceptual representation. And that is precisely the point of immediacy.

D. Reference

Finally, if the representation of space is indeed an intuition, then it must present ("give," in Kant's terms) an object. To be sure, Kant labors nobly to establish that this a priori representation, like the categories, has "objective validity," and is not merely a "play of the imagination." (A239/B298) But that simply means that every empirical intuition must conform to the laws of geometry. If the representation of space is to be an intuition, then Kant must also require those laws to be true by virtue of reference to an existing object.

And that poses a problem. For, in a footnote to the "First Antinomy," Kant says quite explicitly that

> [s]pace is merely the form of outer intuition (formal intuition). It is not a real object which can be outwardly intuited. (A429/B457 n)

And in the "Doctrine of Method," he says that for mathematics in general "there is no question of . . . existence at all." (A719/B747)

These views deal a hard blow to the Leibnizian and Kantian criterion that intuition (in this case, the pure intuition of space) must have a referential role, that it must present an object.

Kant resolves this apparent problem with what I once called his "project centered ontologies,"[48] a notion that displays one of the most significant results of Kant's marriage of Leibnizian principles with human epistemology.

We must distinguish (as Kant most certainly did) between the empirical project and the mathematician's project. When we speak empirically – about boughs for instance, or trees, or planets, or physically occupied space – the truth of our judgments depends upon our finite, receptive perceptions. We have no such perception of mathematical space. And within the empirical project – the project whose limits are explored by the "Antinomy" chapter – the "formal intuition" of space is not an intuition, and its object, "mathematical space," is no object. That is the point of Kant's footnote at A429/B457.

But the rules of evidence for mathematics itself (i.e., carried out as a science in its own right) are different. In the mathematical project, we "abstract" from our perceptual experience. And in mathematics, we substitute imagination for perception. That is why Kant is so optimistic about the solvability of mathematical problems. To be sure, whatever else it may be, mathematical intuition is not empirical perception. But it is still intuition.

Under Kant's assertabilism, this change in style of intuition betokens a corresponding semantic change. As we have seen, Kant's logic of mathematical discourse is consequently classical.

But, once again, the change in epistemology and semantics also draws with it a corresponding change in ontology. That is the Leibnizian legacy: the rarified mathematical intuition *is* also referential. The mathematician has a repertoire of objects: individual numbers, for instance, and mathematical space itself. These are intuited. They ground the truth of our mathematical judgments, and – within the mathematical project – they do exist. This is once again a thin sort of reference (mathematics, for Kant, recall, is not opaque, at least not semantically opaque) and a rarified existence. But mathematics in its own sphere does have its own reference and existence.

E. A Brief Conclusion

It would be nice to end here, for I have shown you how each of the three components of Kant's notion of intuition – singularity, immediacy, and reference – is derived from a Leibnizian counterpart and how the exercise of adapting these Leibnizian aspects of intuition to the special case of human perception engenders some of the central facets of Kant's mature metaphysics. And I hope that you can see how Kant can account for each of these aspects in his contention that the representation of space is a pure intuition. Even the Kantian criterion of reference has its role within what I called the "mathematical project." And these are the points I promised, at the outset, to demonstrate.

But it would be wrong to stop here, for there is an air of relativism about this talk of project-centered ontologies: Leibniz established that intuition, evidence, truth, and objects go together. Kant, we would say, having relativized the notion of intuition, evidence, and truth to empirical and mathematical versions, should and does do the same for objects.

But to accept this relativism would be to paint Kant's adaptation of the Leibnizian paradigm as a sort of mechanical translation, and that would be wrong. Kant is not a relativist. He does not view the empirical and mathematical projects as separate, equivalid enterprises. He believes rather that the empirical project is paramount. It is paramount both practically and ontologically.

We see the practical primacy of the empirical project from Kant's remarks about the objective validity of mathematics. Were it not for the fact that mathematical intuition is a component of empirical intuition, and that mathematical truths govern the empirical world, were it not for these facts, then mathematics would be a "mere play of the imagination" (A239/B296), and would not count as knowledge.[49]

And we see the ontological primacy of the empirical project in the remark I quoted from the "Doctrine of Method." Only empirical – not mathematical – existence is existence in the true sense.

And so we can say that, yes, the pure intuition of space does straightforwardly refer within in the mathematical project. Thinly, to be sure, but it is reference. And we can affirm that geometric truth does rest upon reference to space itself. Nevertheless, only because this pure intuition enables empirical intuition to individuate its own empirical objects do we accept geometry as knowledge. And only because geometric truth regulates empirical truth, do we entitle that pure representation an intuition, and not merely an exercise in imagination.

In the end, then, though we cannot understand Kant without understanding his Leibnizian roots, we must never forget the power and primacy of his empirical, humanizing spirit.

NOTES

I am indebted to Gila Sher and Richard Tieszen, without whose patience and encouragement this essay would not have been written.

1. Passages from the *Critique of Pure Reason* are quoted from the Kemp-Smith translation (New York: St. Martin's, 1961). References are to the first and second edition pagination, as is standard. Passages from the *Critique of Judgment* are quoted from the Meredith translation (Oxford: Oxford University Press, 1952). Page references are to volume V of the *Akademie* edition.
2. See, for instance, "Kant's Philosophy of Arithmetic," in C. Parsons, *Mathematics in Philosophy* (Ithaca, NY: Cornell University Press, 1983); reprinted in C. Posy (ed.), *Kant's Philosophy of Mathematics: Modern Essays* (Dordrecht: Kluwer, 1991).
3. The lattice image is adopted from R. Kauppi's *Über die Leibnizsche Logik* (Acta Philosophica Fennica, Fasc. 12, New York: Garland, 1969).
4. "It is the nature of an individual substance or complete being to have a concept so complete that it is sufficient to make us understand and deduce from it all the predicates of the subject to which the concept is attributed." (*Discourse on Metaphysics*, VIII.)
5. "For if some man were able to carry out the complete demonstration by virtue of which he could prove this connection between the subject, who is Caesar, and the predicate, which is his successful undertaking, he would actually show that the future dictatorship of Caesar is based in his concept or nature and that there is a reason in that concept why he has resolved to cross the Rubicon rather than stop there, and why he has won rather than lost the day at Pharsalus, and why it was reasonable and consequently assured that this should happen. (*Discourse on Metaphysics*, XIII.)
6. I hasten to add that it is the unity and not the existence that is at issue here. The notions of "real" versus "phenomenal" are adjectives describing varieties of unity, not of existence. Indeed, the term "aggregate" does not connote a mind-dependent or imaginary object. Strictly speaking, aggregates are not objects at all and ontology starts only at the level of monads. See M. Mugnai, *Leibniz's Theory of Relations* (Stuttgart: F. Steiner Verlag, 1992) for an interesting discussion of this aspect of Leibniz's metaphysics.
7. "The predicate or consequent therefore always inheres in the subject or antecedent. And, as Aristotle, too, observed, the nature of truth in general or the connection between the terms of a proposition consists in this fact." (G. W. Leibniz, "First Truths," in L. Loemker (ed.), *Philosophical Papers and Letters*, 2nd Ed. (Dordrecht: Reidel, 1956, pp. 267–70).
8. Hints of an assertabilist truth theory are scattered throughout Leibniz's work (and, perhaps more importantly, through the work of the next generation of "Leibnizians"). But here is one explicit place: In "General Inquiries About the Analysis of Concepts and of Truths" [1686] in G. H. R. Parkinson (ed.), *Leibniz Logical Papers* (Oxford: Clarendon, 1966), a work devoted to what we would call semantics, Leibniz says quite explicitly:

That is true, therefore which can be proved, i.e. of which a reason can be given by analysis; . . . [§130] I define "false in general" as that which is not true. For it to be established that something is false, therefore, it is necessary either that it should be the opposite of a truth, or that it should contain the opposite of a truth, or that it should contain a contradiction (i.e., B and not-B), or that it should be proved that, however long an analysis is continued, it cannot be proved that it is true. [§57]

This is a clear assertabilism. Indeed, in the second quoted passage Leibniz preshadows a version of the assertabilist recursion conditions for negations.

9. For us, today, the components of a judgment are words; for Kant (and Leibniz), they were concepts. Leibniz, in fact, objected to Locke's attempt to insert a linguist level between concepts and their objects. See *New Essays*, Book IV, Ch. V. (A. 6. Langley (trans.) Open Court, La Salle, IL, 3rd Ed. 1949).
10. One way to see this difference in kind is to note that monads as "simple" can have no component parts, whereas complete concepts are never simple entities. This is not to deny, however, that Leibniz may have an idealist account of spatially extended objects. R. M. Adams argues forcefully for precisely this view in his *Leibniz: Determinist, Theist, Idealist* (Oxford: Oxford University Press, 1994).
11. See *Discourse on Metaphysics* §8. This notion that the *haecceity* of the object is exhausted by its individual concept is closely related to Leibniz's famous principle of the Identity of Indiscernibles. For, the point of that principle is that nothing outside of their individual concepts can serve to distinguish one monad from another.
12. "Meditations on Knowledge, Truth and Ideas," *Acta Eruditorum*, 1684; translated and reprinted in G. W. Leibniz, *Philosophical Papers and Letters*, ed. L. Loemker (Dordrecht: Reidel, 1956)
13. The structure of a concept can, in Leibniz's view, be even infinitely many layers deep. That is Leibniz's explanation of what Kant comes to call "synthetic" knowledge, that is, knowledge that we are incapable of deriving simply from the analysis of concepts.
14. In *New Essays* (Book IV, Ch. ii) Leibniz elaborates this distinction between *intuitive* and *demonstrative* knowledge. It is developed from Locke's distinction among *intuitive, demonstrative*, and *sensible* knowledge (see *An Essay Concerning Human Understanding*, Book IV, Ch. ii). The *intuitive/demonstrative* distinction is adopted by Hume as well. (See *A Treatise of Human Nature*, Part III, §1, and *An Inquiry Concerning Human Understanding*, §IV, Pt. I.)
15. "When my mind understands at once and distinctly all the primitive ingredients of a conception, then we have intuitive knowledge. This is extremely rare as most human knowledge is only confused or indeed assumed." (*Discourse on Metaphysics*, XXIV)
16. We should recall that it is unity, not existence, that makes empirical objects phenomena.
17. The complete determination of "things" that is emphasized here is contrasted to the merely logical principle of contradiction that is described in Kant's preceding paragraph and that governs concepts. Kant is saying here that objects but not concepts are completely determined. (See note 34.)
18. See Jäsche Logic §10 and *Critique of Pure Reason* A655-6/B683-4. If the faculty of understanding *could* generate an intuition, it would be a Leibnizian intuition, the grasp of an infinitely complex individual concept. Kant admits that in the abstract this is a coherent notion. He calls it "intellectual intuition." So, in this sense Parsons (*Op. cit*, Part I) is right to say that Kant considers the possibility of singular

representations that are not perceptions. But Kant also says that for us humans this is but an empty notion, not part of our human abilities.

19. See Jäsche Logic §1 n.2. To claim that no judgment can be singular simply in virtue of its grammatical structure is in modern terms to claim that there are no singular terms. Manley Thompson has indeed suggested that this is Kant's view. (See "Singular Terms and Intuitions in Kant's Epistemology," *Review of Metaphysics*, 26 (1972), 314–343; reprinted in C. Posy (ed.), *Kant's Philosophy of Mathematics: Modern Essays*, pp. 81–107). Thompson, however, modifies the view by adding that demonstrative expressions ("that floating log I see before me," for instance) are Kant's versions of singular terms. They would be something like Russell's "logically proper names." Thompson is playing on the idea that such expressions are automatically accompanied by intuitions. But in fact they are not. For, a listener with his back turned will not have the accompanying perception. Such a listener could not be taking the judgment in a singular fashion. In the end I think it best to say that, for Kant, singularity is a semantic (or perhaps pragmatic) property of judgments with no "syntactic" counterpart.
20. "Psychologists have hitherto failed to realize that imagination is a necessary ingredient of perception itself. This is due partly to the fact that that faculty has been limited to reproduction, partly to the belief that the senses not only supply impressions but also combine them so as to generate images of objects. For that purpose something more than the mere receptivity of impressions is undoubtedly required, namely, a function for the synthesis of them." (A120, note)
21. "What is actually perceived, and what is more or less clearly co-present and determinate (to some extent at least), is partly pervaded by, partly girt about with a *dimly apprehended depth or horizon of indeterminate reality.*" (Husserl, *Ideen, I*, §27 (W. R. Boyce Gibson (trans.) Collier, New York, 1962)).
22. I explore this a bit more fully in "Where Have All the Objects Gone," *Southern Journal of Philosophy*, 25 (Suppl. 1986): 17–36.
23. This is Kant's distinction (at A193/B238) between subjective and objective time orders.
24. Actually, the matter is more subtle, given Kant's distinction in the "Lecture's on Metaphysics," between productive and anticipatory imagination. The point really must be that Caesar imagines situations in which he views the manifold in a different order, with no assumption about whether or when those situations occur. So, this picture of a projective horizon even characterizes Caesar's perception of the descent of Brutus' dagger on the Ides of March. It matters not that Caesar realizes that he may very soon have no further sensory experiences.
25. The influence, of course, goes in the opposite direction. See my "Kant's Mathematical Realism," *The Monist*, 67 (1984): 115–134, reprinted in C. Posy, *Kant's Philosophy of Mathematics: Modern Essays*, 293–313 for a discussion of Kant's humanized assertabilism.
26. To be sure, these predictions are fallible, and Caesar will recognize them as such. Perhaps, for instance, he is being misled by a passing shadow. That is what makes his judgment a posteriori. But his prediction right now is that, in the fullness of time, all the evidence will support his description of the log. And the question is, what justifies that prediction.
27. "To call an appearance a real thing prior to our perceiving it, either means that in the advance of experience we must meet with such a perception, or it means nothing at all. . . ." (A493/B521)

28. "Now we have the cosmic whole only in concept, never as a whole in intuition." (A518-9/B546-7)
29. Or see it finely enough. For, the same reasoning applies to very small masses.
30. Or at least a causal tie to such an observation.
31. Moreover, this assertabilist reading of the quantifiers suggests a simple intuitionistic Kripke model to simultaneously falsify both (1) and (2). The model consists of an infinite string of nodes, in which the nth node contains the first n integers, and "$>$" is interpreted standardly. In this model, the claim

$$(\forall x) \sim\sim (\exists y)(y > x)$$

is also true, and that intuitionistic double negation captures the sense in which the infinity of space is a regulative force guiding scientific research. Indeed, formulas like this provide a good general interpretation of Kant's notion of regulativity.
32. See, for instance, "*Intuitionistische Betrachtungen über den Formalismus*" in L. E. J. Brouwer, *Collected Works* (Amsterdam: North–Holland, 1975). I have spelled out the connection between assertabilism and these logical issues in "Kant's Mathematical Realism."
33. "What is truth? The nominal definition of truth, that it is the agreement of knowledge with its object, is assumed as granted . . ." (A58/B82). See also A191/B236.
34. This principle is sometimes thought to be a merely regulative requirement. (See, for instance, Gordan Brittan, "The Reality of Reference: Comments on Carl Posy's 'Where Have All the Objects Gone,'" *Southern Journal of Philosophy*, 25 ((Suppl.) 1986): 37–44. That is because Kant does say that

    [t]he complete determination is . . . a concept which in its totality can never be exhibited *in concreto*. It is based upon an idea, which has its seat solely in the faculty of reason – the faculty which prescribes to the understanding the rule of its complete determination. (A573/B601)

    In fact, however, the context of this passage shows that Kant is here concerned not with the principle I have transcribed as (4) but rather with a stronger (second-order) principle of the form $(\forall P)(\forall x)[Px \vee \sim Px]$. He is saying that this principle is only regulatively valid, because we do not have a grasp of the totality of all possible concepts.
35. "The Reality of Reference: Comments on Carl Posy's 'Where Have All The Objects Gone.'" Other authors who have emphasized the referential component of philosophy include Robert Howell, Ralf Merebote, and Paul Guyer.
36. Indeed, in the "Amphiboly" chapter (at A272/B328), when Kant takes up the question of the identity of indiscernables (a principle derived as we have seen from Leibniz's conception of "haecceity"), he quite explicitly replaces the notion of a complete concept with that of a human perception.
37. This is how I interpret the passage at A191/B236, which says

    Since truth consists in the agreement of knowledge with the object, it will at once be seen that we can here enquire only regarding the formal conditions of empirical truth, and that appearance, in contradistinction to the representations of apprehension, can be represented as an object distinct from them only if it stands under a rule which distinguishes it from every other apprehension and necessitates some one particular mode of connection of the manifold.

The rule that Kant speaks of here is the rulelike regularity that is projected in an empirical synthesis. This is what turns the "nominal" definition of truth into a "real" definition.

38. See in particular, S. Kripke, *Naming and Necessity* (Cambridge, MA: Harvard University Press, 1980) and H. Putnam, "Meaning and Reference," *Journal of Philosophy*, 70 (1973): 699–711. Putnam, in his more recent writing, has come to advocate a semantic theory in which reference and assertabilism do cooperate. He has taken pains to acknowledge its Kantian origins and to emphasize those places where he differs from Kant. I shall discuss Putnam's Kantianism on another occasion.
39. In the *New Essays* (Book IV, Ch. ii), Leibniz speaks of our awareness of our own mental states as immediate in this sense. He introduces a distinction between *immediacy of ideas* (an epistemic notion) and *immediacy of feeling* (the direct perception we have of our own existence and of our own inner experiences), but then goes on to suggest that the latter is just a special case of the former.
40. We have already seen Kant state that, when we stay within the limits of what we ourselves have defined, then we can assume a certain omniscience, and there is no room for epistemic opacity.
41. It is tempting to look to Kant's notions of space and time to provide this needed immediacy. Tempting, because space and time are ineliminable from empirical intuitions and because – if we are to believe the "Aesthetic" – their original presentations are not concepts. There are good textual grounds in the "Amphiboly" chapter of the first *Critique* to do so, but this tactic is wrong, because fixing the spatiotemporal position of an object requires objective judgment, and thus the categories. This is one of the central arguments of the "Second Analogy." See my "Transcendental Idealism and Causality" in W. Harper and R. Meerbote (eds.), *Kant on Causality, Freedom and Objectivity* (Minneapolis: University of Minnesota Press, 1981) for a discussion of this argument.

    A more promising approach is to take Kant's arguments from "incongruent counterparts" as evidence for the conceptual opacity of spacial presentation. Kant himself suggests this in the *Prolegomena*. However, this approach is still unsatisfactory for our present purposes. That is because the brute fact that we have experiences of incongruent counterparts does not by itself provide any theoretical explanation within the context of the "Critical Philosophy" for how there could be such experiences. Indeed, it is instructive to note that, although Kant does refer to incongruent counterparts in the *Prolegomena*, whose "analytic" approach is not designed to provide such a theoretical grounding, he makes no mention of them in the more "synthetic" argumentation of the *Critique of Pure Reason*.
42. I should add that this abstraction is not a special rarified mental state in which Caesar contemplates an uncolored cylinder or a shapeless floating object. It is not, as sometimes suggested, a "pure image" in some psychological sense. [See, for instance, Bernard Freydberg's *Imagination and Depth in Kant's Critique of Pure Reason*. Peter Lang, New York: 1994] It is, rather, a decision on Caesar's part to ignore the aspects of color, or shape or motion, etc., in his judgments based upon this mental state and in the general conclusions that he allows himself to draw. See §6 of Kant's Inaugural Dissertation, *On the Form and Principles of the Sensible and Intelligible World* (1770) (AK, II, 385–419). Kant may well be building on a similar point in Berkeley's attack on Locke's notion of abstraction. See, for instance, Berkeley's "Introduction," in *The Principles of Human Knowledge* (London: T. Nelson, 1942) §§13–16.

43. See my "Imagination and Judgment in Kant's Critical Philosophy" in R. Meerbote (ed.), *Kant's Aesthetics* (North American Kant Society Studies in Philosophy, Vol. I, 1991) for an extended discussion of what I called "anchored" and "unanchored" imagination.
44. Kant would say that this characterization amounts simply to Euclidean geometry.
45. This point is made by Philip Kitcher in "Kant and the Foundations of Mathematics" (*Philosophical Review*, 84 (1975): 23–50; reprinted in C. Posy (ed.), *Kant's Philosophy of Mathematics: Modern Essays*.)
46. Indeed, at A792/B820, Kant tells us that only mathematics can support *apagogical* (i.e., indirect) proofs. Today we recognize that these proofs are valid by the classical, but not by the intuitionistic, principles of logic.
47. This is the point of §3 in the "Metaphysical Exposition of the Concept of Space" (A24-25/B39-40).
48. See my "Where Have All the Objects Gone?" Part V.B. I should note one important change from that paper. Here I am emphasizing that Kant does have referential semantics. This is a stronger statement than what I had suggested there.
49. See also A156/B195.

Geometry, Construction, and Intuition in Kant and His Successors

MICHAEL FRIEDMAN

I

I begin with an issue concerning the interpretation of the role of intuition in Kant's theory of geometry that has recently seen myself on one side and Charles Parsons on the other.[1] The interpretation I have defended is a version of what one might call the logical approach to Kantian intuition – an approach first articulated by Evert Beth and Jaakko Hintikka.[2] On this approach the primary role of Kantian intuition is formal or inferential: it serves to generate singular terms in the context of mathematical reasoning in inferences such as we would represent today by existential instantiation. Accordingly, the primary feature that distinguishes Kantian intuitions from purely conceptual representations, on this view, is their singularity – as opposed, that is, to the generality of concepts. Parsons has objected, however, that this formal–logical approach downplays a second feature that Kant also uses to distinguish intuitions from concepts: namely, their immediacy. For Kant, conceptual representation is both general and mediate, whereas intuitive representation is both singular and immediate – that is, it is immediately related to an object. And here, Kant certainly seems to think that the idea of immediacy adds something important – something of an epistemological and/or perceptual character – to the bare logical idea of singularity. Parsons himself suggests that the immediacy in question is to be understood as "direct, phenomenological presence to the mind, as in perception,"[3] and so, this second approach can be characterized as phenomenological. More specifically, the primary role of Kantian geometrical intuition, in this approach, is to acquaint us, as it were, with certain phenomenological or perceptual spatial facts, which can then be taken to provide us with evidence for or to verify the axioms of geometry. In this view, therefore, the question of the origin and justification of the axioms of geometry is prior to that concerning the nature and character of geometrical reasoning from these axioms, whereas on the logical approach, the priority is precisely the reverse.

The particular version of the logical approach to Kantian geometrical intuition that I have defended focuses on the role of Euclidean constructions in the proof procedure actually employed in the *Elements*, the procedure of construction with straight edge and compass articulated in Euclid's first three postulates. The idea is that all the objects introduced in Euclid's reasoning – points, lines, and so on – are iteratively or successively generated by straightedge and compass construction from a given line segment or pair of points. The existence of such objects – and, in particular, of an infinity of such objects – is not simply postulated, as in modern treatments, in existential axioms; rather, it is iteratively or successively generated from an initial object by given initial operations. In this sense, Euclid – again as opposed to modern treatments invoking Dedekind continuity, for example – is concerned only with constructive existence claims and with the potential infinite. Indeed, from this point of view, the existence of an infinity of geometrical objects appears precisely analogous to that of the natural numbers. Moreover, from this point of view, as I have further argued, we obtain a plausible explanation for why Kant thinks that geometrical representation is not purely conceptual. Conceptual representation, as Kant understands it, involves only the logical resources of traditional syllogistic logic. But, with these resources alone, we cannot represent an infinity of objects, not even the potential infinity of the natural numbers. Kant's recognition of this fact, together with his appreciation of the way in which Euclid himself represents the (potential) infinity of geometrical objects by a definite procedure of construction, can thus be taken to explain the Kantian doctrine that construction in pure intuition, and therefore geometrical space, is a nonconceptual species of representation.

On this interpretation, then, the infinity of space is a purely formal–logical feature of mathematical geometry (we would express it today by saying that formal systems of geometry have only infinite models), and the intuitive, nonconceptual character of the representation of space is a consequence of this same formal–logical feature (we would express it today by saying that *monadic* formal systems always have finite models if they have models at all). So, phenomenological or perceptual features of our representation of space play no role whatever here. By contrast, as has been brought out especially clearly in a recent paper by Emily Carson,[4] on the phenomenological approach favored by Parsons the order of explanation is precisely the reverse. The infinity of space is a directly given perceptual fact – it consists of the circumstance that any perceived spatial region belongs within a larger "horizon" as part of a single, uniquely given perceptual space[5] – and it is this perceptual fact that then justifies or explains the use of infinity in mathematical geometry. Perceptual space supplies the framework, we might say, within which geometrical construction takes place and which, accordingly, guarantees that the constructions postulated by Euclid can indeed by carried out.[6] Even if Kant were acquainted with modern, purely logical formulations of mathematical geometry, therefore, he would still need to appeal to spatial intuition – that is, to phenomenological features of our spatial perception – to justify or to verify the relevant axioms.

Now there is an important Kantian text that bears decisively on this issue. It belongs to the dispute with Eberhard in 1790 and occurs in Kant's handwritten notes that were used almost verbatim by his disciple Johann Schulze in the latter's review of Eberhard's *Philosophisches Magazin*. In particular, the passage in question was used by Schulze in his review of essays by the mathematician Abraham Kästner, and it shows, I believe, that the logical approach to Kantian geometrical intuition must, at the very least, be supplemented by considerations congenial to the phenomenological approach. Kant begins by distinguishing space as described by geometry from space as described by metaphysics. The former is generated [*gemacht*] or derivative, and, in this sense, there are many *spaces*. The latter is given or original, and, in this sense, there is only one single *space*. Kant leaves no doubt, moreover, that the infinity of geometrical spaces is grounded in the single, uniquely given metaphysical space:

> [T]he representation of space (together with that of time) has a peculiarity found in no other concept, viz., that all spaces are only *possible* and thinkable as parts of one single space, so that the representation of the parts already presupposes the representation of the whole. Now, when the geometer says that a line, no matter how far it has been extended, can still always be extended further, this does not mean the same as what is said in arithmetic concerning numbers, viz., that they can be always and endlessly increased through the addition of other units or numbers (for the added numbers and magnitudes that are expressed thereby are possible in themselves, without needing to belong together with the previous ones as parts of a whole). Rather, to say that a line can be continued to infinity means that the space in which I describe the line is greater than any line that I may describe in it. Thus, the geometer grounds the possibility of his problem – to increase a space (of which there are many) to infinity – on the original representation of a single, infinite, *subjectively given* space. This agrees very well with the fact that geometrical and objectively given space is always *finite*. For the latter is only given in so far as it is generated [*gemacht*]. To say, however, that the metaphysical, i.e., original but merely subjectively given space – which (because there are not many) cannot be brought under any concept capable of construction but which still contains the ground of all possible constructions – is *infinite* means only that it consists of the pure form of the mode of sensible representation of the subject as a priori intuition. Therefore, the possibility of all spaces, which proceeds to infinity, is *given* in this space as a singular representation. (Ak. 20, pp. 419–21)[7]

The explicit contrast that Kant draws here between the infinity of space and that of the natural numbers is made even sharper a few lines later when Kant endorses the idea that mathematics considers only the potential infinite [*infinito potentiali*] and that "an infinity in act [*actu infinitum*] (the metaphysically-given) is not given on the side of the object, but on the side of the thinker" – where the latter "infinity in act" nevertheless "lies at the basis of the progression to infinity of geometrical concepts."

It is clear, therefore, that the analogy between the infinity of space and that of the natural numbers can only apply to space as described by the geometer.

Geometry deals with a successively generated sequence of spaces (spatial objects) that is potentially infinite as a whole and thus necessarily finite at every stage. By contrast, space as described by the metaphysician – as described in the Metaphysical Exposition of the Concept of Space in the Transcendental Aesthetic of the first *Critique* – has quite a different character. And it is precisely this special character of metaphysical space, moreover, which, for Kant, grounds or explains the possibility of the infinity of space as described by the geometer. So far, then, the basic ideas of the phenomenological approach appear to be vindicated. The crucial question, however, concerns exactly how metaphysical space – "the pure form of the mode of sensible representation of the subject as a priori intuition" – is supposed to accomplish this grounding. Is the given infinity of space as a pure form of sensible intuition supposed to be directly seen, as it were, in a simple act of perceptual or quasi-perceptual acquaintance? Are we supposed to have direct perceptual or quasi-perceptual access to such infinity entirely independently of geometry – which access we can then use to justify or to verify the possibility of Euclidean constructions? Both of these ideas appear to be very doubtful. For we are certainly not perceptually presented with an infinite space as a single given whole; and, since the visual field is itself always finite, it does not even appear to be true that any perceived spatial region is directly given or perceived as part of a larger such region. The idea of independently given phenomenological facts capable of somehow grounding or justifying the possibility of geometrical construction can quickly appear to be absurd.

Several pages earlier in the same notes for Schulze's reply to Kästner, Kant himself discusses the question of explaining or justifying the possibility of geometrical construction as follows:

> [I]t is very correctly said [by Kästner] that "Euclid assumes the possibility of drawing a straight line and describing a circle without proving it" – which means without proving this possibility *through inferences*. For *description*, which takes place a priori through the imagination in accordance with a rule and is called construction, is itself the proof of the possibility of the object. Mechanical *delineation* [*Zeichnung*], which presupposes description as its model, does not come under consideration here at all. However, that the possibility of a straight line and a circle can be proved, not *mediately* through inferences, but only immediately through the construction of these concepts (which is in no way empirical), is due to the circumstance that among all constructions (presentations determined in accordance with a rule in a priori intuition) some must still be *the first* – namely, the *drawing* [*Ziehen*] or describing (in thought) of a straight line and the *rotating* of such a line around a fixed point – where the latter cannot be derived from the former, nor can it be derived from any other construction of the concept of a magnitude. (Ak. 20, pp. 410–11).

What grounds or explains the possibility of geometrical construction, then, is simply the immediate activity of our a priori imagination by which we draw or describe a straight line in thought and then rotate such a line around a fixed point. Indeed, Kant had already made it clear in the first *Critique* that it is precisely such imaginative activity that grounds the axioms of geometry:

I can represent no line to myself, no matter how small, without drawing it in thought, that is gradually generating all its parts from a point, and thereby first registering this intuition. . . . On this successive synthesis of the productive imagination in the generation of figures is based the mathematics of extension (geometry), together with its axioms, which express the conditions of a priori sensible intuition under which alone the schema of a pure concept of outer appearance can arise. (A162–3/B203–4)

And, as this passage intimates, the axioms of geometry are capable of no further proof, because it is only via the imaginative activity in question that the relevant geometrical concepts can be thought (compare A234–5/B287).

This last idea is given special emphasis in §24 of the second edition Transcendental Deduction, which also further articulates the imaginative activity in question. After characterizing the activity of the productive imagination as "figurative synthesis" or "transcendental synthesis of the imagination," and explaining that figurative synthesis is a "transcendental action of the imagination" expressing the "synthetic influence of the understanding on inner sense," Kant illustrates his meaning as follows:

We always observe this in ourselves. We can think no line without *drawing* it in thought, no circle without *describing* it. We can in no way represent the three dimensions of space without *setting* three lines at right angles to one another from the same point. And we cannot represent time itself without attending, in the *drawing* of a straight line (which is to be the outer figurative representation of time), merely to the action of synthesis of the manifold, through which we successively determine inner sense, and thereby attend to the succession of this determination in it. Motion, as action of the subject (not as determination of an object*), and thus the synthesis of the manifold in space – when we abstract from the latter and attend merely to the action by which we determine *inner* sense in accordance with its form – [such motion] even first produces the concept of succession. (B154–5)

And in the footnote, Kant explicitly links motion in the relevant sense with the imaginative description of space underlying the axioms of geometry:

*Motion of an *object* in space does not belong in a pure science and thus not in geometry. For, that something is movable cannot be cognized a priori but only through experience. But motion, as the *describing* of a space, is a pure act of successive synthesis of the manifold in outer intuition in general through the productive imagination, and it belongs not only to geometry, but even to transcendental philosophy.

Thus motion in the relevant sense – the pure act of successive synthesis in space as transcendental activity of the subject – grounds or underlies geometry by also belonging to the metaphysical consideration of space characteristic of transcendental philosophy. As Kant puts it in §26 in the footnote at B161, it is "through it [i.e., the transcendental synthesis of the imagination] (in that the understanding determines sensibility) that space or time are first *given*" as intuitions.

In what sense is the motion in question an "action of the subject"? Kant states in §24 that the understanding, as active subject, exerts the transcendental synthesis of the imagination on the "passive subject [i.e., inner sense] whose faculty it is" (B153). But it is also the case, as Kant explains at the very beginning of the Metaphysical Exposition of the Concept of Space, that the subject of outer sense is itself in space. Space as the form of outer sense enables us to represent objects as outer precisely by representing them as spatially external to the perceiving subject:

> Space is no empirical concept that has been derived from outer experiences. For, in order that certain sensations are related to something outside me (that is, to something in another place in space than the one in which I find myself), and, similarly, in order that I be able to represent them as outside of and next to one another – and thus not merely as different but as in different places – the representation of space must already lie at the basis. Therefore, the representation of space cannot be obtained from the relations of outer appearance through experience; rather, this outer experience is itself only possible in the first place by means of the representation in question. (A23/B38)

Space as the form of outer sense contains the point of view of the subject, from which the objects of outer sense are perceived and around which, as it were, the objects of outer sense are arranged.[8] And it follows that by changing this point of view – by moving in and through space – the subject can change its perspective on the objects of outer sense.

Let us suppose, then, that by the fundamental transcendental action of the imagination that Kant calls "figurative synthesis," the subject imaginatively locates itself in space at a definite position (a particular point of view) and with a definite orientation (a particular perspective on the spatial world as perceived from this point of view). Such an orientation is established, for example, by choosing three particular line segments set at right angles from a common point.[9] The objects of outer sense then appear as arranged around this subjective point of view and thus capable of being seen from it in accordance with the chosen orientation or perspective. And this much, moreover, belongs to the a priori structure of *pure* spatial intuition. In particular, *empirical* spatial intuition or perception is necessarily conceived as taking place within this already-established formal structure. Empirical spatial intuition occurs when an object spatially external to the point of view of the subject affects this subject – along a spatial line of sight, as it were – so as to produce a corresponding sensation in it; and it is in this sense, therefore, that the pure form of (spatial) sensible intuition expresses the manner in which we are affected by (outer) objects. It is in this sense, too, I believe, that the immediacy of pure spatial intuition, in contrast to the mediate character of merely conceptual representation, is to be understood. For, by virtue of the formal or a priori structure of spatial perception, pure spatial intuition contains causal, indexical, and demonstrative elements not present in merely conceptual representation. Pure spatial intuition thereby expresses the a priori form, we might say, by which the perceiving subject can be immediately related to (outer) objects.[10]

By the same fundamental action of the transcendental imagination through which the subject imaginatively locates itself at a given point of view with a given orientation, this subject can also, as noted above, imaginatively change the given point of view and orientation by imaginatively moving in and through space. In particular, the subject can imaginatively change the given point of location by a translation through space and imaginatively change its given orientation by a rotation around this point. Moreover, by an appropriate combination of such translations and rotations the subject can thereby imaginatively put itself in position to perceive *any* outer object located *anywhere* in perceptual space. It is in this sense, I believe, that perceptual space is necessarily both singular or unitary and infinite or unbounded. Perceptual space is singular or unitary because any outer object must be perceivable by the same perceiving subject, and thus all outer objects must be located within the same formal structure of possible perceptual relations: all outer objects must be reachable via translation and rotation, as it were, from a single initial given point of view. By the same token, perceptual space is infinite or unbounded because, although any particular momentary visual field is indeed bounded or finite, by moving in and through space and thereby changing its perspective, the subject changes its visual field so as to embrace successively more and more regions of the single, unitary perceptual space. It is in this sense – that is, kinematically – that any given spatial region is perceived within an "horizon" eventually comprising all possible perceptual spatial objects. And this also clarifies the sense, it seems to me, of the otherwise puzzling idea that metaphysical space – that is, the formal structure of perceptual space described in the Metaphysical Exposition – involves an "infinity in act [*actu infinitum*] (the metaphysically-given) [that] is not given on the side of the object, but on the side of the thinker."

Finally, this same formal structure of perceptual space can be seen as grounding or explaining the constructive procedure expressed in the axioms of Euclidean geometry. As we have seen, Kant takes the possibility of straight-edge and compass construction to be grounded in the imaginative activity of drawing a straight line via the rectilinear motion of a given point (here, in connection with B154–5, see especially B292) and describing a circle via the rotation of a given line. The construction of a straight line, in other words, is executed by a translation, and the construction of a circle is executed by a rotation. (In modern terminology, straight lines and circles appear as orbits of the Euclidean group of motions.) The possibility of translational and rotational motion is primary, therefore, because it is given in the pure formal structure of perceptual space. Geometrical space is then iteratively or constructively generated within the formal structure of perceptual space by successively applying the fundamental operations of drawing a straight line and describing a circle, and this is the precise sense, I believe, in which the possibility of mathematical geometry is grounded in or explained by the formal structure of perceptual space.[11] It does not necessarily follow, however, that the structure of perceptual space can be taken to provide an independent epistemological justification for the axioms of geometry – still less that there is an independently accessible realm of phenomenological facts capable of providing such a justification through some kind of quasi-perceptual direct acquaintance. The spatial intuition grounding the axioms of geometry is fundamentally kinematical, in my view, and it is expressed in the formal structure of translational and rotational motion (in modern terminology, the structure of the group of Euclidean motions). That perceptual space in fact has or embodies this formal structure can in no way be simply read off of our perceptual experience, as it were, independently of our knowledge of geometry. On the contrary, the only way in which we know that perceptual space in fact embodies this structure is precisely through our knowledge that geometry is applicable to it.[12] Kant's theory of pure spatial intuition provides an explanation of the possibility of geometry – an explanation, in particular, of the nonconceptual and intuitive or perceptual character of geometry. But it does not provide, and does not attempt to provide, an independent epistemological foundation.[13]

II

Giving motion – understood in terms of the possible changes in position and orientation of the subject's point of view – a central role in Kant's doctrine of pure spatial intuition raises a variety of interpretative questions concerning, on the one hand, the scope of geometrical construction and, on the other, the involvement of the understanding and the transcendental unity of apperception in the characteristic features of sensibility.

With respect to the first set of questions, it is noteworthy that Kant himself is forced explicitly to reconsider the nature and scope of geometrical construction in the course of the dispute with Eberhard – during the very same period, therefore, when our initial texts on the infinity of space were written. Eberhard had appealed to Apollonius's treatise on conic sections in order to argue against the Kantian doctrine that geometrical concepts require the construction of a corresponding intuition. For Apollonius uses the defining planar characterization of a parabola, $y^2 = ax$, without showing how to draw or delineate such a curve in the plane. Indeed, although every individual point on the curve can be constructed by straight edge and compass, the curve itself cannot be continuously traced thereby; the latter task requires more complicated means of construction, which are often, in contradistinction to "geometrical" constructions with straight edge and compass, termed "mechanical" constructions. Yet, since Apollonius is able to develop the entire theory of conic sections without considering such mechanical constructions at all, it cannot be true that the successful use of geometrical concepts always requires the construction of a corresponding intuition.[14]

Kant's reply is twofold. He objects, in the first place, that Apollonius does indeed provide a construction in intuition – not, to be sure, in the plane but rather in space:

Apollonius first constructs the concept of a cone, that is, he presents it a priori in intuition (this is now the first action whereby the geometer verifies beforehand the objective reality of his concept). He cuts it in accordance with a determinate rule, e.g., parallel to a side of the triangle that cuts the base of the cone (*conus rectus*) through its apex at right angles, and proves a priori in intuition the properties of the curved line that is generated by means of this section on the surface of the cone. He thus brings forth a concept of the ratio in which the ordinates of this curve stand to the diameter [i.e., the relation $y^2 = ax$], which concept, namely (in this case) of the parabola, is thereby given a priori in intuition; and therefore its objective reality – that is, the possibility that a thing with the properties in question can be given – is proven in no other way *except that one supports it with the corresponding intuition*. (Ak. 8, p. 191)

Thus, in his very first definition, Apollonius generates a cone by rotating an infinite straight line in space, fixed at a given point, around the diameter of a given circle (the base of the cone). In Proposition I.11, he generates the parabola from a section whose diameter is parallel to one of the sides of the axial triangle, and proves thereby that the characteristic equation, $y^2 = ax$, where a is the so-called latus rectum or parameter, then holds.[15] The curve is thus derived or constructed in space, and for this reason, the ancients termed problems involving conic sections, in contradistinction to "plane" problems constructible by straight edge and compass, "solid" problems.[16]

Kant further objects, in the second place, that mechanical constructions in the plane are completely irrelevant:

[Apollonius's editor, Borelli[17]] speaks of mechanical *construction* of the concepts of the conic sections (except for the circle) and says that mathematicians teach the properties of the latter without mentioning the former – which is certainly a true observation but a very insignificant one; for instructions on how to *delineate* [*zeichnen*] a parabola in accordance with the prescriptions of the theory are only for the artist, not for the geometer.*

And the footnote then clarifies the distinction that Kant has in mind here:

*The following may serve to secure against the misuse of the expression, the *construction* of concepts, of which the *Critique of Pure Reason* frequently speaks and by which it has first precisely distinguished the procedure of reason in mathematics from its procedure in philosophy. In the more general meaning, all *presentation* of a concept through the (spontaneous) production of an intuition corresponding to it can be called construction.

If this takes place through the mere imagination in accordance with an a priori concept, it is called *pure* construction (which the mathematician must lay at the basis of all his demonstrations. . .). If, however, it is exerted on any kind of matter, it could be called *empirical* construction. The former can also be called schematic, the latter technical. The latter type of construction, which is actually only improperly so-called (because it belongs not to science but to art and is achieved with instruments), is now either *geometrical* construction by means of compass and ruler or *mechanical* construction, for which other instruments are necessary – as, for example, the delineation [*Zeichnung*] of the remaining conic sections besides the circle. (Ak. 8, pp. 191–2)

In particular, then, Kant himself is perfectly aware that the general conic section is not constructible with straight edge and compass.

In this same passage, however, Kant classifies ruler and compass constructions as "empirical" or "technical" as well. Does this mean that even the constructions of elementary Euclidean geometry are also mathematically irrelevant? That this is emphatically not the case is clear from a footnote to §I of the *First Introduction to the Critique of Judgement* (which is thus written shortly before the reply to Eberhard):

*This pure and for precisely this reason noble science [i.e., geometry] seems to compromise its dignity when it admits that, as elementary geometry, it uses *instruments*, although only two, for the construction of its concepts – namely the compass and the ruler, which constructions alone it calls geometrical, while those of higher geometry, by contrast, it calls mechanical since for the construction of the concepts of the latter more composite [*zusammengesetztere*] machines are required. But one also understands by the former, not the actual instruments (*circinus et regula*), which could never give these figures with mathematical precision; rather, they should mean only the simplest modes of presentation of the a priori imagination, which no instrument can imitate. (Ak. 20, p. 198)

And it is clear from this passage, together with the passage on the transcendental synthesis of the imagination at B154–5 and the passage from the reply to Kästner at Ak. 20, pp. 410–11 cited earlier, that the "simplest modes of presentation of the a priori imagination" in question are just the two activities (in thought) of drawing [*ziehen*] a straight line and describing a circle (by rotating a line segment about a fixed point in a plane). Construction of straight lines and circles, when performed via figurative synthesis in the a priori imagination rather than with real draftsman's instruments on real pieces of paper, thus continues to be paradigmatic of "properly so-called" pure or schematic mathematical construction.

There are therefore two different distinctions at play here. The first is a distinction, within mathematics, between those curves or figures constructible via straight lines and circles and more complex figures such as the conic sections. The former are constructible by straight edge and compass in the idealized mathematical sense, and this question is entirely independent of the capabilities of any actual draftsman's instruments. The second distinction, by contrast, is between pure mathematical construction, whether belonging to "elementary" or to "higher" geometry, and actual empirical delineation. In this sense, even constructions with straight edge and compass, when it is a matter of actual draftsman's instruments rather than idealized mathematical operations, lie outside the concerns of the geometer. Kant's dichotomy between schematic and technical construction precisely corresponds to the second distinction, but his use of "mechanical construction" tends to blur the two. In the passage from the reply to Kästner at Ak. 20, pp. 410–11, for example, a priori mathematical "description" is opposed to "mechanical delineation," and a few lines later in the above reply to Eberhard Kant sets up an opposition between "pure, merely schematic construction" and "*mechanical* [construction]" (Ak. 8, 192). This latter usage of "mechanical" thus corresponds to "empirical" or "technical" and does not involve the contrast between ruler and compass and more complicated forms of construction.[18]

Lying behind this ambiguity is an important issue of principle concerning the precise scope of admissible geometrical or mathematical operations – an issue made particularly acute by the investigation of a large variety of new curves in the seventeenth and eighteenth centuries. The *locus classicus* for this issue is Descartes's *Géométrie*, which, as is well known, develops a novel version of the distinction between "geometrical" and "mechanical" curves. The former include all algebraic curves – not only lines, circles, and the conic sections but also curves defined by algebraic equations of third and higher degree – whereas the latter comprise the nonalgebraic or transcendental curves. Moreover, the algebraic curves, according to Descartes, are all constructible by appropriate generalizations of the straight edge and compass, by idealized instruments that arise by iteration, as it were, of the most elementary ones. We can construct lines and circles and then rotate and translate them to produce new curves (like the conic sections); we can then rotate and translate these new curves to produce further curves (like the so-called Cartesian parabola, which is of the third degree); and so on. Transcendental curves, by contrast, do not find a place in this iterative hierarchy of possible constructions. They may exist "mechanically" in actually given empirical nature, but they forever exceed our precise mathematical grasp – our capacity clearly and distinctly to proceed step-by-step via intuitively evident rules.[19]

Unfortunately, there is not enough evidence to determine where Kant himself stands on this issue. He never, to my knowledge, considers curves more complex than the conic sections, and here his viewpoint appears to be entirely traditional. Conic sections are intuitively presentable through the ancient "solid" constructions on a cone, which itself arises through the rotation of a line with a fixed point in space. For Kant, as we have seen, what is primary are the basic operations – the "simplest modes of presentation of the a priori imagination" – by which the subject can execute translations of and rotations around a given point of view in space, and he appears to hold that only constructions that can arise thereby are geometrically and mathematically admissible. Yet some delimitation of what "can arise thereby" actually means is necessary if a notion of admissible construction is to be at all well defined, and Kant unfortunately says nothing to suggest such a delimitation. If arbitrary combinations of translations and rotations are allowed, we can clearly construct any continuous curve, and then there is no reason, in particular, to dismiss "mechanical" constructions of the conic sections as mathematically irrelevant.[20] So what is needed, then, is some iterative extension of a set of basic operations analogous to Descartes's. From a fundamental logical and mathematical point of view, however, the issue proves to be a deep one indeed. For it eventually leads, via the need to assimilate transcendental as well as algebraic curves, to the free use of infinitesimal methods and thus, in the end, to the realization that a radically new type of iteration essentially involving limit operations is required.

In any case, the relationship that Kant does set up between geometrical construction, on the one hand, and motion in space (i.e., translations and rotations), on the other, raises significant questions about his own doctrine of sensibility. In particular, if, as we argued above, the two key features of intuitive space – its unity and infinity – depend on the possible motions of the subject's point of view, then these features appear to depend on the unity and identity of the subject – and thus, in the end, on the transcendental unity of apperception – rather than on independently given features of space as a form of sensibility. Space is unitary because every possible object therein must be reachable from a given initial point of view by an appropriate combination of translations and rotations; and space is infinite or unbounded because any initial perceptible region is thereby extendible without limit to any other perceptible region. These two key features of intuitive space therefore directly depend on the requirement that every spatial region be accessible via continuous motion by a single perceiving subject, and without this requirement there would simply be no guarantee whatever that all possible spatial regions belong to a single, unitary and unbounded, comprehensive system of such regions.[21]

This dependence of key features of sensibility on the transcendental unity of apperception, and thus on the understanding, is closely related, in turn, to the well-known distinction that Kant makes between space as "form of intuition" and as "formal intuition" in §26 of the second edition Transcendental Deduction:

*Space represented as *object* (as is actually required in geometry) contains more than the mere form of intuition – namely, [it contains] *uniting* [*Zusammenfassung*] of the manifold in accordance with the given form of sensibility in an *intuitive* representation, so that the *form of intuition* gives [a] mere manifold but the *formal intuition* gives unity of representation. In the Aesthetic I counted this unity [as belonging] to sensibility, only in order to remark that it precedes all concepts, although it in fact presupposes a synthesis that does not belong to the senses but through which all concepts of space and time first become possible. For, since through it (in that the understanding determines sensibility) space or time are first *given*, the unity of this a priori intuition belongs to space and time, and not to the concept of the understanding (§24). (B160–1)

The reference to geometry and to §24 implies, I believe, the conception of motion in space that was first suggested above by the passage at B154–5. And that geometrical motion in this sense is a direct expression of the transcendental unity of apperception is explicitly stated in §17:

Therefore, the first pure cognition of the understanding, on which its entire remaining use is grounded, and which is also, at the same time, entirely independent of all conditions of sensible intuition, is the principle of the original *synthetic* unity of apperception. Thus, the mere form of outer sensible intuition, space, is not yet any cognition at all; it gives only the manifold of a priori intuition for a possible cognition. But to cognize anything in space, for example, a line, I must *draw* [*ziehen*] it and therefore bring about synthetically a determinate combination of the given manifold – so that the unity of this action is, at the same time, the unity of consciousness (in the concept of a line), and thereby alone is an object (a determinate space) first cognized. (B137–8)

The unity of the intuitive representation in question therefore depends directly on the unity of consciousness and thus, in the end, on a conceptual unity.

Why, then, does Kant also assert, in the last sentence of the footnote at B160–1, that "the unity of this a priori intuition belongs to space and time, and not to the concept of the understanding"? The point, I think, is that the relationship between the understanding and sensibility effected by the transcendental synthesis of the imagination is a reciprocal one. To be sure, space would not be unitary in the relevant sense without the "action of the understanding on sensibility" (B152) manifested in figurative synthesis. Nevertheless, the unity thereby produced is not itself a conceptual unity, whereby a number of representations (subordinate concepts) are contained *under* a given representation (superordinate concept); it is, rather, a distinctly intuitive unity, whereby a number of representations (spatial regions) are contained *in* a given representation (that of a single space) (see B39–40 and compare B133 n). All spatial regions belong to a single space in that they all must be reachable from here, as it were, but reachable-from-here is not a conceptual relation. By the same token, although the transcendental synthesis of the imagination is a realization or embodiment of the "pure intellectual synthesis" contained in the synthetic unity of apperception (B150–2), it must also go beyond pure intellectual synthesis since the latter in fact requires pure sensibility if it is to succeed in unifying a given manifold (B153–4). There would thus be no unity in the relevant sense without the mutual cooperation of understanding and sensibility, without that interaction between the two faculties "through which the categories, as mere forms of thought, then acquire objective reality, that is, application to objects that can be given to us in intuition" (B150–1).

Even in pure mathematical synthesis in the pure imagination, therefore, the categories are necessary to bring unity into the intuitive manifold – a point Kant makes explicitly in §20 of the *Prolegomena*:

> Even the judgements of pure mathematics in its simplest axioms are not exempt from this requirement [i.e., subsumption under a pure concept of the understanding]. The principle that the straight line is the shortest between two points presupposes that the line is subsumed under the concept of magnitude, which is certainly no mere intuition but has its seat solely in the understanding and serves to determine the intuition (the line) with respect to the judgements that may be made of it in relation to their quantity, namely [in relation to] plurality . . . (Ak. 4, pp. 301–2)

And in the Table of Pure Concepts of the Understanding following in §21, Kant lists the categories of quantity in the form: "unity (the measure), plurality (the magnitude), totality (the whole)" (p. 303). So, it is clear, then, that the categories involved in pure mathematical synthesis, and thus in the mathematical unity of space as an object of geometry, are the categories of quantity.[22]

In the terminology of the important footnote added at B201, we are therefore involved with a *mathematical* synthesis of "*composition* (*compositio*)" rather than a *dynamical* synthesis of "*connection* (*nexus*)." We are involved with the mathematical categories (here the categories of quantity) rather than the dynamical categories of relation and modality – where the former ground the possibility of the principles of mathematics, whereas the latter ground the possibility of "general (physical) dynamics" (A162/B202). And this distinction has fundamental implications for the nature and status of the pure geometrical motion (viz., translation and rotation), which, according to our interpretation, first embodies mathematical synthesis. In particular, since the dynamical categories and thus "general (physical) dynamics" are not yet at issue, we are not concerned here with the questions about distinguishing true from apparent motion and establishing a privileged frame of reference arising in the context of Newtonian dynamics.[23] The motion with which we are concerned here is purely relative or, perhaps better, purely mathematical, in that we abstract from all questions of speed, acceleration, duration, and so on, and attend only to its character as a continuous transformation.[24]

III

The above interpretation of Kantian spatial intuition attempts to build a bridge between the phenomenological and logical approaches by viewing the relevant formal structure of intuitive or perceptual space as fundamentally kinematical: it is a matter of the possible translational and rotational motions (in modern terms, the group of rigid motions) by which the perceiving subject can move in and through space so as to put itself in potential perceptual contact with all possible spatial objects. This structure then grounds the formal procedure of geometrical construction underlying pure mathematical geometry by generating the two basic operations of drawing straight lines and describing circles (in modern terms, as orbits of the group of motions). From this point of view, therefore, Kant's own conception of spatial intuition is not so far from that developed in the nineteenth century by Hermann von Helmholtz. Indeed, it is of course Helmholtz who first explicitly articulates a program for founding geometry on the formal structure of perceptual space based, via the condition he calls free mobility, on the group of rigid motions.[25] Nevertheless, it is well known that Helmholtz presents his position as anti-Kantian, and this in two central respects.

The first and most obvious respect in which Helmholtz presents his conception as anti-Kantian is that Helmholtz explicitly attacks the idea that the specific structure of *Euclidean* space is grounded in our spatial intuition or is in any way a priori. From Riemann's work and his own mathematical investigation of the "space-problem"[26] Helmholtz has learned that the relevant formal structure of possible motions in and through perceptual space (the structure characterized by the condition of free mobility) does not yield specifically Euclidean space, but rather the three classical cases of spaces of constant curvature: Euclidean space, spherical or elliptic space, and hyperbolic space. By imagining a mobile perceiver located in one or another of the non-Euclidean spaces of constant curvature, we can then make it perfectly evident that the formal structure of spatial intuition alone (i.e., the possibility of free mobility) does not uniquely single out the Euclidean case:

> This will suffice to show how one can, in the way suggested, derive from the known laws of our sensible perceptions the series of sensible impressions that a spherical or pseudo-spherical world would give us if such a world existed. We thereby never come upon an inconsistency or impossibility, any more than in the calculative treatment of metrical relationships. We can picture to ourselves the appearance of a pseudo-spherical world outwards in all directions, just as well as we can develop the concept of such a world. We therefore cannot grant that the axioms of our geometry [i.e., Euclidean geometry] are grounded in the given form of our faculty of intuition or are somehow implicated in such a form.[27]

And it is in this sense, therefore, that Helmholtz defends an empiricist conception of geometry. The axioms of specifically Euclidean geometry are neither necessities of thought (because we can consistently develop the more general concept of Riemannian metrical manifold) nor necessities of intuition (because the formal structure of spatial perception leaves all three classical cases of constant curvature still open). Specifically Euclidean geometry thus can be obtained only from the observed facts governing the behavior of mobile rigid bodies in the actual world. (Helmholtz himself has no reason to doubt, of course, that the observed facts do indeed support specifically Euclidean geometry.)

This does not mean, however, that the Kantian idea of spatial intuition and its a priori structure are wholly erroneous. On the contrary, Helmholtz's famous assertion that "space can be transcendental without the axioms being so" is intended precisely to underscore the fundamental truth that he finds in the Kantian doctrine:

> Kant's doctrine of the a priori given forms of intuition is a very happy and clear expression of the situation. Yet this form must be contentless and free enough in order to take up every content that can ever enter into the form of perception in question. But the axioms of geometry [i.e., Euclidean geometry] limit the form of intuition of space so that every thinkable content can no longer be taken up therein, if geometry is to be at all applicable to the actual world. However, if we leave these axioms aside, then the doctrine that the form of intuition of space is transcendental is free from any stumbling block. Here Kant was not critical enough in his critique. Certainly, however, we are here concerned with propositions of mathematics, and this piece of the critical work must be taken care of by the mathematician.[28]

Since our spatial intuition in fact has an a priori formal structure expressed mathematically in the condition of free mobility, Kant's doctrine is, so far, unobjectionable. Only the later mathematical discovery of the classical non-Euclidean geometries and the fact that precisely the *three* classical cases are given by the condition of free mobility allow us to correct the one flaw in Kant's original doctrine (which discoveries, we might add, Kant could in no way have been expected to anticipate).[29]

The second, and, in the present context, perhaps even more interesting respect in which Helmholtz presents his conception as anti-Kantian concerns the idea of spatial intuition itself. For Helmholtz presents his kinematical picture of spatial intuition expressed in the condition of free mobility – the picture of spatial intuition as involving the formal structure of the possible changes in point of view and orientation of the perceiving subject – as explicitly opposed to the "popular" or "older" concept of intuition (which he sometimes attributes to Kant himself, but sometimes only to the "Kantians of strict observance") according to which spatial intuition is a simple and unanalyzable momentary psychological act providing us with direct "evidence in a flash [*blitzähnliche Evidenz*]." And it is only by invoking what he takes to be his new, kinematical picture of spatial intuition that Helmholtz is then able to argue against the claims of contemporary Kantians that, although non-Euclidean geometries may be mathematically thinkable, they are not spatially intuitable and therefore are not possible candidates for describing the structure of our spatial intuition.[30] If the above interpretation of Kant's own doctrine of spatial intuition is at all correct, however, it turns out that at least the germ of Helmholtz's kinematical conception is already present in Kant himself. In this sense, Kant's explanation of the a priori status of Euclidean geometry in terms of the necessary structure of our pure form of spatial intuition contains the seeds of its own destruction.

Here, however, it is imperative to note a third respect in which Helmholtz's conception is very definitely anti-Kantian – a central difference between the two conceptions of spatial intuition that Helmholtz, because of his naturalistic transformation of the meanings of "a priori" and "transcendental," does not himself emphasize at all. For what Helmholtz considers as belonging to the a priori or transcendental structure of spatial intuition involves, from a Kantian perspective, *empirical* rather than *pure* intuition. Helmholtz constructs the relevant group of rigid motions expressing the free mobility of the perceiver in and through perceptual space from the muscular and kinesthetic sensations of the subject as it voluntarily initiates such motions, which motions are essentially considered, therefore, as movements of the subject's body. The sense in which the structure of these bodily sensations constitutes an a priori or transcendental form of intuition, then, is simply that this structure belongs on the side of the subject and does not simply picture or mirror an external realm of "things in themselves."[31] For Kant, by contrast, the relevant group-theoretical structure involves only the motions of a disembodied point of view and has nothing to do, therefore, with any bodily sensations. Kant is concerned only with that "action of the understanding on sensibility" (B152) whereby the (transcendental) subject locates itself in space at a definite point of view and with a definite orientation.[32] Indeed, Kantian pure, as opposed to empirical, intuition can, of course, involve no sensations or actual perceptions at all.[33] Kant's doctrine of space as a *pure* form of outer intuition is in this sense entirely unique, and it cannot be satisfactorily understood, I believe, except by invoking the basic ideas of the logical interpretation of this doctrine. It is only because there is no room in Kant's own conception of logical, conceptual, or analytic thought for anything corresponding to pure mathematical geometry that there is a place, accordingly, for a wholly nonconceptual faculty of pure spatial intuition. For Helmholtz, by contrast, there is no difficulty at all in formulating pure mathematical geometry conceptually or analytically with no reference to spatial intuition whatsoever (via the Riemannian conception of metrical manifold), and an appeal to spatial intuition or perception is only then necessary to explain the psychological origin and empirical application of the pure mathematical concept of space.[34]

This necessary separation, in Helmholtz's conception, of the pure mathematical concept of space from perceptual or intuitive space implies, moreover, that there is a fundamental gap between the precision and exactitude of the mathematical concept and the rough and approximate character of the relevant perceptual or intuitive experience. And it is precisely by emphasizing and exploiting this gap that Henri Poincaré develops his contrasting conventionalist interpretation of geometry. For Poincaré entirely agrees with Helmholtz that the psychological origin and empirical application of mathematical geometry is to be explained by the structure of the group of rigid motions of the perceiver in perceptual space expressed in the condition of free mobility. Poincaré also entirely agrees that the structure of this group (i.e., the structure of perceptual space) is based on our motor or kinesthetic bodily sensations as we voluntarily move or displace our body in and through perceptual space. Accordingly, that aspect of perceptual space most relevant to geometry, for Poincaré, is what he calls motor space – the space generated by the group of bodily displacements. Finally, Poincaré also entirely agrees with Helmholtz that, precisely because the condition of free mobility leaves all three classical cases of geometries of constant curvature still open, specifically Euclidean geometry is neither a necessity of thought nor an a priori product of our form of spatial intuition. For Poincaré, however, an empirical explanation of the origin of Euclidean geometry is not the only remaining alternative. On the contrary, precisely because the exact mathematical concept of continuous group can only be an *idealization* of our rough and approximate experience of bodily displacements, the group we end up with must inevitably reflect our own free choice, which choice is guided but not constrained by the rough and approximate experience with which we begin.

For Poincaré, therefore, our choice of specifically Euclidean geometry (which he, like Helmholtz, has no reason to question) is based primarily on its mathematical simplicity: on the circumstance, namely, that only the Euclidean group of motions contains a normal subgroup of translations:

> Geometry is not an experimental science; experience forms merely the occasion for our reflecting upon the geometrical ideas which pre-exist in us. But the occasion is necessary; if it did not exist we should not reflect; and if our experiences were different, doubtless our reflections would be different. Space is not a form of our sensibility; it is an instrument which serves us not to represent things to ourselves, but to reason upon things.
>
> What we call geometry is nothing but the study of formal properties of a certain continuous group; so that we may say, space is a group. The notion of this continuous group exists in our mind prior to all experience; but the assertion is no less true of the notion of many other continuous groups; for example, that which corresponds to the geometry of Lobachevsky. There are, accordingly, several geometries possible, and it remains to be seen how a choice is made between them. Among the continuous mathematical groups which our mind can construct, we choose that which deviates least from that rough group, analogous to the physical continuum, which experience has brought to our knowledge as the group of displacements.
>
> Our choice is therefore not imposed by experience. It is simply guided by experience. But it remains free; we choose this geometry rather than that geometry, not because it is more *true*, but because it is more *convenient*.
>
> ... We choose the geometry of Euclid because it is the simplest.... [I]t is simpler because certain of its displacements are interchangeable with one another, which is not true of the corresponding displacements of the group of Lobachevsky.[35]

Poincaré's conventionalism is thus based, in the end, on the traditional Platonic gap between pure mathematical ideas and the rough perceptual experience from which they arise and to which they are to be applied. More precisely, when one combines this traditional gap with the Helmholtz–Lie solution to the "space-problem," one sees that there are three and only three mathematical counterparts to our intuitive experience of perceptual space – among which, therefore, a conventional choice must be made. By contrast, as we saw earlier, the entire point of Kant's own doctrine of *pure* spatial intuition is precisely to overcome this traditional Platonic gap – the gap, in Kantian terms, between reason and the understanding on one side and sensibility on the other. Indeed, as we argued at the end of §II, Kant's doctrine of the transcendental synthesis of the imagination is intended precisely to unite the understanding and sensibility once and for all, so that, in particular, "pure mathematics, *in its full precision*, [is made] applicable to objects of experience" (A165/B206, my emphasis).

Nevertheless, although Poincaré's conception of the role of spatial intuition in geometry is, in this respect, quite antithetical to the Kantian doctrine of pure spatial intuition, Poincaré's conception of the role of intuition in arithmetic is closely analogous to Kant's. As is well known, Poincaré vehemently opposes the logicist doctrine that arithmetic is a part of logic and hence a product, in Kantian terms, of the pure understanding. He holds instead that arithmetic is based on an irreducible intuition of succession or indefinite iteration, by which the mind is immediately aware of its own capacity indefinitely to repeat any given operation. It is this immediate awareness that grounds both the potential infinity of the number series and the characteristically mathematical form of reasoning expressed in mathematical induction. And this awareness is intuitive rather than conceptual precisely because neither fundamental property of our arithmetical thought is reducible, for Poincaré, to purely logical thinking – even if we widen our conception thereof to include the new mathematical logic. In this sense, Poincaré's defense of the idea that arithmetic is synthetic a priori is indeed genuinely Kantian, and Poincaré's conception of what we might call pure arithmetical intuition is in fact closely analogous to Kantian pure intuition. The one difference is that Poincaré's arithmetical intuition is not so directly and explicitly tied to sensibility, to the idea of time as the form of inner sense.[36]

What is perhaps not so well known is that Poincaré also emphasizes the importance of arithmetical intuition (the intuition of succession or indefinite iteration) in geometrical reasoning. As we have seen, Poincaré holds that the object of geometry is a group of rigid motions and thus, in accordance with the Helmholtz–Lie theorem, that space, although not necessarily Euclidean, must nonetheless have constant curvature. And it is precisely in this context that he appeals to arithmetical intuition:

[*S*]*pace is homogeneous and isotropic*. It may also be said that a movement which has once been produced may be repeated a second and a third time, and so on, without its properties varying. In the first chapter, where we discussed the nature of mathematical reasoning, we saw the importance which must be attributed to the possibility of repeating indefinitely the same operation. It is from this repetition that mathematical reasoning gets its power; it is, therefore, thanks to the law of homogeneity, that it has a hold on the geometric facts.[37]

This passage seems to be closely connected, in turn, with the circumstance that Poincaré, again in accordance with the Helmholtz–Lie theorem, explicitly excludes from consideration the more general theory of Riemannian manifolds including spaces of variable curvature:

If therefore the possibility of motion is admitted, there can be invented only a finite (and even a rather small) number of three-dimensional geometries. Yet this result seems contradicted by Riemann, for this savant constructs an infinity of different geometries, and that to which his name is ordinarily given is only a particular case. All depends, he says, on how the length of a curve is defined. Now, there is an infinity of ways of defining this length, and each of them may be the starting point of a new geometry.

That is perfectly true, but most of these definitions are incompatible with the motion of a rigid figure, which in the theorem of Lie is supposed possible. These geometries of Riemann, in many ways so interesting, could never therefore be other than purely analytic and would not lend themselves to demonstrations analogous to those of Euclid.[38]

Thus Poincaré appears to be perfectly clear (unlike Helmholtz, for example) that Riemann has indeed shown how to introduce the notion of distance or measurability into a manifold without relying on the motion of rigid bodies and hence on free mobility. Poincaré's claim is rather that nonhomogenous manifolds of variable curvature are not susceptible to Euclidean-style systems of demonstration, so that, therefore, they are not in the same sense synthetic.[39]

Now it is by no means clear precisely what Poincaré means by this assertion. Nevertheless, the way in which he juxtaposes geometrical and arithmetical intuition here may suggest a connection between the group of motions in a space of constant curvature and an iterative procedure of geometrical construction, a connection that would generalize what we argued earlier in the case of Kant's theory of geometrical construction and, specifically, Euclidean geometry.[40] In that case we argued that the Euclidean group of motions (the group of Euclidean translations and rotations) constitutes the basis, for Kant, of the procedure of construction with straight edge and compass underlying the proof structure of the *Elements*. This procedure iteratively generates the domain of Euclidean geometry so that, in particular, we are concerned here only with constructive existence claims and the potential infinite. We are concerned, that is, with a domain precisely analogous, in this respect, to the natural numbers. Moreover, as is well known, this feature of the domain of elementary Euclidean geometry can be expressed analytically by the circumstance that a Cartesian space over the entire field of real numbers is by no means necessary for representing the existence claims implicit in Euclid's postulates. On the contrary, the domain of elementary Euclidean geometry is represented precisely by a Cartesian space over the much smaller Euclidean subfield of the reals, which results by closing the rationals under the operation of extracting real square roots. (And it is this representation, of course, that we use to prove the impossibility within elementary Euclidean geometry of various "higher" constructions such as the trisection of an angle.)

It is interesting, then, that something closely analogous is true in all three classical cases of spaces of constant curvature. In all three cases, one can formulate an elementary geometry where, in place of the Dedekind continuity axiom, one simply has an axiom of intersection for straight lines and circles. The domains of these elementary geometries therefore consist of precisely those points generated by straight-edge and compass constructions in the sense of each of the geometries in question, and each of these domains is analytically representable by an appropriate space over a Euclidean subfield of the reals – the familiar Cartesian space in the case of Euclidean geometry, certain Klein spaces in the cases of hyperbolic and elliptic geometry. Moreover, in this more general analytic treatment, all three cases are viewed in accordance with the Cayley–Klein program as embedded within projective geometry (in the elliptic case, the embedding is trivial), so that, in particular, the three different groups of motions appear as subgroups of the projective group. The analytic representations in question are then induced by corresponding analytic representations of projective geometry. In this sense, there does indeed seem to be a general connection between groups of rigid motions and geometrical construction subsisting in all three classical cases of constant curvature.

The most interesting case of this situation occurs in hyperbolic or Bolyai–Lobachevsky geometry, a central feature of which is the existence of limiting or asymptotic parallel lines.[41] It is not only the case that, given a line l and a point P not on l, there are an infinity of lines through P that do not intersect l (and are in this sense parallel to l), but among all such nonintersecting lines, there are exactly two distinguished ones, the limiting or asymptotic parallels to l through P, that precisely divide the set of all lines through P into two classes – those that lie within the angle determined by the two asymptotic parallels on the same side as l (i.e., within the region $\mathcal{R}$ in Figure 1) and intersect l, and those that do not lie within this angle and do not intersect l.

The pair of asymptotic parallel lines through P therefore determines a Dedekind cut in the set of all lines through P (more precisely, in the set of all rays or half-lines originating at P) with respect to the property of intersecting line l, and, accordingly, the existence of such lines is traditionally justified by a Dedekind continuity axiom. It was Hilbert in 1903 who showed that one could,

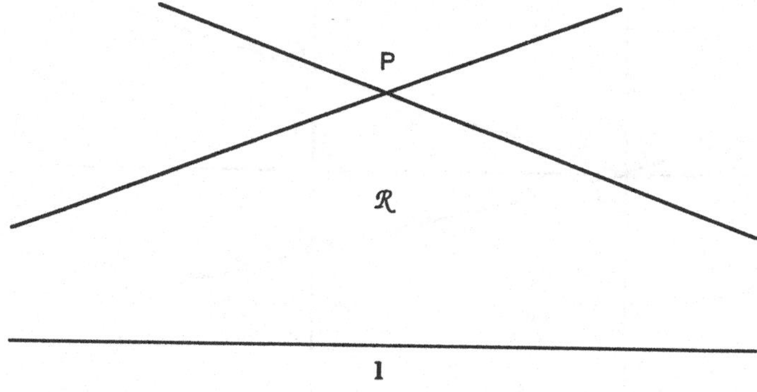

Figure 1

instead, simply add an axiom asserting the existence of asymptotic parallels to his axioms of incidence, order, and congruence characterizing absolute geometry, and one thereby obtains all the usual theorems of hyperbolic geometry without needing to invoke full Dedekind continuity.[42] The key part of Hilbert's approach is the construction of a field based on equivalence classes of asymptotic parallels (the so-called end calculus), which then can be used to coordinatize the set of points so as, in effect, to embed the geometry in question within projective geometry. From this embedding within projective geometry, the usual formulas of hyperbolic geometry then follow. It turns out that the field thus constructed by Hilbert is precisely a Euclidean field. In particular, the axiom of intersection for lines and circles is a consequence of Hilbert's axiom of asymptotic parallels.

It is natural to ask whether the converse also holds. That is, given Hilbert's axioms of incidence, order, and congruence characterizing absolute geometry, the negation of the Euclidean parallel postulate, and the axiom of intersection for lines and circles, can we then derive the existence of asymptotic parallels? If so, we would, in effect, have constructed the asymptotic parallels with straight edge and compass within hyperbolic geometry. And it is in fact the case that Bolyai himself gave a construction with straight edge and compass of the asymptotic parallels already in 1832[43] (where, in Figure 2, PX and PY are congruent to QR).

It turns out, however, that to complete the derivation in question, one also needs to invoke the Archimedean axiom. This follows from work of Hessenberg, Hjelmslev, and Bachmann, which yields an embedding of any geometry satisfying the axioms of absolute geometry into projective geometry so as thereby to construct a canonical coordinatization by a field K. Moreover, the axiom of intersection for lines and circles holds if and only if K is Euclidean. But, if the

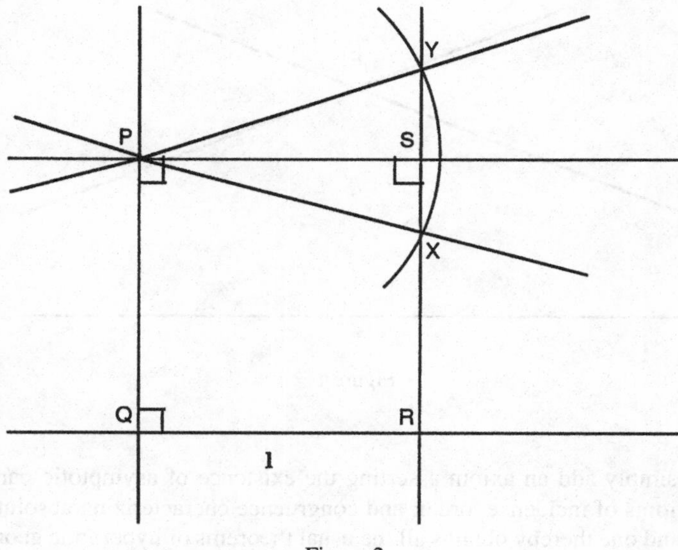

Figure 2

negation of the Euclidean parallel postulate holds and K is non-Archimedean, then Bolyai's construction yields lines through a point P having a "common perpendicular at infinity" with the given line l, which lines, however, are not necessarily asymptotically parallel to l. Each such line through P meets l in an ideal point at infinity, but the ideal points at infinity do not necessarily have a minimum or limiting value. If K is Archimedean, by contrast, then Bolyai's construction does yield asymptotic parallels, and, in fact, the resulting geometry must be isomorphic to the usual Klein model for hyperbolic geometry over an Archimedean Euclidean field (where, in Figure 3, the points in our model are all interior to the bounding circle of ideal points, the polar constructions outside the bounding circle depict the perpendicularity relations of Figure 2, and the interior circle with center P and radius equal to QR appears as a conic). (Without the Archimedean axiom, by contrast, we can construct models in which the interior points are all infinitesimal, the bounding "circle" of ideal points consists of all finite points, and the ultra-ideal points outside the bounding circle are infinite; Bolyai's construction then yields ideal points on the bounding circle, but there is no limiting or minimum value.) In this sense, by taking the Archimedean axiom as an additional constructive constraint, we can give a constructive treatment of Bolyai–Lobachevsky geometry analogous to Euclidean geometry.

As noted earlier, an analogous treatment can be given of elliptic geometry, although here, we of course need to generalize the axioms of absolute geometry as well. In all three classical cases of constant curvature, the connection we

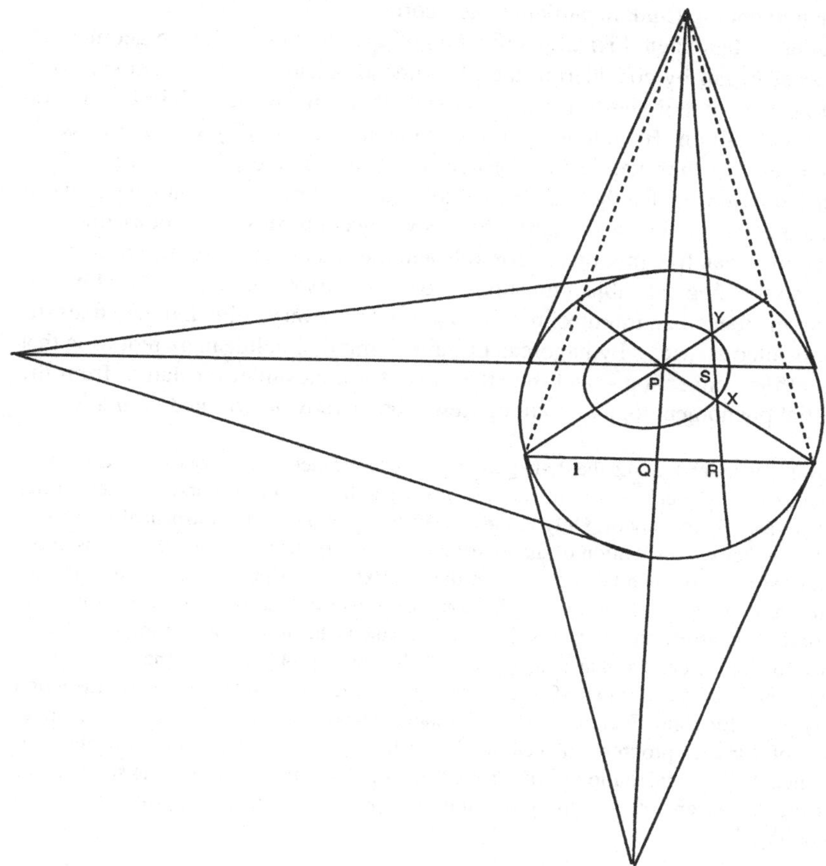

Figure 3

have attributed to Kant between space as a form of intuition or outer perception given by a group of motions and space as an object of geometrical construction therefore appears to hold.[44] Yet it is also interesting to note, finally, that if we do not insist on such a connection with geometrical construction, it is still possible to envision a "phenomenological" foundation, based on a consideration of the possible motions of the perceiving subject in and through perceptual space, of the more general class of Riemannian manifolds, including spaces of variable curvature. This, in fact, was precisely the philosophical motivation behind Hermann Weyl's generalization of the Helmholtz–Lie solution to the "space-problem." Weyl developed his analysis in the context of Husserlian phenomenology and, in particular, as a correction to Oskar Becker's (Husserl's student) phenomenological justification of specifically Euclidean geometry.[45] Becker began by considering the phenomenological subject as embedded in space at a point of view with respect to which it can change both its orientation and its position. By imposing a condition of free mobility on such possible changes, we then derive the constant curvature of the given space, and by further requiring that the translations constitute a normal subgroup, we arrive at specifically Euclidean geometry. For Weyl, by contrast, we do not assume that Helmholtzian free mobility is possible and thus that the perceptual space of the phenomenological subject must have constant curvature. Instead, Weyl begins with the idea of an infinitesimal rotation group at every point and then fixes the associated metric as Pythagorean or infinitesimally Euclidean by requiring that an affine connection – and thus the idea of infinitesimal translation from the initial phenomenological point of view – be thereby determined *uniquely*:

> A way for understanding the Pythagorean nature of the metric expressed in the Euclidean rotation group precisely on the basis of the separation of a priori and a posteriori has been given by the author: Only in the case of this group does the intrinsically accidental quantitative distribution of the metric field uniquely determine in all circumstances (however it may have been formed in the context of its a priori fixed nature) the infinitesimal parallel displacement: the non-rotational progression from a point into the world. This assertion involves a deep mathematical theorem of group theory which I have proved. I believe that this solution of the space-problem plays the same role in the context of the Riemann-Einstein theory that the Helmholtz-Lie solution (section 14) plays for rigid Euclidean space. Perhaps the postulate of the unique determination of "straight-progression" can be also justified from the requirements of the phenomenological constitution of space; Becker would still like to ground the significance of the Euclidean rotation group for intuitive space on Helmholtz's postulate of free mobility.[46]

Although Weyl's group-theoretical solution to the generalized "space-problem" of course retains its mathematical interest entirely independently of this connection with Husserlian phenomenology, the philosophical motivations behind Weyl's approach attest, nevertheless, to the enduring fascination of the idea that geometry is to be based on a consideration of space as a (kinematical) form of intuition.[47]

In Weyl's work, therefore, the program, begun by Kant, of considering the geometry of space as given in our form of outer intuition has in a sense come full circle. In Kant, as we have seen, this program has both a logical or constructive and a phenomenological or perceptual dimension. The logical or constructive side of Kant's conception is grounded in the proof procedure of Euclid's *Elements*, where we iteratively generate the objects of geometry by a definite procedure of construction, so that the objects in question constitute a *potentially* infinite totality. The phenomenological or perceptual side of Kant's conception is expressed by the "action of the understanding on sensibility" (B152), whereby the subject imaginatively locates itself in space at a definite point of view and with a definite orientation, so that, by virtue of the resulting formal structure of pure outer intuition, the space in which we perceive outer objects is necessarily both singular or unitary and infinite or unbounded. In Euclidean geometry, moreover, we find mathematical counterparts to both sides of Kant's conception, in that the translations and rotations at the basis of the Euclidean group of rigid motions generate the two fundamental elements of Euclidean construction – straight lines and circles – as their orbits. In the nineteenth century, Kant's conception is both generalized and radically transformed in the work of Helmholtz and Poincaré, where a group-theoretical and perceptual/kinematical interpretation of the foundations of geometry is extended also to the classical non-Euclidean geometries of constant curvature. Since, however, this nineteenth-century generalization now has an entirely conceptual model of geometry given by Riemann's theory of manifolds, spatial perception as it figures in the foundational conceptions of Helmholtz and Poincaré is now, in Kantian terminology, empirical as opposed to pure intuition. Nevertheless, it is still possible, from a mathematical point of view, to connect this generalized group-theoretical conception of geometry with an appropriate generalization of Euclidean construction. In the case of the variably curved Riemannian manifolds employed in the general theory of relativity, by contrast, the approach taken by Helmholtz and Poincaré must itself be radically transformed. And the import of Weyl's reaction to this situation, from the present point of view, is that a group-theoretical and perceptual/kinematical interpretation of the foundations of geometry can, in a new sense, be sustained, whereas, at the same time, the connection with a definite procedure of construction is abandoned. In Weyl's work, we might say that we find a definitive divorce between the logical or constructive and the phenomenological or perceptual dimensions of Kant's original doctrine.

NOTES

Earlier versions of this paper were presented at a Workshop on Modern Mathematical Thought, University of Pittsburgh–Carnegie Mellon University; at the University of Western Ontario; at a History of the Philosophy of Science (HOPOS) Workshop; at a conference in Honor of William Tait at the University of Chicago; and at a Kant Conference at the University of St. Andrews. I am indebted for comments and discussion to John Bell, Henk Bos, Graciela De Pierris, Michael Dickson, George Gale, Robert Hanna, Ulrich Majer, Kenneth Manders, Onora O'Neill, Charles Parsons, Alan Richardson, Simon Saunders, Howard Stein, Mark Wilson, and Allen Wood. I am especially indebted to Robert DiSalle, without whom this paper would not have been written (see note 25).

1. M. Friedman, "Kant's Theory of Geometry," *Philosophical Review*, 94 (1985): 455–506, reprinted, with revisions, as Chapter 1 of *Kant and the Exact Sciences*

(Cambridge, MA: Harvard U. Press, 1992); C. Parsons, "The Transcendental Aesthetic," in P. Guyer (ed.), *The Cambridge Companion to Kant* (Cambridge UK: Cambridge U. Press, 1992). Parsons' original discussion of this issue, to which I was responding in 1985, is "Kant's Philosophy of Arithmetic," in S. Morgenbesser, P. Suppes, and M. White (eds.), *Philosophy, Science, and Method: Essays in Honor of Ernest Nagel* (New York: Cornell U. Press, 1969), 568–94, reprinted, with a postscript, in *Mathematics in Philosophy* (Ithaca, NY: Cornell University Press, 1983), 110–49.

2. E. Beth, "Über Lockes 'Allgemeines Dreieck'," *Kant-Studien* 49 (1956–7): 361–80; J. Hintikka, "On Kant's Notion of Intuition (Anschauung)," in T. Penelhum and J. MacIntosh (eds.), *Kant's First Critique* (Belmont, Calif.: Wadsworth, 1969); "Kant's 'New Method of Thought' and His Theory of Mathematics," *Ajatus* 27 (1965): 37–43; "Kant on the Mathematical Method," *The Monist*, 51 (1967): 352–75. These last two are reprinted in *Knowledge and the Known* (Dordrecht: Reidel, 1974).
3. "The Transcendental Aesthetic," p. 66; this refers back to "Kant's Philosophy of Arithmetic," p. 112 of *Mathematics in Philosophy*.
4. E. Carson, "Kant on Intuition in Mathematics," presented to the Canadian Philosophical Association in the spring of 1994. This paper has appeared more recently as "Kant on Intuition in Geometry," *Canadian Journel of Philosophy*, 27 (1997): 489–512.
5. See Parsons, "The Transcendental Aesthetic," p. 70: "[T]here is a phenomenological fact to which [Kant] is appealing: places, and thereby objects in space, are given in a [single] space, therefore with a 'horizon' of surrounding space."
6. See Parsons, "The Transcendental Aesthetic," pp. 77–8: "[Euclidean constructions] are constructions *in* intuition; space is, one might say, the field in which the constructions are carried out; it is by virtue of the nature of space that they *can* be carried out."
7. All translations from Kant's German are my own and are cited, except for the *Critique of Pure Reason*, by volume and page numbers of the Akademie edition of *Kant's gesammelte Schriften* (Berlin: Reimer (later Walter de Gruyter), 1902–); the *Critique of Pure Reason* is cited by the standard A and B pagination of the first (1781) and second (1787) editions, respectively. Kant's reply to Eberhard, *On a Discovery According to Which Any New Critique of Pure Reason Has Been Made Superfluous by an Earlier One* (1790), is translated, together with valuable supplementary materials, in H. Allison (ed.), *The Kant-Eberhard Controversy* (Baltimore: Johns Hopkins U. Press, 1973): the present passage can be found on pp. 175–6.
8. I thus disagree with Henry Allison's contention, in *Kant's Transcendental Idealism* (New Haven: Yale U. Press, 1983), pp. 83–6, that "outside [*außer*]" has a nonspatial meaning here – meaning "distinct from me (the self)" in the first clause and "(numerically) distinct from one another" in the second clause – so that, in the second clause, Kant is referring to space's role as a principle of individuation. It seems to me, on the contrary, that the parenthetical insertion in the first clause, as well as the "and next to" added (in the second edition) to the second clause, make it clear that "outside" has a spatial meaning. Kant is not outlining an abstract principle of individuation here but rather articulating an a priori perceptual structure: "outside me" is perceptually indexical, and has the force of "outside of (and thus capable of being seen from) this point of view."
9. Compare the above-quoted passage from B154–5 on "representing the three dimensions of space" with the discussion of orientation in *On the First Ground*

of the Distinction of Regions in Space (1768) at Ak. 2, pp. 378–9. This latter discussion begins with the construction, "because of the three dimensions [of space]," of "three surfaces . . . that all intersect one another at right angles."

10. Allison's main ground for rejecting a spatial reading of "outside" in the passage at A23/B38 (note 8, above) is that this makes Kant's claim appear tautological and thus analytic. However, whereas it is indeed tautological that all "outer" objects are in space, it does not follow that the articulation of the a priori structure of this perceptual space is itself analytic. On the contrary, this formal structure is described precisely by the *synthetic* a priori science of geometry; and the task of the transcendental philosopher is then to describe the human cognitive faculties that make this possible. In the end, therefore, the difference between the two readings rests, I believe, on Allison's attempt (which, in fact, is common to most interpretations of the Transcendental Aesthetic) to make the argument of the Metaphysical Exposition entirely independent of the consideration of geometry (which is then supposed to be confined to the Transcendental Exposition): see *Kant's Transcendental Idealism*, pp. 81–2, 98–9. I am indebted to Graciela De Pierris for discussion of this point.
11. The priority of motion and the circumstance that construction is essentially kinematical rather than instantaneous implies that *points* are not independently constructible – they emerge only as products of the process of drawing lines and describing circles (as intersections, endpoints, and so on). This clarifies Kant's claim, made in the context of a discussion of continuous, "flowing" quantities, that "[p]oints and instants are only limits, that is, mere places of their [space's or time's] limitation; but places always presuppose those intuitions that they limit or are to determine" (A169/B211).
12. How do we know, in particular, that Euclidean constructions can indeed be carried out – and also iterated indefinitely? On the logical interpretation, since there is no purely logical or conceptual representation possible, the only way we can even think of or represent, say, the proposition that a circle is always constructible with a given center and radius, is by actually *possessing* the construction in question (as a Skolem function, as it were, for the existential quantifier); and, if we have the construction, the proposition is then automatically true. The proposition is thus a priori true, because its truth is a condition of its mere possibility. On the phenomenological interpretation, by contrast, the truth of such geometrical axioms is not already settled by their mere possibility: geometrical intuition and perceptual spatial "facts" are then called in to settle this question. See *Kant and the Exact Sciences*, Ch. 1, p. 66, and especially Ch. 2, §IV. (This note and the preceding one were prompted by queries from Michael Dickson.)
13. Parsons has suggested to me in conversation that an emphasis on the importance of directly perceptible phenomenological facts in grounding or explaining the axioms of geometry need not involve a commitment to an epistemological foundation for geometry given from outside this science itself; on the contrary, it may simply involve the attempt to articulate the significance of intuitively spatial evidence *within* the science of geometry. This is an important suggestion, which I will touch upon briefly later.
14. Eberhard's appeal to Apollonius appears in his *Philosophisches Magazin*, I (1789): 158–9. See Ak. 20, p. 505. Kant orchestrated a full-scale counterattack, which included (besides his own *On a Discovery*) a review essay by Reinhold (Ak. 20, pp. 385ff). See Allison's *The Kant-Eberhard Controversy*, for further details and a selection of relevant materials.

15. See T. L. Heath (ed.), *Apollonius of Perga: Treatise on Conic Sections* (Cambridge UK: Cambridge U. Press, 1896), pp. 1–9 (Prop. I.11 is Proposition 1 in Heath's numbering). The notation "$y^2 = ax$" is used by Kant.
16. This terminology derives from Pappus's commentary on Apollonius's two books on plane loci. Loci (curves) that are neither plane nor solid – including both higher algebraic curves and transcendental curves – are here termed "curvilinear." See A. Jones (ed.), *Pappus of Alexandria: Book 7 of the Collection* (Berlin: Springer, 1986), pp. 104–7.
17. Eberhard explains in a correction published in his *Philosophisches Magazin,* III (1790/91): 205–7, that he had mistakenly cited Borelli's 1661 edition of Books V–VII; the edition in question is actually that of Books I–IV by Claudius Richardus (1655).
18. In the *Metaphysical Foundations of Natural Science* (1786), Kant distinguishes between "geometrical" and "mechanical construction" (Ak. 4, 493) and, accordingly, between the "*mathematical construction*" and the "*mechanical* execution [*Ausführung*]" of the composition of velocities – where the latter shows "how it can be *brought forth* through nature or art by means of certain instruments and forces" (494).
19. Descartes's new version of the distinction between "geometrical" and "mechanical" curves, which is explicitly intended as a correction to the traditional classification of Pappus (see note 16), occurs at the beginning of Book II of the *Géometrié*, in a section entitled "What curved lines are admitted in geometry" – see D. Smith and M. Latham (trans.), *The Geometry of René Descartes* (La Salle, II: Open Court, 1925), pp. 40–9. For discussion, see A. Holland, "Shifting the Foundations: Descartes's Transformation of Ancient Geometry," *Historia Mathematica*, 3 (1976): 21–49, and especially H. Bos, "On the Representation of Curves in Descartes' *Géométrie*," *Archive for History of Exact Science*, 24 (1981): 295–338 and "The Structure of Descartes' *Géométrie*," in G. Belgioioso (ed.), *Descartes: il Metodo e i Saggi* (Rome, Istituto della Enciclopedia Italiana, 1990). I am indebted to Bos, and also to Mark Wilson, for urging me to consider this question of the scope of geometrical construction.
20. Newton gives a well-known mechanical construction of conics (which may very well have been familiar to Kant), by rotating lines and thereby describing intersections, in Lemma XXI of *Principia*, Book I.
21. As noted earlier, I therefore reject the idea – characteristic of the phenomenological interpretation (compare note 5) – that the unity and unboundedness of space can be directly and immediately given as some kind of quasi-perceptual fact.
22. Compare Kant's remarks about the relationship between the "category of *magnitude*" and the perception of the spatial figure of a house at B162. I am indebted to Parsons for prompting this discussion of the connections among intuitive spatial unity, the transcendental unity of apperception, and the categories of quantity.
23. These latter questions are central to Kant's *Metaphysical Foundations of Natural Science*. See Chapters 3 and 4 of *Kant and the Exact Sciences*, as well as my "Causal Laws and the Foundations of Natural Science," in P. Guyer (ed.), *The Cambridge Companion to Kant*, pp. 161–99.
24. From a modern, four-dimensional point of view, we are concerned only with continuous transformations within a single plane of simultaneity, and thus the dynamical question of how different planes of simultaneity are related to one another over time is not relevant here. Kant himself, in the first chapter on Phoronomy of the *Metaphysical Foundations of Natural Science*, says that, in phoronomy, "motion can be considered solely as *describing of a space*, but still in such a way that I attend not

merely, as in geometry, to the space that is described, but also to the time in which and thus the speed with which a point describes the space" (Ak. 4, p. 489). And in §24 of the second edition Deduction, in the passage from B154–5 cited earlier, after giving examples of geometrical synthesis, Kant describes how we represent "time itself" by attending, not to the space described in drawing a straight line, but to the act of successive determination by which we thereby determine *inner* sense and thus "first produce the concept of succession." (The concept of succession is of course a component of the *dynamical* category of causality – more precisely, of its schema: A144/B183 and compare B291–2.) I am indebted to Howard Stein for prompting this discussion of the nature and status of geometrical motion.

25. The basic idea of using Helmholtz's kinematical conception of spatial intuition to build a bridge between the phenomenological and logical approaches to interpreting Kant's own theory of spatial intuition is due to Robert DiSalle – in comments on the paper by Emily Carson cited in note 4. My own work on the present paper grew directly out of conversations with DiSalle.
26. "Über die Tatsachen, die der Geometrie zugrunde liegen," *Nachrichten von der königlichen Gesellschaft der Wissesschaften zu Göttingen*: no. 9, (1868): 39–71, translation in R. Cohen and Y. Elkana (eds.), *Hermann von Helmholtz: Epistemological Writings* (Dordrecht, The Netherlands: Reidel, 1977), Ch. II.
27. "Über die Ursprung und die Bedeutung der geometrischen Axiomen," first given as a lecture in 1870 and published in *Populäre wissenschaftliche Vorträge*, Vol. 2 (Braunschweig: F. Vieweg, 1871). I cite from H. Hörz and S. Wollgast (eds.), *Philosophische Vorträge und Aufsätze* (Berlin: Akademie Verlag, 1971), p. 214 – this corresponds to the translation in R. Cohen and Y. Elkana (eds.), *Hermann von Helmholtz: Epistemological Writings*, p. 23.
28. Appendix III to the address of 1878, "Die Tatschen in der Wahrnemung," first published in *Vorträge und Reden* (Braunschweig: F. Vieweg, 1884). I cite from H. Hörz and S. Wollgast (eds.), *Philosophische Vorträge und Aufsätze*, p. 299, which corresponds to Cohen and Elkana (eds.), *Hermann von Helmholtz*, pp. 162–3. "Space can be transcendental without the axioms being so" is the title of Appendix II.
29. I thus interpret Helmholtz as holding that the general, "transcendental" form of spatial intuition includes the condition of free mobility (and thus constant curvature) but not the properties of specifically Euclidean geometry (zero curvature). As Howard Stein, in particular, has emphasized to me, this is certainly not the only possible interpretation: one might also take the most general, "transcendental" form of space to include, for example, only topological and manifold properties. The present reading is supported by the following passage from "On the Origin and Meaning of the Axioms of Geometry," which initiates the criticism of the specifically Kantian theory of intuition:

> We will now have to ask further where those particular determinations come from that characterize our space as plane space, since these, as has been shown, are not included in the general concept of an extended magnitude of three dimensions and free mobility of the structures contained therein. They are not *necessities of thought*, which flow from the concept of such a manifold and its measurability or from the most general concept of a rigid structure contained therein and its freest mobility. (Hörz and Wollgast (eds.), *Philosophische Vorträge*, pp. 206–7; Cohen and Elkana (eds.), *Hermann von Helmholtz*, p. 17)

One should also note that whenever Helmholtz explicitly states what he calls "the axioms of geometry," these always characterize specifically Euclidean space.

30. See "The Facts in Perception" – Hörz and Wollgast (eds.), *Philosophische Vorträge*, pp. 262–5; Cohen and Elkana (eds.), *Hermann von Helmholtz*, pp. 128–31.
31. See Hörz and Wollgast (eds.), *Philosophische Vorträge*, pp. 256–8; Cohen and Elkana (eds.), *Hermann von Helmholtz* pp. 122–4.
32. Kant's discussion of orientation cited in note 9 uses the human body to determine the relations above–below, right–left, and forward–backward. This amounts to using one's body to pick out a particular triad of mutually perpendicular line segments corresponding, respectively, to these three relations. This procedure should be viewed, I believe, as an account of how we *apply in experience* the purely geometrical notion of orientation – which notion is itself given by an entirely arbitrary construction of three mutually perpendicular line segments in *pure* intuition. It would be interesting to apply these ideas to Kant's conception of incongruent counterparts, but this will have to wait for another occasion. (This note was prompted by comments and suggestions from Robert Hanna, Onora O'Neill, and Allen Wood.)
33. This is connected with the circumstance, remarked in note 24, that the purely geometrical motion involved in figurative synthesis does not yet involve the questions of *time*-determination characteristic of the dynamical as opposed to the mathematical categories. Compare A160/B199:

    In the application of the pure concepts of the understanding to possible experience the use of their synthesis is either *mathematical* or *dynamical*. For it applies partly to mere *intuition*, partly to the *existence* of an appearance in general. But the a priori conditions of intuition are necessary throughout in relation to a possible experience, those of the existence of objects of a possible empirical intuition are in themselves only contingent.

    The dynamical categories essentially involve the conditions for transforming perceptions or empirical intuitions into law-governed experience, and thus, as I have argued elsewhere (see note 23), they also involve the conditions for transforming apparent motions into true motions – where the (Newtonian) laws of motion here realize or embody the Analogies of Experience. The purely mathematical synthesis expressed in pure geometrical motion, by contrast, has nothing to do with the laws of motion. Once we follow Helmholtz in basing geometry on real bodily motion, however, we simply cannot avoid entangling geometry with the laws of motion: we cannot avoid facing the fact, in modern terms, that the four-dimensional structure of space-time is primary.
34. The crucial point is that, on the phenomenological interpretation, the truths of geometry appear as brute "perceptual facts" – even if they are conceived as intuitively evident truths internal to the (supposedly) a priori science of geometry as in note 13. On this kind of interpretation, there is no particular difficulty in thinking or conceiving the truths of geometry independently of spatial intuition, and the latter is then called in only to establish that some particular set of axioms (the Euclidean axioms) is in fact true. On the logical interpretation, by contrast, the truths of geometry function as a priori preconditions without which it would be impossible even to think of or to conceive spatial structures in the first place: without the truths of geometry there would simply be no "spatial facts" (see note 12). If one now asks where the "spatial facts" invoked by the phenomenological interpretation come from, this interpretation is then vulnerable, from a philosophical point of view, to naturalism and empiricism – according to which such "facts" must rest, in the end, on contingent conditions of our perceptual apparatus and/or contingent

characteristics of physical space (see Parsons, "The Transcendental Aesthetic," pp. 72–5). Here I am again indebted to De Pierris – compare her discussion of Parsons's paper in her "Review of *The Cambridge Companion to Kant*," *Ethics*, 104 (1994): 655–7.

35. "On the Foundations of Geometry," *The Monist*, 9 (1898), pp. 41–3.
36. For a discussion of Poincaré's theory of arithmetic, see J. Folina, *Poincaré and the Philosophy of Mathematics* (New York: Macmillan, 1992).
37. *La Science et l'Hypothèse* (Paris: Flammarion, 1902), Ch. IV – I cite from the translation of G. Halsted in *The Foundations of Science* (Lancaster, Pa.,: The Science Press, 1913), p. 75.
38. *La Science et l'Hypothèse*, Ch. III, citing from G. Halsted (trans.), *Foundations of Science*, p. 63. In the popular Dover edition of *Science and Hypothesis* (New York, 1952), p. 48, the last sentence is incorrectly translated as: "These geometries of Riemann, so interesting on various grounds, can never be, therefore, purely analytical, and would not lend themselves to proofs analogous to those of Euclid" – thereby entirely reversing its sense (and the preceding sentence on p. 47 incorrectly has "variable figure" instead of "invariable figure").
39. Since Poincaré thus rules out nonhomogeneous manifolds of variable curvature, his conventionalism is entirely incompatible with the general theory of relativity – a circumstance that has led to considerable confusion among his followers: see my "Poincaré's Conventionalism and the Logical Positivists," in J.-L. Greffe, G. Heinzmann, and K. Lorenz (eds.), *Henri Poincaré: Science and Philosophy* (Berlin and Paris: Blanchard and Akademie Verlag, 1996), pp. 333–44. Indeed, since Poincaré (like Helmholtz) bases geometry on the free mobility of real physical bodies, his conception is also incompatible with the space-time structure of special relativity – where, despite the fact that each individual plane of simultaneity is Euclidean, there is still no free mobility of rigid bodies in space-time (compare notes 24 and 33).
40. Poincaré's explicit discussion of "The Reasoning of Euclid" in "On the Foundations of Geometry," pp. 32–4, focuses on proofs that proceed by translating and rotating figures rather than on Euclidean constructions. Yet, as we will see, it is nonetheless possible to forge a connection between these two ideas. Helmholtz suggests such a connection in "On the Origin and Meaning of the Axioms of Geometry" – Hörz and Wollgast (eds.), *Philosophische Vorträge*, pp. 190–1; Cohen and Elkana (eds.), *Hermann von Helmholtz*, pp. 4–5.
41. For this and the next three paragraphs, see M. Greenberg, "Euclidean and Non-Euclidean Geometries Without Continuity," *American Mathematical Monthly*, 86: 757–64.
42. "Neue Begründung der Bolyai–Lobatschefsckyschen Geometrie," *Mathematische Annalen*, 57 (1903): 137–50. This appears as Appendix III to *Foundations of Geometry* (La Salle: Open Court, 1971).
43. Section 34 of J. Bolyai, *Scientiam Spatii Absolute Veram*, published as an Appendix to W. Bolyai, *Tentamen Juventutem Studiosam in Elementa Matheseos Purae*, translated in G. Bonola, *Non-Euclidean Geometry* (New York: Dover, 1955), pp. 37–8.
44. For the elliptic case, see W. Schwabhäuser, "On Models of Elementary Elliptic Geometry," in J. Addison (ed.), *The Theory of Models* (Amsterdam: North-Holland, 1965). If one wants an elementary geometry going beyond straight-edge and compass constructions to include all algebraic curves in the manner of Descartes (note 19), then one can, by using a first-order continuity schema, also construct

geometries over real closed fields in all three cases of constant curvature: see W. Schwabhäuser, "Metamathematical Methods in Foundations of Geometry," in Y. Bar-Hillel (ed.), *Logic, Methodology and Philosophy of Science* (Amsterdam: North-Holland, 1965).

45. Weyl explains the background of his mathematical investigations in Husserlian phenomenology in the Introduction to *Raum, Zeit, Materie* (Berlin: Springer, 1918), translated as *Space-Time-Matter* (New York: Macmillan, 1952). For Becker, see "Beiträge zur phänomenologischen Begründung der Geometrie und ihrer physikalishen Anwendung," *Jahrbuch für Philosophie und phänomenologische Forschung*, 6 (1923): 385–560. For further discussion, in the context of Rudolf Carnap's contrasting (nonkinematical) conception of geometrical intuition developed in his dissertation of 1921, see my "Carnap and Weyl on the Foundations of Geometry and Relativity Theory," *Erkenntnis*, 42 (1995): 247–60.
46. *Philosophie der Mathematik und Naturwissenschaft* (Berlin: Leibniz Verlag, 1927), §18, pp. 99–100, translated as *Philosophy of Mathematics and Natural Science* (Princeton, NJ: Princeton University Press, 1949), p. 137.
47. In "Die Einzigartigkeit der Pythagoreischen Maßbestimmung," *Mathematische Zeitschrift*, 12 (1922): 114–46, where Weyl first proves his group-theoretical theorem, he begins by stating that the infinitesimally Euclidean nature of the metric is "*characteristic of space as form of appearance*" (p. 116). (One should note that Weyl's "purely infinitesimal" approach involves an extended conception of metrical structure as well, where a "Weyl structure" on a manifold consists of a *class* of conformally equivalent Riemannian metrics, each paired with an accompanying "gauge factor.")

Parsons on Mathematical Intuition and Obviousness

MICHAEL D. RESNIK

I. Introduction

In a series of important papers, Charles Parsons has argued that we have mathematical intuitions of certain "quasi-concrete," mathematical objects, such as numerals or geometric shapes. With Kant and Hilbert, he maintains that these intuitions constitute some of our evidence for certain elementary mathematical truths, and that they can be used to explain the obviousness of these truths. I focus on his account of elementary number theory and attempt to clarify his views. I have no objection to the thesis that we intuit quasi-concrete objects, provided that we use more muted notions of intuition and quasi-concreteness than Parsons apparently does. I also raise doubts about using intuition to explain the obviousness of elementary mathematical truths. Let us begin with an exposition of his views.

II. Quasi-concreta and Intuitions of Them

According to Parsons, a quasi-concrete object is an abstract entity such as a symbol, a word, a color, or a sound, understood of course, as something over and above its utterances or instances, and yet also understood as capable of being given to us through the perception of them. As he puts it, "what they [quasi-concreta] are is determined by some instantiation or representation in the concrete," and thus they differ from "pure" abstract objects, such as numbers, which have "no intrinsic concrete representation."[1] Unlike Penelope Maddy, whose views those of Parsons bring to mind, he does not maintain that quasi-concreta are located with their instances or have causal properties; they are like "pure" abstract entities in being outside of space-time and causally inert. Thus Parsons also contrasts with Maddy in denying that we perceive these objects as we perceive their instances. Instead we *intuit* quasi-concrete objects through perceiving or imagining their instances.

Parsons treats quasi-concreta as individuals rather than as properties or universals; they fall within the range of first-order rather than higher-order variables. Since he seeks intuitive models for elementary mathematical theories, treating types and shapes as individuals is the natural and simpler option for him. But it is also in keeping with the Kantian roots of his view, since for Kant we intuit objects rather than concepts.

As I stated earlier, for Parsons, the natural numbers are pure abstract objects, and we cannot intuit them. Yet we can intuit elements and initial segments of omega sequences of quasi-concreta forming models of number theory. Some of these models represent the numbers more perspicuously than others. One of these is a model that Parsons derives from Hilbert. It is an omega sequence of stroke-strings, where each succeeding stroke-string is obtained by adding one stroke to its predecessor:

/, //, ///, ////, /////, ...

Now compare this with a permutation of this sequence, for example,

/////, //, /, ///, //////, ////, ...

or with an omega sequence of line segments. Although each can serve as a model for number theory, only the first consists of quasi-concreta whose internal structures are isomorphic to the initial segments of the model they end, and thus also isomorphic to the initial segments of the number sequence that they represent. This feature of the first model not only makes it more perspicuous than the others but also allows us to represent certain arithmetic operations directly as operations on the stroke-strings. Addition is a familiar case. We simply represent the summing of two numbers by the operation of juxtaposing their stroke numerals. The commutativity and associativity of addition are an immediate consequence of this representation and the corresponding features of juxtaposition of stroke-strings.

This is not so for the second model. Although juxtaposition of its elements remains associative and commutative, juxtaposition can no longer represent addition. Concatenating "//" with "/" yields the fourth element of the model rather than the third. The Arabic numerals fare no better in making the commutativity of addition obvious. Nothing in the look of "7 + 5" and "7 − 5" reveals why $7 + 5$ equals $5 + 7$ but $7 - 5$ does not equal $5 - 7$.

According to Parsons, we can obtain intuitive knowledge about quasi-concrete symbol types by perceiving their tokens. For example, by inspecting the first sequence written above, we can see that the type "//" is part of the type "///," and that the latter is the successor of the former, provided we define the successor of a stroke-string as the result of concatenating a stroke to it.

Some of the passages in "Ontology and Mathematics" suggested to me that Parsons also believed that we can know the stroke versions of the first four Peano axioms on a purely intuitive basis.[2] However, his later works clearly indicate that this is not so. For in responding to James Page's criticism[3] that we do not intuitively know that "/" is not a successor or that stroke-successor is one-one, Parsons writes:

> I can imagine its being argued that . . . elementary truths of this kind are analytic and that for that reason knowledge of them does not rest on intuition. In the case of the concept of string of strokes, where intuition is needed even to understand the concept, even if one accepted the claim that these statements are analytic in some sense it would not follow that they are independent of intuition.[4]

Thus, although we use and need intuition to know that "/" has no proper parts, we might not use it alone to know that "/" is not a successor. Still, intuition is necessary for knowing the stroke versions of the first four Peano axioms, and it can be sufficient for knowledge about particular stroke-strings.

I was misled by the discussion between Parsons and Page concerning the axiom that every stroke-string has a successor. Parsons had argued that we can know that every stroke-string has a successor by imagining ourselves extending the token of a given stroke-string with one more stroke. Page then objected that the images generated in such thought experiments are not sufficiently definite to know that the new stroke-string is the old one with one more stroke. Parsons conceded that the experiments cannot assure us that extending a string always leads to one of a new type or that the successor operation can be iterated indefinitely. Yet he still maintained that we can use such images to know that every stroke-string has a successor.[5] It is likely that the reason he took this position is that, contrary to what Page (and I) had been assuming, the successor of a stroke-string need not be distinct from the string in a model of just the first four Peano axioms. Indeed, as Parsons noted in his reply to Page,[6] we need mathematical induction to prove that the successor of something is always distinct from it, and Parsons has never claimed that we have intuitive knowledge of induction. On the other hand, a model in which some number is identical to its successors must be a nonstandard extension of a standard model.[7] This is because the axioms that 0 is not a successor and successor is one-one allow us to prove that $s(0)$ is distinct from 0, $s(1)$ is distinct from 1, and so on for each number using just *modus ponens* and *universal instantiation*. Of course, proving that we can do this for each number requires the use of mathematical induction in the meta-language. That Page and I were so easily confused is a good indication that our presystematic conception of a string implies that strings are subject to inductive reasoning.

Parsons does not present an explicit explanation of how intuition of stroke types accounts for the obviousness of certain number theoretic truths. But we can see how explanations consistent with his views would go. One of these would note that since the stroke progression is isomorphic to the natural number sequence, all the purely structural properties of the former hold for the latter. Assuming that this is itself obvious, it follows that any fact about the numbers that corresponds to an obvious fact about stroke-strings is obvious too – provided, of course, that the inference in question preserves obviousness.[8]

Another possibility for explaining the obviousness of parts of elementary number theory arises from Parsons's suggestion that we construe each number as a functional type. The idea is that the tokens of the numerals "///," "iii," and "3," for example, count as tokens of the number three, since each is a token of a numeral having the same function in counting.[9] On this approach, just as the intimate relation between inscribed tokens and their types makes certain truths about types obvious, so would the relationship between the numerical tokens and numbers (construed as functional types) make obvious certain truths of number theory.

Underlying these explanations is the assumption that anything we know by intuition is obvious. Intuition is supposed to put us into a direct cognitive relationship with an object. So if to be obvious is to be something that will be immediately acknowledged by anyone in a position to know, intuition would yield obvious truths.

III. Illusiveness of Intuitions of the Quasi-concrete

We only see the individual token and not the type, so one wonders how a single observation can furnish us with reliable information about the type or about all tokens of the same type. Since symbol types are quasi-concreta, information about the type is in some sense also information about its possible tokens, or – if we do not want to countenance *possibilia* – about what it is possible to token. The problem here is the old Kantian problem of generalizing from a supposedly paradigmatic instance, but it is independent of the ontology of quasi-concrete types. For the nominalist has the analogous problem of projecting from one token to all tokens of the same type. It is also independent of the issue of whether we can intuit types, since even realists who admit types but no intuitions must explain how we can obtain knowledge about types from tokens.

The solution is to add something extra that overrides the presumption against generalizing from an instance; we must appeal to the circumstances in which the observation is made or to suppressed premises that are presumed to be part of the observer's background knowledge. Parsons writes:

> Of course a perception of a string of stroke-tokens is not by itself an "intuition" of a stroke-string type. One has to approach it with the *concept* of a type, first of all to have the capacity to recognize other tokens as of the same type or not. Something more than the mere capacity is involved, which might be described as seeing something *as* the type.[10]

This suggests that for Parsons the something extra needed to justify the inference from one token instance to the type or to other tokens of the same type is that one see the instance as representing features of the type or of all tokens of its type.

What is it to see some token of a type "*as* the type" itself? I will take this to mean that in addition to the ability one must (ordinarily) have to construct tokens of the type and to recognize tokens of the type, one must also believe that there are types and that they can be *directly presented to us* when we perceive or imagine their tokens. I have inserted "ordinarily" because I do not want these conditions to be overly rigid. Someone who was unable to write, for example, could have the concept of a symbol type. Also I do not see that one's belief in types or the type/token relationship must be explicit or clearly articulated for one to intuit types. For Parsons holds that intuiting types is "quite spontaneous and natural" when we identify the words (word types) in the course of understanding utterances in a natural language.[11]

Parsons compares intuition of types to ordinary perception of physical objects. Let us elaborate a bit on this analogy. We perceive ordinary objects through perceiving surface features of their timeslices. We never take them in their spatiotemporal entireties. For example, at sunset on a particular day I might see just the north face of Table Rock Mountain and not the whole mountain. Similarly, though we never see types, we might think of seeing their appearances or aspects in seeing their tokens. Furthermore, just as we can discern features of a material object (e.g., that it is a chair) by looking at one side of it at a particular time, we can discern characteristics of type through perceiving a few of its tokens. This seems to nicely solve the Kantian problem. Just as examining a chair today can inform us that it is, was, and always will be made of oak, examining a single stroke-string token can tell us whether the stroke-string is a successor string. Finally, just as it can be simply obvious to certain people that the chair is oak, it can be obvious that the stroke-string is a successor.[12]

Whether this is the correct way to read Parsons, an important disanalogy between types and perceivable objects raises doubts about the plausibility of this account of intuition. Material objects are composed of their spatiotemporal aspects; thus we can perceive them by perceiving their parts. But types are not composed of their tokens – at least not on Parsons's account that allows for uninstaniated types – and we do not perceive types in perceiving their parts. Moreover, although we causally interact with material objects by interacting with their parts, we do not causally interact with types at all.[13]

Consider instead the closely related view of someone who believes that types exist and that we must be able to construct and recognizes tokens of types in order to know the types, but who goes on to deny that types are presented to us via tokens. The contrast between the previous view and this one is somewhat similar to that between two theses one finds in the philosophy of science. One of these is that, given the appropriate training and circumstances, we can literally perceive subatomic particles by watching an experimental apparatus or a computer readout. The other thesis, while taking an equally realist stance toward subatomic particles, holds that in looking at the experiment or computer screen we only observe things that reliably indicate features of subatomic particles. When this issue is presented outside of any further context it seems of no consequence whether we say that we only figuratively perceive subatomic particles or that we literally perceive them in an evolving sense of perception. Yet the same issue is quite significant when it arises within the context of the debate over the reality of the distinction between theoretical and observable entities.[14]

Now if all that Parsons means by someone intuiting a type is that the person possesses the appropriate abilities, believes (perhaps merely tacitly) that under certain circumstances tokens can reliably indicate features of types, and is observing a token under those circumstances, then I have no quarrel with his talk of intuiting types. But, as I have indicated, it seems clear that Parsons means more than this, namely, that in intuition, quasi-concrete objects can stand in "a direct presence . . . to the mind."[15] Perhaps we can get a better grip on this idea by considering the nature of the quasi-concrete objects that are supposed to be directly present to the mind.

What distinguishes a symbol type from a number is that the first but not the second is supposed to have an intrinsic concrete representation in tokens. I take this to mean that it is part of our conception of the type that it "looks" or "sounds" a certain way – the way its tokens look or sound. Thus, for example, an italicized "*A*" has a certain look, and it is different from the look of a plain letter "A." Now I find this no less puzzling than the idea of intuiting types, because we cannot perceive quasi-concrete objects. Furthermore, the "look" or "sound" of an expression type varies with context and background assumptions. Under our default assumptions, we can token English expressions in different typefaces and handwriting styles or utter them in different accents. Thus a wide variety of marks and sounds can look like or sound like a particular English expression. Other assumptions replace the default ones when, for instance, we are supposed to use a certain typeface or mimic somebody's accent. This has the effect of changing or specifying more fully the "look" or "sound" of the expressions.

Something similar is true of stroke-strings. The point Parsons made with the stroke-strings depended upon their having certain structural features, and some "look" or other. But within fairly wide limits their particular "look" made no difference; a sequence of star-strings, viz.,

*, **, ***, . . . ,

would have served him just as well. As Eric Heintzberger pointed out to me, even mixing strokes and stars could be made to work, given an appropriate set of background assumptions.

Thus our conception of quasi-concrete strings contains two quite different components, one pertaining to their structural features and the other pertaining to their "look." The former states that there is an initial symbol x, that x standing alone is an x-string, that concatenating x to the end of an x-string is an x-string, and that all x-strings can be generated from the initial x-string by a finite number of such concatenations. To this we add that x and concatenation look like____, where we fill in the blank with a stroke or a star or whatever symbol that distinguishes our x-strings from other kinds and something else to illustrate the look of concatenation. It is this extra step that distinguishes the quasi-concrete strings from the more abstract strings of pure string theory, to which we attribute only structural features.

Parsons has made an important contribution to the epistemology of mathematics in emphasizing that we have conceptions of strings that go beyond the purely structural conception of a string and that the study of these extra elements may be key to understanding our grasp of elementary mathematics. I do not intend to undermine this. Nor do I want to deny that in ordinary language we talk of the look of expressions and say things like "Gothic letters look quite different from Roman ones." On the other hand, I do not see why we must take such talk literally or invoke intuitions of types (in the sense in which they are directly present to the mind) in order to explicate the conceptions of strings that Parsons has highlighted. For instead of saying that a stroke looks like "/" and that a star looks like "*" we can say a stroke *is tokened* by making a mark that looks like "/" and that a star *is tokened* by making a mark that looks like "*"; instead of saying that we see the "/", we can say that we see a token of it, and so on.

Unlike the nominalist, however, I am not claiming that all reference to types can or should be replaced by reference to tokens. Consequently, if all Parsons means by saying that quasi-concrete types have intrinsic concrete representations via tokens is that it is part of our conception of them that their tokens have a certain look or sound, then I have no objection to his view. I simply see neither the gain nor the need for a conception in which the types themselves have a look to them.

If I am correct then the difference between purely abstract strings and quasi-concrete ones lies not in the objects themselves, but in the theories referring to them. Our theories of the quasi-concrete have ties to observation (or acts of tokening) that our theories of the purely abstract strings achieve only by virtue of their intuitive interpretations as quasi-concreta. A consequence of this is that, from the structuralist perspective, to which both Parsons and I subscribe, there may be no fact of the matter as to whether the quasi-concrete strings are identical to the purely abstract ones. For that matter, there may be no fact of the matter as to whether a quasi-concrete stroke is identical to a quasi-concrete star, despite the difference between tokening-a-stroke and tokening-a-star.[16]

We are still left with the problem with which we began this section, namely, that of determining which features of the token we are to take as representative. Parsons is clearly right in thinking that part of possessing the concept of a type and seeing the token as a token of that type involves knowing which features of a token are representative of its type. But I do not see that thinking of the type as something that has a certain "look" or that can be directly present to the mind does any work in solving this problem in addition to that done by the more tempered readings of Parsons's view that I have proposed.

I suspect that Parsons hopes that his theory of mathematical intuition may help us understand the apriority of mathematics.[17] I think that a convincing case can be made that people who know how to use the stroke system or some similar system of numeration can use thought experiments to acquire knowledge about symbol types that they could not acquire through perception, or at least could not acquire as well. For example, by reflecting on how one constructs stroke-strings, one can learn that any string of the form . . . / precedes one of the form . . . ////. It is quite unlikely that one might come to believe this by inducing it from experience, and even if one did, doing so would not provide as strong a warrant for one's belief as the thought experiment would. This knowledge is related to our knowledge that adding 3 to a number produces a greater number. Thus, the apriority of the knowledge of strokes based upon the thought experiment would yield the apriority of the corresponding mathematical knowledge.

I am not endorsing this view. And, as Parsons' work has demonstrated again and again, a very careful analysis would be necessary both to determine the role of intuition in this process and to establish the absence of empirical hypotheses. In any case, I cannot see that it would be essential to the process that we intuit stroke-strings in any sense stronger than that of imagining that we have tokened them.

I would also like to have a better understanding than I have been able to obtain from reading either Kant or Parsons of the exact role that visualizing the stroke-strings plays. If we actually infer properties of the strings by inspecting our images or drawings, then there seems to be an irreducible uncertainty built into the process of intuition. For while we may be confident that we see or imagine a token with paradigmatic feature *X* and generalize from it, we might mistake *X* for another feature *Y* or may mistakenly identify *X* as paradigmatic. On the other hand, if it is built into our concept of intuiting that we cannot make such mistakes, then why is the thought experiment not a disguised deduction? And if so, does it follow that intuition in these cases is dispensable after all? Or is it for us humans an indispensable substitute for deduction?[18]

Notice, by the way, that in the examples that I have been considering lately, we generalize from the intuition. The dispensability of intuition in these cases would not entail its dispensability for establishing the existence of quasi-concrete models of arithmetic. Parsons has emphasized the importance of this role for intuition.

IV. Explaining the Obviousness of the Obvious

Before discussing Parsons' views on the obviousness of mathematics, it is worth clarifying the notion of obviousness itself. The obvious is supposed to be open to view or knowledge; it is supposed to be plain, manifest, or evident. This suggests that when something is obvious to us, we feel certain of it, that we do not feel the need for further evidence for it. Self-evident truths are supposed to be obvious to us upon our considering them, I take it. But the obvious need not be self-evident since considering something other than the statement itself may be needed to make it obvious to us. For example, a proof can make a previously unobvious theorem obvious. (Of course, it should be obvious to anyone with experience reading proofs that not all proofs do this.) If something makes a statement obvious to us, then it is already good evidence for that statement.

Everything I have said so far can be interpreted as concerned with the state of mind we are in when we find something obvious, but calling something obvious has a public function as well. When we report that some theorem is an obvious consequence of some others, we do not simply report our state of mind. We also indicate that you should find the consequence so too. The step is supposed to be obvious to any qualified reader; that is why we see no need to spell out our reasoning.

This should remind us of the jokes about what is obvious to various famous mathematicians – and, of course, opaque to us. What is obvious to someone is a function of context, their background knowledge and training, as well as their interest and attention. To me, the fellow at the bar next to me may appear to be an amusing character, but to the bartender it may already be plain that he is a combatative drunk and to the psychiatrist watching us that he is a paranoid, and perhaps to someone else that he is a talented poet. Something can make something obvious that would not be otherwise. It can become obvious that a person is quite plain when they are stripped of their fancy clothes or makeup. More to the point of this paper, a notation or method of proof can make a mathematical claim obvious. It is obvious to us that $10150 + 1793$ cannot be 18111 because in our notation the latter "ends" in 1 rather than 3. It probably would not be obvious if the numbers were given in unary or binary notation.

Although the obviousness of something qualifies it as evidence for something else, obviousness is not a sure guide to the truth. The consequences that Saccheri deduced from negating the parallel postulate were obviously false to him. Of course, one might distinguish the obvious from the seemingly obvious, and allow obviousness to guarantee truth, but I will not. I think I am with Parsons here, because he does not try to connect the apriority or necessity of mathematics to the obviousness of its elementary parts.

Let us now turn to Parsons's views on intuition and the obvious. The most explicit thing he says in this connection is that the need to explain the obviousness of elementary truths of mathematics is a reason for introducing intuition into the epistemolgy of mathematics. He continues by criticizing Quine's holism for being unable to offer such an explanation and states that, in Quine's view, the truths of elementary arithmetic should be "bold hypotheses, about which a prudent scientist would maintain reserve, keeping in mind that experience might not bear them out. . . ."[19] I take this to indicate that Parsons thinks we can be certain of obvious elementary arithmetical truths and may use them as evidence for the less obvious ones.

We have already noted that the obvious depends upon the context, the background knowledge of the person judging, and so on. Thus, whether looking at tokens of stroke-strings makes elementary truths about stroke-strings obvious will depend upon who is doing the looking, and so on. Parsons does hold that whether someone looking at a token intuits its (or a) type depends upon his or her conceptual framework, but he does not make clear whether someone can intuit a type without finding elementary truths about it obvious. Offhand, I would think that it would depend upon the truths in question and the circumstances of the intuiting. Intuiting "///" and "/////" may make it obvious to a person that the latter is the second successor of the former without making it obvious to the same individual that every stroke-string of the form . . . /// is the second successor of one of the form . . . /.

In any case, it seems clear that Parsons thinks that the string theoretic versions of the first four Peano axioms can be made obvious though appropriate intuitions. As we have seen, however, Parsons concentrates on the intuition of stroke-strings and not on the intuition of numbers, which, on the most straightforward reading of his view, are not intuitable at all. (If we construe numbers as the functional types of equivalent numeral tokens, then we might be said to intuit numbers in an extended sense.)

We have already speculated on how intuiting stroke-strings might make the corresponding truths about numbers obvious. One explanation took it to be obvious that the stroke-strings are isomorphic to the natural numbers, the other that the numbers are the functional types of numerals. Assuming that we want to explain why elementary truths of arithmetic were obvious to the ancients and are obvious to unsophisticated members of our own culture, we can reject this explanation as depending upon the possession of concepts that they did not have. I think it is more likely that numbers entered the ken of the ancients and continue to enter the ken of our young through speaking of numerals as numbers. Rightly or wrongly, we continue to engage in talk that it takes someone with Fregean compunction to foreswear, and we freely speak of writing *numbers* in books or on chalkboards. This explanation is fine if all we are trying to explain is our states of mind and our social practices. On the other hand, if obviousness is supposed to ensure truth, then the explanation must be found elsewhere. Then, our second explanation based upon Parsons' suggestion that numbers are functional types may be the best to pursue. I am not optimistic about its prospects, however, since, like the first explanation, it presupposes a much richer conceptual scheme than that needed for intuiting the numerals of one or two symbol systems.

Another way in which things become obvious to us is through "seeing" patterns. Parsons does not mention this, but it is clearly related to his intuiting, since types are symbol patterns. It seems to me that when we recognize the pattern of extending stroke-strings, it becomes obvious to us that every stroke-string has a successor. If this is so, then induction, or at least a weaker form of it, can be made obvious too. For we can be shown the pattern of proving Fn for a specific number n by first proving $F0$, $F1$, and so on, in proofs taking the form:

1. $(x)(Fx \rightarrow Fx + 1)$	Premise
2. $F0$	Premise
3. $F1$	1, 2 UI, MP
4. $F2$	1, 3, UI, MP
.	
.	
.	
$n + 2$. Fn	1, $n + 1$, UI, MP.

Of course, this makes it obvious to us that every number is an F if we confuse being able to prove that each number is an F with being able to prove that all of them are.

In conclusion, I think Parsons is correct in pointing out the importance of our knowledge of symbol systems to the epistemology of mathematics. I have no objection to his speaking of intuiting quasi-concrete objects, provided that these objects are not understood as distinguished from pure abstract ones through having looks or sounds, and provided we understand intuiting a type as a way of speaking of the beliefs we acquire on the basis of our conceptual setup and our perceptions of or imaginations of tokens. Furthermore, I take the difference between quasi-concrete and pure abstract objects to arise from our theories about them, and not in theory-independent features of the objects themselves. When it comes to explaining the obviousness of the obvious, I think that Parsons has offered us the beginnings of an account. I am much less sure that it can be developed far enough to explain why most of the obvious arithmetic truths are obvious.

NOTES

I would like to thank Eric Heintzberger and Richard Tieszen for helpful comments on an earlier draft of this paper.

1. Charles Parsons, "Intuition and Number" in *Mathematics and Mind* (Oxford: Oxford University Press, 1994), p. 143.
2. Parsons has always denied that the induction axiom can be known intuitively.
3. James Page, "Parsons on Mathematical Intuition," *Mind*, 102 (1993): 223–32.
4. Charles Parsons, "On Some Difficulties Concerning Intuition and Intuitive Knowledge," *Mind*, 102 (1993): 233–46. See p. 241.
5. Parsons, "On Some Difficulties Concerning Intuition," p. 245.
6. Ibid.
7. The simplest model is one with an extra "number" that is its own successor. The standard numbers behave as usual. Since induction is need to prove that every number but 0 has a predecessor, the model needs only one nonstandard number.
8. In "Intuition and Number," Parsons remarks that logical inference need not preserve intuitive evidence when "the domain of quantification has a nonintuitive character" (p. 152). Since the quantification is over structural properties of omega sequences, his remark clearly applies. But logic might still preserve obviousness even if it does not preserve intuitive evidence.
9. Charles Parsons, "Mathematical Intuition," *Proceedings of the Aristotelian Society*, New Series, 80 (1979–80): 145–68. See p. 163.
10. "Mathematical Intuition," p. 154.
11. Ibid.
12. I am grateful to Richard Tieszen for emphasizing the importance of the perceptual analogy in Parsons' thinking, and for suggesting a version of the interpretation presented in this paragraph.
13. I am not denying that we speak of the seeing of words or messages as causing actions. But this talk should be interpreted in terms of tokens just as talk of writing down numbers should be interpreted in terms of tokening numerals.
14. Richard Tieszen cautioned me that the analogy to perceiving subatomic particles is too far removed from ordinary perception. This is because we must construct and interpret scientific instruments to observe subatomic particles, whereas tokens present us with types immediately. Because writing is an invention, and we must learn to read and write, I do not find the analogy as stretched as Tieszen does.
15. Parsons, "Mathematical Intuition," p. 147. Although he uses these words in the course of paraphrasing Kant, Parsons takes himself to be explicating the Kantian notion of intuition. We can leave open the question of whether intuiting a type is more like perceiving a material object or perceiving a theoretical entity in the literal sense.
16. I have inserted hyphens to emphasize that I am talking about two kinds of act rather than two kinds of object.
17. In the introducing *Mathematics in Philosophy*, Parsons states that he is "inclined

to think that a defense [of the apriority of mathematics] is possible." See Charles Parsons, *Mathematics in Philosophy* (Ithaca, NY: Cornell University Press, 1983), p. 18.

18. In "Computation and Mathematical Empiricism," *Philosophical Topics*, 17 (1989): 129–144, I argue that, for us, concrete computation is an indispensable substitute for deduction.
19. "Mathematical Intuition," pp. 151–2.

Gödel and Quine on Meaning and Mathematics

RICHARD TIESZEN

Charles Parsons (1995b, p. 309) has noted that Gödel never discussed the deeper issues about meaning that are addressed by Quine. Parsons says that the only place where Gödel even begins to approach these issues is in a paper entitled "The Modern Development of the Foundations of Mathematics in the Light of Philosophy" (Gödel *1961/?). In this paper, Gödel argues that a foundational view is needed that would allow us to cultivate and deepen our knowledge of the abstract concepts that underlie formal or "mechanical" systems of mathematics. It should be a viewpoint that is favorable to the idea of clarifying and making precise our understanding of these concepts and the relations that hold among them. Gödel says that phenomenology offers such a method for clarification of the meaning of basic mathematical concepts. The method does not consist in giving explicit definitions but consists instead "in focusing more sharply on the concepts concerned by directing our attention in a certain way, namely, onto our own acts in the use of those concepts, onto our powers in carrying out our acts, etc." It is through such a methodological view that we might hope to facilitate the development of mathematics and to gain insights into the solvability of meaningful mathematical problems.

The current philosophical climate is certainly not favorable to this idea. Gödel wrote a number of other papers (e.g., *1951, *1953/1959-III and -V, 1947/1964, 1972a) in which, in effect, he criticized views of mathematics that were not favorable to it. His criticisms were based on his ideas about the incompleteness theorems, consistency proofs, and the solvability of mathematical problems, and he applied these ideas to the views of Carnap, Hilbert, and others. As far as I know, Gödel never wrote about Quine's philosophical views. In this paper, I want to extend some of Gödel's arguments to Quine's view of mathematics. Quine's view has been very influential and it stands as a major rival position to Gödel's. I shall argue that, like Carnap and other empiricists, Quine has no place for the kind of nonreductive meaning clarification that is needed to facilitate the development of mathematics and to gain insights into the solvability of meaningful mathematical problems. Quine's view, like the other views criticized by Gödel, fails to reconcile mathematics with empiricism. It fails to achieve the kind of balance between empiricist and rationalist views that Gödel argues for in the 1961 paper. These are the central claims I want to argue for in this paper. I do not attempt to give detailed support to all of the ideas that are part of Gödel's alternative view of mathematics (e.g., on rational intuition), although I do think it is possible to support some of them. Furthermore, I do not discuss the kinds of extrinsic or a posteriori grounds for developing mathematics that Gödel mentions in some of his papers. Gödel does not mention these in the 1961 paper. My focus will be on the intrinsic, meaning-theoretic grounds that are part of Gödel's view of conceptual intuition.

I agree with Parsons that it is only in the 1961 paper that Gödel begins to enter the circle of ideas in which Quine's discussion of meaning moves. Accordingly, I will start with a few observations on the 1961 paper and on the papers from the fifties (*1951, *1953/1959-III and -V) that are part of its immediate background. These observations are intended to set the stage for the argument that follows.

I. The Call for Meaning Clarification (1961)

In the papers from the fifties, we see that Gödel thinks that mathematical expressions have their own content and that they refer to objects and facts in a way that is analogous to perceptual reference. Gödel thinks that there is a distinction between empirical science and mathematics but he holds that there is an analogy between our experiences in these two domains. Carnap makes a sharp distinction between empirical science and mathematics and holds that the two are not analogous. Quine, of course, thinks it is not possible to make such a sharp distinction between the two. Empirical science and mathematics are continuous.

According to Gödel, we learn about mathematical concepts, objects, and facts by a kind of rational perception or intuition that is analogous to sensory perception. Gödel thinks the analogy holds on several counts: in each case, our intuitions are forced or constrained in certain respects; there is a kind of inexhaustibility in each case; and we can be under illusions in each case. The notion of rational perception is construed so broadly that it may include the perception of concepts themselves (e.g., the general concept of set, or the concept of the natural numbers) and, in some cases, the objects (e.g., sets, natural numbers) or facts to which mathematical sentences may refer. Evidently, there can be an awareness of the concepts even if we do not or cannot intuit individual objects falling under the concepts (see, e.g., Gödel 1947/1964, p. 258). We can, as it were, grasp the meanings of mathematical terms that express systems of concepts independently of knowing whether there are individual objects falling under the concepts.

Gödel argues, against Carnap, Hilbert, and others, that the second incompleteness theorem suggests that we must reflect on the meanings or "thought contents" of mathematical expressions. We must try to cultivate and deepen our knowledge of the abstract concepts that underlie formal or mechanical systems of mathematics. It is through such a methodological view that we might hope to facilitate the development of mathematics and to gain insights into the solvability of meaningful mathematical problems. The incompleteness theorems show that we cannot adequately capture mathematical concepts in formal systems. Instead, we refer through a formal system to what is not presently enclosed in that system, which only shows that our mathematical intentions or concepts were not adequately captured in the first place. As was noted earlier, Gödel states that this "clarification of meaning consists in concentrating more intensely on the concepts in question by directing our attention in a certain way, namely, onto our own acts in the use of those concepts, onto our own powers in carrying out those acts, etc." The phenomenological view to which Gödel is appealing holds that it is by virtue of the contents or "meanings" of our conscious acts that we are referred to objects and facts, whether in mathematics or in other domains of experience (see Tieszen 1992, 1996). Meaning or content in mathematics determines extension (if there is one). Generally speaking, the extension of a mathematical expression underdetermines its meaning or content. The same object or fact can be given from a number of different perspectives, or under a number of different contents. I will also say that objects can be given under different "concepts," and I will henceforth think of concepts as that part of the intentional content expressed by predicates. The viewpoint thus recognizes both intensional and extensional aspects of mathematics and logic.

It is usually possible to identify at least some of the concepts that are at work in a given domain of experience. This much is not usually an epistemological mystery, even if we do not understand everything about the concepts we are using. Simply consider the contents of the 'that'-clauses we use to express our experience in the domain. Although an actual or imagined object can be given under different concepts, it cannot be given under just any concept. Concepts under which an object might be given can, for example, contradict or be consistent with one another, where the only way to determine this is to reflect on the meaning of the terms that express the concepts. Concepts can be of different categories or types, thus making category mistakes possible. There are bounds on what can fall under a concept, although there may be a wide range of variation. This is all part of the phenomenological view of intentionality and, according to this view, the logic of concepts is not to be identified with purely formal logic. If reference in mathematics is a function of meaning or content and reference can be indeterminate in various respects, then there is a need for meaning clarification. We do not have a grasp of the precise boundaries of all of the basic concepts we use in mathematical thinking. Thus, we should try to determine the boundaries of our basic mathematical concepts, what is compatible with them and what is not, and what will fall under a given concept and what will not. Note that a purely *formal* theory of concepts could not be sufficient for this epistemological task. One just has to plunge in and acquaint oneself with the concepts in a given domain. Thus, I think that it is not an objection that Gödel's view only makes sense if we have or can devise an "adequate" formal theory of concepts. This is not, however, to say that we do not need a better account of concepts.

According to the thesis of intentionality, consciousness is always consciousness of something or other. The fact that there is directedness implies that there is categorization in our experience. Mathematical reason, by virtue of exhibiting intentionality, is responsible at any given stage for the nonarbitrary categorization and identification of mathematical objects and facts. There can be different meaning categories or categories of concepts and corresponding regional ontologies within mathematics itself. To see this categorization, we only need to look to the existing science of mathematics, and to the different domains of mathematical practice. It follows, for example, that there need not be only one true concept of set. There could be different concepts of set (e.g., predicative, intuitionistic, maximal iterative), and then we simply develop our intuitions with respect to these different concepts. Mathematical propositions are true or false relative to different concepts. Indeed, there are different sets of axioms for different concepts.

The content of mathematical acts is therefore a condition for the possibility of the science of mathematics. This content is seen in the different areas of mathematics. It is also seen in differences in the intended meanings of formal mathematical theories. Sometimes we attempt to clarify or make precise the intended meanings of mathematical theories. When we do not discern the intended meaning of a formal system, we often attempt, quite automatically, to supply one.

I also note that a phenomenological view allows for more immediate and more theoretical aspects of experience in both mathematical and physical experience. Propositions that are more or less obvious can be found in both cases. Empirical theories are built up through reflection, generalization, and abstraction, as are the more theoretical parts of mathematics. In mathematics, however, there is even greater generalization. Husserl (1913) also argues that mathematics depends upon a type of "formal generalization" that should not be confused with empirical generalization. I return to this point later.

II. Meaning Clarification and Reductionism

One of the most important consequences of these ideas is that we should be critical of programs that are reductionistic about mathematics. I think this is a point that Gödel makes in many of his philosophical papers (see especially *1951, *1953/1959-III and -V, *1961/?). The point is put succinctly in a remark of Gödel's recorded by Hao Wang (1996, p. 167):

> Some reductionism is right: reduce to concepts and truths, but not to sense perceptions. . . . Platonic ideas [what Husserl calls "essences" and Gödel calls "concepts"] are what things are to be reduced to. Phenomenology makes them clear.

Meaning clarification is supposed to amount to the (more or less systematic and conscious) analytic unfolding of the content of mathematical concepts. But the methods of "meaning clarification" offered by Carnap's program, Hilbert's formalism, nominalism, and psychologism misconstrue mathematical content. The point applies to a wide range of "isms": empiricism, naturalism, fictionalism, instrumentalism, pragmatism, and even mechanism and logicism (see Tieszen 1994, 1996). The methods offered by these philosophical views are simply not adequate to the task of facilitating the development of mathematics and helping us gain insights into the solution of meaningful mathematical problems. All of these viewpoints try in one way or another to eliminate or modify the kind of act/content/object structure involved in the intentionality of mathematical experience. In particular, they try to substitute other kinds of content for the given mathematical content. In some of his early work, Carnap tried to substitute syntax for mathematical content. But I think Gödel (*1953/1959) has given us good reason to believe that this effort fails. Hilbert's original program, which also fails, contains a variation on this theme: substitute finitary, concrete content, which may be understood in terms of syntax, for abstract, infinitary mathematical content. The lines are drawn distinctly in Hilbert's method. We are to start with the part of mathematics that is finitary, concrete, real, contentual, meaningful, and available to immediate intuition, and this can be set off from what is infinitary, abstract, ideal, merely formal, "meaningless," and a product of thought without intuition. The incompleteness theorems show that the proposed substitution is unworkable.

Generally speaking, these programs insist on a substitute for the given mathematical content, where the substitute involves notions of possibility, necessity, and generality that are more restricted than or different in type from the given or intended content. It is as if all mathematical content *must* somehow be conservative over the favored substitute and if it turns out that it cannot be so construed, then it is to be neutralized or downplayed. The incompleteness theorems tell us, however, that in mathematics we are sometimes faced with genuine extensions of content. Even intuitionism substitutes a different (albeit mathematical) content for the sentences of higher set theory. We can see this, for example, in the fact that the continuum hypothesis splits into many different statements in intuitionism (which may be quite interesting in their own right).

Gödel's point is that we cannot smuggle in such substitutions unless we are willing to solve mathematical problems in terms different from those in which the problems are put (see especially the comments in the early part of 1947/1964). Philosophers of these persuasions do not, in effect, make the phenomenological reduction when they come to concepts in sciences such as mathematics. They do not set aside their ideological prejudices (see Husserl 1911, 1913). This surely explains Gödel's remark (Wang 1987, p. 193) that we might be able to see concepts more clearly if we practiced the phenomenological epoché. This is an important part of what it means to practice the epoché. It is not the silly, quasi-mystical undertaking that some commentators have made it out to be. I will consider in the next section how Quine's philosophy distorts mathematical content.

III. The Contrast with Quine

It is clear from the 1961 paper that Gödel means to reject a wide range of viewpoints about mathematics. This includes the views of Carnap and Hilbert. By Gödel's sights, Quine's view of mathematics would also not strike the appropriate balance between empiricist and rationalist views. Quine is an empiricist about mathematics but he wants an empiricism without the two dogmas of logical positivism. If we use the classification of worldviews given in the 1961 paper, then Quine's empiricism can be seen in his skepticism about concepts, meaning, intensions, intentionality, and even the parts of mathematics that are not applied. Indeed, Quine is skeptical of a notion of reason that would depend on these ideas. In his critique of the analytic–synthetic distinction, we see that Quine "reconciles" the a priori nature of mathematics with empiricism by attempting to undercut this division with a pragmatic holism.

Quine's holism, as is well known, is quite broad in scope. It encompasses the fields of empirical science, mathematics, and logic. Of course it could be even broader were it to encompass additional fields such as ethics. Quine does not say much about ethics but there are holists who also would want to include ethics in the mix (see Føllesdal's [1988] discussion of these matters). Others have gone even further. Richard Rorty for example, criticizes Quine for marking off the whole of science from the whole of culture. Rorty's determination to avoid such a division issues in a rather forlorn characterization of science as solidarity. In the other direction, one might be a holist within the fields of empirical science, mathematics, and logic separately, without attempting to combine them. Thus, it seems that in principle one could be a more localized holist or one could try to extend the view to an unbounded holism.

Quine's characterization of his holism toward the end of "Two Dogmas of Empiricism" is still worth quoting:

> Total science is like a field of force whose boundary conditions are experience. A conflict with experience at the periphery occasions readjustments in the interior of the field. Truth values have to be redistributed over some of our statements.... Having reevaluated one

statement we must reevaluate some others, which may be statements logically connected with the first or may be statements of logical connections themselves. . . . No particular experiences are linked with any particular statements in the interior of the field, except indirectly through considerations of equilibrium affecting the field as a whole. (Quine 1951, pp. 42–3)

Quine says that if his extended form of holism is correct then it is folly to to seek a boundary between synthetic statements, which hold contingently on sense experience, and analytic statements, which hold come what may. Any statement can be held true come what may if we make drastic enough adjustments elsewhere in the system. Conversely, no statement is immune to revision, including statements of mathematics and logic.

The view that there is no interesting philosophical distinction between mathematical and empirical truths reappears in many of Quine's writings (see, e.g., Quine 1960, 1966, 1970, 1974, 1992). There are only gradations of abstraction and remove from the particularities of sense experience in these truths, but no sharp boundaries and no qualitative differences or differences in type. Granted, mathematical content cannot be understood along the lines of earlier, cruder forms of empiricism (e.g., Mill's). It cannot, for example, be understood in terms of empirical induction. Instead, mathematical content will have to be more like the content of the theoretical hypotheses of natural science. It will be more centrally located in the holistic web of belief, less likely to be revised in the face of recalcitrant experience but, in principle, revisable. In this scheme, there is no clear demarcation point of the analytic, the a priori, or the "necessary." Notions of mathematical necessity, possibility, and generality are assimilated to natural necessity, possibility, and generality. Similarly, there is no clear demarcation point of "certainty," and the alleged certainty of mathematics will have to be understood accordingly. (I leave aside discussion of the certainty of mathematics in this essay since I think the issue is complicated by a number of factors.) In this context it will also be difficult to understand the other rationalist element of mathematics that Gödel mentions in 1961, which is the idea that a kind of "meaning clarification" based on conceptual intuition might play an important role in facilitating the development of mathematics and in helping us to solve open mathematical problems. In Quine's work, there is certainly no notion of meaning clarification based on conceptual intuition. The very idea of such a type of meaning clarification would be met with skepticism.

In contrasting Quine's pragmatic holism with Gödel's view, I will argue that it is the breadth of Quine's holism that leads to problems. A holism that attempts to encompass the fields of empirical science, mathematics, and logic does not do justice to mathematical meaning or mathematical content. Quine's deflationary remarks on analyticity depend upon this overarching holism. On a more localized holism within mathematics, or within mathematics and logic, the deflation may not be possible. One would not be forced to hold that mathematical or logical statements are revisable on the basis of what happens in empirical science (e.g., to simplify empirical theory), but one could hold that they may be revisable on the basis of what happens inside mathematics itself. That is, they may be revisable on the basis of uniquely mathematical evidence.

IV. Analyticity

It will be useful to establish first a few simple reference points about the notion of analyticity. We know that Gödel distinguishes a narrow from a wide notion of analyticity and rejects the claim that mathematics is analytic in the narrow sense. Gödel (1944, p. 139) says that "analyticity" may have the purely formal sense that terms can be defined (either explicitly or by rules for eliminating them from sentences in which they are contained) in such a way that axioms and theorems become special cases of the law of identity and disprovable propositions become negations of this law. In a second sense, he says, a proposition is called "analytic" if it holds owing to the meaning of the terms occurring in it, where meaning is perhaps undefinable (i.e., irreducible to anything more fundamental). Gödel makes a similar distinction in a later paper (Gödel *1951, p. 321) except that in this context he is thinking more specifically of the notion of analyticity in positivism and conventionalism. In this later paper, he says that to hold that mathematics is analytic in a broad sense does not mean that mathematical propositions are "true owing to our definitions." Rather, it means that they are "true owing to the nature of the concepts occurring therein." This notion of analyticity is so far from meaning "void of content" that it is possible that an analytic proposition might be undecidable. Gödel claims that we know about propositions that are analytic in the broad sense through rational intuition of concepts. Analyticity is thus linked to a kind of concept description and analysis.

Quine's arguments in "Two Dogmas of Empiricism" and related works are focused on a much narrower notion of analyticity than Gödel's preferred notion. Quine looks to the views of Carnap and the logical positivists, and to the tradition that stems from accepting Hume's distinction between matters of fact and relations of ideas. Gödel could perhaps even agree with these arguments. Indeed, we should keep in mind Gödel's comments on what the incompleteness theorems show about the analyticity of mathematics, viz., mathematics could not be analytic in the narrow senses that he indicates. But if mathematics is analytic in the wide sense, then relations between mathematical concepts must be of a rather substantial nature. Determining relations between concepts, as we described it earlier, must be different from determining purely formal relations, relations of synonymy, explicit definition, convention, and "semantical rules" of the type that Quine considers in "Two Dogmas." At the same time, it must not be based on sense experience and it is not simply a function of rounding out our theories of sensory objects.

It is very important to keep in mind Gödel's examples of propositions that are analytic in the wide sense. In particular, I think we must start with examples in mathematics (and possibly logic) before we begin to worry about whether there are wide analytic propositions containing terms drawn from other domains of our experience. Gödel says in various places that there exist unexplored series of axioms that are analytic in the sense that they only explicate the content of the concepts they contain. An example from foundations is provided by the same phenomenon Gödel uses to refute Carnap's idea that mathematics is syntax and is void of content: the incompleteness theorems. On the basis of the incompleteness theorems, an unlimited series of new arithmetic "axioms," in the form of Gödel sentences, could be added to the present axioms. These axioms become evident again and again and do not follow by formal logic alone from the previous axioms. Here we might say that, by way of a series of independent rational perceptions, we are only explicating the content of the concept of the natural numbers. The Gödel sentences we obtain are compatible with this concept and do not overstep its bounds. In the procedure for forming Gödel sentences, we do not diagonalize out of this concept, even though we do step outside of the given formal system. Moreover, new propositions or axioms may help us solve problems that are presently unsolvable or undecidable. One can already look at the undecidable Gödel sentence G for a formal system F in this way. We see that F will prove neither G nor $\neg G$. But metamathematical reasoning shows us that G is true if F is consistent, and we can thus "decide" G on these grounds. By adding this sentence to F, we can create a new formal system that will solve (albeit trivially in this case) a problem that was previously unsolvable. This idea is related to Gödel's comments on finding a viewpoint that is conducive to solving meaningful mathematical problems through a clarification of concepts in which we ascend to higher forms of awareness.

In the case of the construction provided by the incompleteness proof, it is not clear that we reach a "higher" state of consciousness in a significant way, for these "axioms" are not mathematically interesting. However, the incompleteness results for arithmetic do open up the possibility of finding interesting results like those of Paris and Harrington (1977). The Paris–Harrington theorem is a genuinely mathematical statement (a strengthening of the Finite Ramsey theorem) and is undecidable in Peano arithmetic (PA). The Finite Ramsey theorem itself is provable in PA. The Paris–Harrington theorem refers only to natural numbers but its proof requires the use of infinite sets of natural numbers. This is a good example of Gödel's idea of having to ascend to stronger, more abstract (in this case, set-theoretic) principles to solve lower-level (number-theoretic) problems.

Gödel has the same model in mind for other parts of mathematics. He notes how we can solve problems that were previously unsolvable, by ascending to higher types, and how we can also obtain various kinds of "speedup" results in this way (see especially the trenchant conclusion that Boolos [1987] reaches). An example that Gödel presents in many papers is based on axioms of infinity in set theory. These axioms assert the existence of sets of greater and greater cardinality, or of higher and higher transfinite types, and Gödel (e.g., in 1972a, p. 306) says they only explicate the content or meaning of a general iterative concept of set. In other words, these axioms do not overstep the bounds of this concept, but are in fact compatible with it. So here we have axioms that are different from one another, but that appear to explicate the same concept(s). We have a kind of unity through these different axioms. Such a series may involve a very great and perhaps even an infinite number of actually realizable independent rational perceptions. Gödel says this can be seen in the fact that the axioms concerned are not evident from the beginning, but only become so as the mathematics develops. To understand the first transfinite axiom of infinity, for example, one must first have developed set theory to a considerable extent. Once again, we are able to solve previously unsolvable problems with these new axioms. Gödel even thought we might find an axiom that would allow us to decide the continuum hypothesis (CH).

One might think of CH as an example in the following way: CH is known to be independent of first-order ZF (ZF^1). One can argue, however, that ZF^1 is not really adequate to the intended interpretation. ZF^1 does not come close to the goal of describing the cumulative hierarchy with its membership structure because ZF^1 has many nonstandard models. For the intended interpretation, we do better to look to second-order ZF (ZF^2) (compare, e.g., Kreisel 1965, 1971). (The possibility of viewing set theory in this manner may be closed to Quine, given his strictures about higher-order logic.) Although ZF^2 is not itself categorical, it is known that its models are isomorphic to an inacessible rank. Now, on the basis of the first incompleteness theorem, one might suppose that the concept of set-theoretic truth is richer than the concept of set-theoretic provability. There will be truths of ZF^2, for example, that are not provable in ZF^2. In particular, CH should have a truth value in the universe of ZF^2, even if we do not presently know what it is. The truth or falsity of CH depends on the breadth of the set-theoretic hierarchy and not on its height. The relation of $\aleph_1$ and $2^{\aleph_0}$ is determined by the internal structure of the stages of the hierarchy. It can therefore be argued that the truth value of CH is fixed by the contents of an initial segment of the hierarchy. By stage ω, sets of cardinality $\aleph_0$ appear. By stage $\omega + 1$, sets of cardinality $2^{\aleph_0}$ appear. And by stage $\omega + 3$, the pairing functions necessary for the truth of CH will have appeared. In other words, there is some reason to believe that CH should have a truth value under the *intended interpretation* of the axioms of this theory. The intended interpretation is to be understood in terms of the comments made earlier about intentionality. That is, it is to be understood in terms of the idea that we are directed toward a domain or universe by virtue of the meanings or contents of our acts, and we can then further explore this domain. In this directedness, we have an example of what Gödel calls "rational intuition." As we said earlier, rational intuition need not always be fully determinate. That is precisely why meaning clarification is needed. We need not have and usually do not have a fully determinate understanding of a domain from the outset. Moreover, there are a variety of ways in which the intended interpretation might be corrupted. We must be careful not to substitute some other content for the given or intended content, even though this is what reductionist views (such as empiricism and naturalism) would have us do.

V. Rational Intuition and Analyticity

Gödel gives examples in which there is a conceptual or meaning-theoretic link between some given axioms and new axioms that are logically or formally independent of the given axioms. The link cannot therefore be an analytic link in the sense of being formally analytic. Thus, it cannot be a link that excludes intuition in the way that formal logic is supposed to exclude intuition. Something outside of the given logical formalism must be involved. But we also do not learn about the link on the basis of sense experience. This is why Gödel says we learn about it through "rational intuition." There must be a (partial) intuition of a concept (intention) whereby the earlier axioms are related to the new axioms. There must be a grasp of the common concept or concepts (intentions) under which the axioms are unified. We can then begin to explore additional concepts that consistently extend a given concept.

The notion of intuition of concepts is not mysterious if one is prepared to recognize the fact that mathematical awareness exhibits intentionality. As we said earlier, the fact that there is directedness implies that there is categorization in our experience. For an act to be directed in a particular way means that it is not directed in other ways. Our beliefs are always about certain *types* or *categories* of objects and it is just these categories that we are referring to as "concepts." It is safe, for example, to say that we *know* that certain things are not instances of the concept "natural number" and that other things are instances of this concept. Thus, we must have some grasp of this concept even if our grasp is not fully precise and complete. Instead of saying that we have a partial "grasp" of a concept like this, we might as well say that we "intuit" the concept. The term intuition is used because the concept is immediately given as a datum in our mathematical experience once we adopt a reflective attitude toward this experience. It is given prior to further analysis of the concept and to the consideration of its relation to other concepts. The objections raised to Quine's views in this paper seem to me to point toward such a notion of rational intuition. If it is possible to show that there are serious flaws in Quine's view of mathematics and if the exposure of these flaws seems to presuppose the notion of rational intuition, then we have all the more reason to take the notion of rational intuition seriously.

The view of analyticity just sketched is far from the tight little circle of preserving or obtaining truths by synonym substitution or by the semantical rules that Quine considers in "Two Dogmas." The relation of one axiom to another cannot be one of synonymy, or of "truth by semantical rules." Indeed, I shall suggest later that wide analytic truths share some (but not all) features with the theoretical hypotheses of natural science to which Quine generally wishes to assimilate mathematics.

Recognizing wide analytic truths is, in a sense, just a way of making room for notions of meaning and the a priori that are needed to account for mathematical developments that have taken place since the late nineteenth century. It is a way of making room for the meaning of modern mathematics. When Quine says there is no difference in principle between empirical science and mathematics, we must keep in mind that this claim derives what plausibility it possesses from a purely extensionalist or truth-functional view of the sameness and difference of the meaning of predicates and sentences. Quine's "meaning holism" is a thoroughly extensionalist viewpoint. On such a viewpoint it is possible to say that sets of sentences cannot be self-contained or autonomous in terms of their meaning or content because the concepts associated with predicates are ignored and only extensions of predicates or truth values of sentences are considered. If we consider the concepts, then we will find various compatibilities and incompatibilities, concepts with their own horizons and boundaries, categories of concepts, and so on.

VI. Mathematical Content and Theoretical Hypotheses of Natural Science

Gödel's notion of wide analyticity yields something more like what Quine obtains by assimilating mathematical content to theoretical hypotheses about nature. (Quine has acknowledged this in response to Parsons' "Quine and Gödel on Analyticity." See Leonardi and Santambrogio [1995b, p. 352]) Mathematical content is like the content of theoretical hypotheses in that it is more general or universal. The content of theoretical hypotheses is farther removed from the particularities of sense experience. Neither mathematics nor natural science is content-neutral. Also, Gödel would want to say that neither type of content is arbitrary. We are forced or constrained in some respects. Mathematical sentences, like theoretical hypotheses, are not true by definition, linguistic convention, or synonymy. In both cases, we go far beyond a narrow notion of the analyticity of mathematics. Both have real content.

There are, however, some important disanalogies. Charles Parsons has pointed out some of the most important differences. Parsons (1983b, p. 195) states that theoretical and experimental physics are about the same subject matter, that experiments are carried out to verify or falsify the theories of theoretical physics, and so on. There is no similar unity of subject matter or of purpose between mathematics and physics. Mathematical truth does not depend on the tribunal of sense experience, but theoretical hypotheses do, even if only indirectly. Propositions of mathematics are not falsified when physical theories are abandoned or modified. If the mathematics changes in applications, it changes in the sense that a particular structure from the mathematician's inventory is replaced by another. There do not seem to be competing theoretical hypotheses in core parts of mathematics that are replaced with the passage of time.

Parsons' remark that theoretical and experimental physics are *about* the same subject matter should, in my view, be read in terms of the concept of intentionality. When he says that there is no similar unity of subject matter or purpose between mathematics and physics, this just signifies that the mind is directed in different ways in these fields and toward different goals. Different sets of properties and relations are applicable to the objects or concepts in each case.

Parsons notes that, according to Quine, there is no higher necessity than physical or natural necessity. Set theory is supposed to be on par with physics in this respect. The problem is that there is tension in Quine's own view of this. It conflicts with his view of mathematical existence. Not only does Quine treat mathematical existence and truth as independent of the possibilities of construction and verification, but he also treats them as independent of the possibilities of representation in the concrete (Parsons 1983b, p. 186). Quine is a platonist about set theory. The notion of object in set theory, however, and the structures whose possibility it postulates, are much more general than the notion of physical object and spatiotemporally or physically representable structure. But then how can Quine maintain that these possibilities are "natural" and that the necessity of logic and mathematics is not "higher"?

Although it is sensible to hold that neither mathematical propositions nor the theoretical hypotheses of natural science are topic-neutral, mathematics is nonetheless closely connected with logic in that its potential field of application is just as wide (Parsons 1979–80, p. 152). Quine does mention the applicability of mathematics to other sciences. Parsons argues that this indicates that mathematics has greater generality, and he notes that we might think of the generality of mathematics and logic as different in kind from that of laws in other domains of knowledge. He (Parsons 1983a, p. 18) cites with approval Husserl's idea that there is a difference in kind between the "formal generalization" involved in mathematics and logic and the generality of the laws of particular regions of being, such as the physical world.

This is related to another difference noted by Parsons: elementary mathematical truths do not seem to be more rarefied and theoretical than the theoretical hypotheses of natural science. On the contrary, they seem quite obvious. Quine's view cannot explain the obviousness of elementary mathematics and parts of logic (Parsons 1979–80, p. 151). In fact, there seem to be very general principles that are universally regarded as obvious, whereas in an empiricist view, one would expect them to be bold hypotheses about which a prudent scientist would maintain reserve. For an empiricist such as Quine, obviousness should not typically accompany general hypotheses about nature. Think of how these hypotheses have to be built up over time, and of how they are refined and adjusted in various ways in the process. The generalizations of physical theory are not typically elementary and obvious, but formal generalization may be elementary and obvious. Formal generalization need not be involved only in highly theoretical assertions.

Finally, Parsons, Dummett, and many others have suggested that differences such as those between classical and intuitionistic mathematics are naturally explained as differences about *meaning* or content. The truth of mathematical propositions in these different areas is based, broadly speaking, on different notions of meaning.

VII. Application of Ideas on Incompleteness, Consistency, and Solvability

All of this suggests that mathematical content differs in significant ways from the content of theoretical hypotheses about nature. Mathematical content has different dimensions in terms of the type or degree of generality, obviousness, possibility, and necessity involved. As was mentioned earlier, the second incompleteness theorem indicates that if we are interested in consistency proofs then, generally speaking, we cannot substitute content involving narrower notions of possibility and generality for content involving wider notions of possibility and generality. For example, there is a clear sense in which mathematical induction in PA involves a narrower notion of mathematical possibility than is involved in transfinite induction on ordinals $<\varepsilon_0$. There is a clear sense in which the introduction of primitive recursive functionals of finite type involves a notion of possibility or generality that goes beyond what we find in PA.

The point is that we cannot substitute less general for more general content, contents involving different types of generalization, or contents involving a "lower" necessity for those involving a "higher" necessity. Gödel would say that we cannot eliminate rational intuition. Rational intuition is needed to find consistency proofs, and it is generally involved in the cognition of consistency. Appeals to more general principles, and wider conceptions of possibility are needed to justify the consistency of principles that involve less general, narrower notions of possibility. Of course Quine could say that he is not interested in proof-theoretic consistency proofs anyway, since all we really need to worry about is what works in or is indispensable to natural science. But there are a number of difficulties with this response. Here I mention only the fact that consistency problems in the practice of mathematics and logic are perfectly legitimate and interesting. Quine also wants to help himself to the reduction of mathematics to set theory. Which set theory should we chose? Quine's NF or ML? Or ZF? Considerations about the likelihood of the consistency of these theories presumably should have some bearing on how we answer this question. It is not clear how appeals to natural science could help.

In any case, Gödel's work has shown us how these considerations about incompleteness and consistency are related to the matter of finding new axioms, solutions to mathematical problems, speedup results, and to the development of mathematics itself. On Quine's naturalistic holism, we always want to stay within the narrowest set of possibilities and generalities needed to round off the mathematics required by natural science. This conflicts with the need to go to wider notions of generality and mathematical possibility to obtain solutions to *mathematical* problems (or to obtain consistency proofs). If we want a solution to a mathematical problem in the form 'P or not P' which is not now solved, it is not a good idea to insist that we substitute narrower or different ideas (or even ideas at the same level) of generality and possibility for the mathematical content given in the sentences that form the background of the problem. (See the examples of the Paris–Harrington theorem and CH in Section IV, and of $V = L$ in Section IX.) What we need is a new principle that is independent of the existing principles but consistent with them, and that we see (by rational intuition) to be true, given that we are willing to take the other principles to be true. A kind of informal rigor is involved here. If, in particular, we substitute something that is supposed to be assimilable to *theoretical hypotheses about nature* for the given mathematical content of P, then we are not going to see that P is true (or, e.g., that a particular theory is consistent), and thus find a solution to 'P or not P'. We would simply not recognize the possibility of a sentence that could be true but that lies outside of these narrower generalities or possibilities, because the latter are officially the criteria of truth. It is conceptual intuition that accounts for the fact that we are not restricted by these bounds. This is similar to the problem of recognizing the possibility of a true but unprovable formula for a formal system, given that one is supposed to remain in the narrower sphere of immediate concrete intuition of the type that is supposed to accompany finitary mathematics (see Tieszen 1994). The difference is that in Quine's case it is natural science, not finitist or constructive mathematics, that is supposed to provide the most secure and reliable basis of knowledge. Natural science is epistemically privileged. The problem is that it is privileged to the extent that it can blind us to finding consistency proofs, new axioms, solutions to open problems, and to the general development of mathematics.

To avoid these objections, Quine could try to revert to his set-theoretic platonism. The problem with this maneuver, as we have seen, is that he is then in a bind with his own views on natural necessity. He does in fact make some effort to inactivate unapplied parts of set theory, as we will see later. This suggests that he wants to backtrack on his platonism, but without embracing constructivism. Another problem with the maneuver is that Quine's view of set theory does not embody a notion of content that would allow us to avoid the basic objection.

VIII. Mathematical Content and Quine's Conception of Set Theory

It is telling that, in practice, Quine actually assimilates set theory more to logic than he does to theoretical hypotheses about nature. In this guise, Quine thinks of sets as extensions of predicates. The axioms of set theory attribute extensions to certain predicates (see, e.g., Quine 1937, 1940, 1969, 1974, 1992). As Parsons has noted (Parsons 1983b, p. 198), this logic-like conception of sets is rather Fregean. This makes Quine's views "deviant" relative to most contemporary thought on the subject. We need to keep in mind how the ideas associated with the iterative or "mathematical" conception of set differ from Quine's views.

Now mathematical content could not be like the content of theoretical hypotheses about nature and at the same time be assimilable to logic or Quinean set theory. But mathematical content differs from what we find in logic, and it differs from what we find in set theory as construed by Quine. It differs from logic because logic is supposed to be content-neutral but mathematics is not. Mathematical content also cannot be assimilated to Quinean set theory. If we suppose otherwise, we are saddled with a very impoverished notion of mathematical content. Mathematicians do not actually do mathematics from within NF or ML (or even from within other forms of set theory). To hold otherwise is to ignore mathematical practice. Indeed, what is the intended interpretation of NF or ML? These are syntactically motivated theories. It is not clear that there is a concept behind them. Quine's view of set-theoretic content does not yield the proper directedness and regulation.

Quine has been very slow to warm to the idea of the cumulative hierarchy as an interpretation of the axioms of ZF. His view has not included the idea of advancing to higher sets/types or to new axioms of infinity as a natural and even intrinsic extension of set theory. Hence, the ideas about new axioms and problem solving that are described by Gödel are simply missing in Quine's work. Gödel's examples of the analytic unfolding of the content of the concept of set are always built around the general, "mathematical" concept of set. In Quine's conception of set theory, we would have nothing like the idea of solving problems by advancing to higher types or to new axioms of infinity. We simply would not be able to obtain this idea from Quine's conception, but, given the concept with which Gödel works, such ascension suggests itself. Why not look into it? It is in the horizon of the concept, so to speak, and it has been fruitful and led to many new developments and results that would be overlooked or downplayed if Quine's views were taken seriously.

These reflections on set theory are related to another problem for Quine's view: the problem of accounting for unapplied parts of mathematics.

IX. Mathematical Content and Unapplied Parts of Mathematics

As we noted, one of the problems with Quine's view is that the elementary, more obvious parts of mathematics must be assimilated to highly theoretical hypotheses about nature. On the other hand, Quine has difficulties accounting for the theoretical parts of mathematics that do not have applications in natural science. This is the fate of Quine's view. Pragmatic holism allows us to do justice neither to elementary, obvious parts of mathematics nor to advanced, theoretical parts of mathematics. That is, it does not allow us to do justice to mathematics. The view I favor runs orthogonal to this. On a more localized holism, we could have the distinction between general and specific, or between the theoretical and the applied in empirical science as well as in mathematics. Unapplied parts of mathematics do seem to have content or meaning and we have already indicated how this is possible. In Quine's view, this is a serious problem.

Many of Quine's most explicit remarks on the problem have come in response to Parsons' promptings. Here are two examples:

> So much of mathematics as is wanted for use in empirical science is for me on a par with the rest of science. Transfinite ramifications are on the same footing insofar as they come of a simplificatory rounding out, but anything further is on a par with uninterpreted systems. (Quine 1984, p. 788)

> I recognize indenumerable infinities only because they are forced on me by the simplest known systematizations of more welcome matters. Magnitudes in excess of such demands, e.g., Beth$_\omega$ [the cardinal number of $V_\omega(N)$ and of $V_{\omega+\omega}$] or inaccessible numbers, I look upon only as mathematical recreation and without ontological rights. (Quine 1986, p. 400)

In a recent book, Quine (1992) reexamines objections to his views on set theory. He says that sentences like CH and the axiom of choice, although not justified by their applications in natural science, can still be submitted to considerations of simplicity, economy, and naturalness that contribute to molding scientific theories generally. Such considerations, he says (1992, p. 95), support Gödel's axiom of constructibility, $V = L$. This axiom inactivates the more gratuitous flights of higher set theory. On the basis of the argument in Sections I, II, IV, V, and VII, however, we should have the strong sense that the point has been missed. Should these really be the reasons, or the only reasons, for *deciding* that $V = L$? It is certainly not clear that we should close off investigation for these reasons. To do so would be to ignore what is in the horizon of the general mathematical or iterative concept of set.

It seems to me that we are at least owed an account of the meaningfulness of higher set theory, even if there is reason to be skittish about its ontology or its radical platonism about set-theoretic objects and facts. On this matter, the problems with Quine's view are manifold. First, what constitutes a "simplificatory rounding out" of the type cited in the first passage quoted above? Perhaps we should just accept all of first-order ZF, but the other remarks cited run counter to the idea that this is the simplest known systematization of more welcome matters. In any case, if a theory such as ZF turns out to provide the simplest systematization, then Quine's stipulations about simplicity could conflict with his stipulations about natural necessity. Given the ontological commitments of ZF, the bind with a pragmatic holism that assimilates mathematics to natural science emerges once again. It is also questionable whether research in higher set theory should be discouraged on the grounds of what are at a given time the simplest *known* systematizations of more welcome matters. The simplest known systematization today may not be the simplest known systematization tomorrow.

It is also not clear how the idea of a simplificatory rounding off could be compatible with advancing to the more robust notions of mathematical possibility and necessity that are required in the face of the incompleteness theorems. Advancing to higher types or sets is a natural and intrinsic extension of Gödel's concept of set, but it is not clear that it answers to Quine's concerns for simplicity and economy with respect to natural science. In short, it is not clear what the appeal to simplicity and economy could amount to in the presence of the incompleteness phenomena and, more generally, of the *inexhaustibility of mathematics*. The kind of ascent needed to solve open mathematical problems takes us further and further from natural science and its problems and has nothing to do with rounding out the mathematics needed for the natural sciences. Empiricist accounts of mathematics tie mathematical concepts more closely to acts of sense perception in one way or another and thus place constraints on mathematical thinking instead of fostering its expansion. On the other hand, acts in which there is a free *imagination* of possibilities, when placed in the service of reason, lead to a far less constrained view of mathematical possibility.

One has to ask how to proceed with research about open mathematical problems. For Quine, the "simplificatory rounding out" is always a rounding out of what is needed for natural science. What cannot be handled in this way is treated as on a par with uninterpreted systems or mathematical recreation and, as we see, is regarded as gratuitous. But what is gratuitous or simplest with respect to the objective of solving *mathematical* problems? One could argue that with such an objective in mind it would be gratuitous and perhaps even contrary to canons of simplicity to decide in favor of $V = L$.

In a sense, the problem goes beyond Quine's view of higher set theory. The problem is that Quine has no place at all for the intended meanings of mathematical theories. Quine has on occasion said that a serious divergence over logic or set theory is just a confrontation of rival formal stipulations or postulates. He has said that we make deliberate choices and set them forth unaccompanied by any attempt at justification other than appeals to elegance or convenience. Although this may be true of Quine's attitude toward mathematics and set theory, it does not gibe with the usual attitude toward these subjects. It is not accurate as a report on mathematical practice.

It is worthwhile to consider the meaning and motivation of Quine's set theories ML and NF in light of these comments on rival formal stipulations. In point of fact, ZF has enjoyed far more popularity in mathematics than Quine's ML or NF. This is most likely because there is some sense of what the intended interpretation of ZF is, even if it is not perfectly clear. This yields more confidence in the consistency of ZF than in ML or NF. Given the choice between ZF, NF, and ML as the set theory to which mathematics is to be reduced, it is clear that ZF is preferable. Considerations of this type show that we do and must worry about the consistency of higher set theory. This concern cannot be overridden by or collapsed into a pragmatic or instrumentalist criterion that focuses only on what works in or is indispensable to natural science.

As Wang (1986, p. 162) puts it, Quine tries to combine an emphasis on formal precision with a gradualism that tends to blur distinctions and emphasize relativity or difference in degree. The drive for precision gives preference to reference over meaning, extensional objects over intensional objects, language over thoughts and concepts. But science generally possesses far less formal precision than classical predicate logic or Quine's ML or NF. In contrast, I am urging that much of science depends on informal rigor, that is, on rigorous but informal analyses of concepts. Mathematical content has the function of determining research and making it meaningful even if this research is not applied to or "justified" by what is needed in natural science. Mathematical content is not exclusively a function of finding the simplest systematizations of the mathematics required by the natural sciences. It is underdetermined (and possibly even corrupted) by such a process. The meaningfulness of the propositions involved in stating an open mathematical problem, for example, must be autonomous in this sense. There is evidence that is unique to mathematics because there is content or meaning that is unique to mathematics. It is very difficult to see how, in Quine's view, we should explain problem solving in those parts of mathematics that have no applications in natural science, for we are not provided with a way to understand the meaningfulness of the propositions involved in these problems.

I am arguing that these mathematical intentions or contents are indispensable to research in mathematics. They are conditions for the possibility of the science of mathematics. It is possible, although not recommendable, for theoretical mathematicians to turn their backs on natural science and to get along quite well. Natural science is dispensable to the work of theoretical mathematics, as it is to the work of a logician. However, mathematics is indispensable to natural science. Quine's indispensability arguments do not speak to the question of what is required for the practice of mathematics itself.

X. Mathematics and (Propositional) Attitude Adjustment

Note how these problems about mathematical content are related to the unstable place that the concept of intentionality occupies in Quine's philosophy. Quine generally wants to be a behaviorist about mental phenomena. In his flight from intentionality and intension he does not wish to recognize concepts or other intensions as abstract entities, nor does he wish to recognize mental entities. At the same time, he wants to somehow recognize intentionality and its irreducibility, but to downplay it (see, e.g., Quine 1960, pp. 219–21). However, he has recently recommended Dennett's work on these matters (Quine 1992, p. 73), and so he would perhaps accept a pragmatic or instrumentalist view of intentionality and of the need for intentional content. The situation is rather unclear.

It is a generally accepted fact about intentionality that the content or meaning of intentional states determines what those states are about. The notion of meaning or content is thus inserted into an account of our thinking in various domains. I argue that an appeal to concepts is needed to explain the kind of directedness, categorization, and regulation that is involved in research in mathematics, and that such an appeal is not available in a behaviorist account of mental phenomena. As Gödel suggests, what we really need to do is to focus "more sharply on the concepts concerned by directing our attention in a certain way, namely, onto our own acts in the use of those concepts, onto our powers in carrying out our acts, etc." Content or meaning determines what our research is about in various domains of mathematics and it can do so independently of applications in natural science. The content required for this function is simply not the kind of content that Quine's view prescribes. If Quine were to accept this claim, on Dennettian grounds, then he could do more justice to mathematics. He would then have to change his thinking about mathematics. Quine's present position on mathematics indicates that he does not want to take intentional content seriously and that he wants to be a behaviorist. Thus, there is tension between Quine's views on intentionality and his view on mathematics.

XI. Conclusion

If the arguments in this paper are correct, then Quine fails to reconcile mathematics with his pragmatic brand of empiricism. The attempt to extend holism across the fields of empirical science, mathematics, and logic does not do justice to mathematical content. Thus, he fails to arrive at a position with the appropriate balance of the empiricist and rationalist views that Gödel advocates in various papers. In surveying Quine's work, it appears that we must think of mathematical content in terms of either (i) the content of theoretical hypotheses of natural science; or (ii) extensions of predicates, as in Quine's view of set theory; or (iii) formal stipulation that is governed by considerations of elegance and convenience. Quine invariably substitutes one or another of these for mathematical content, depending on the issue at hand. It is clear that mathematical content could not be like the content of theoretical hypotheses about nature and at the same time be assimilable to logic or Quinean set theory. As I have indicated, mathematical content cannot be like the content of theoretical hypotheses in the relevant ways. It is also not exhausted by logic or Quinean set theory. Finally, mathematics cannot just be formal stipulation governed by considerations of elegance and convenience. Reflection on the incompleteness theorems and related ideas on consistency and solvability helps us to establish these facts.

In attempting to reconcile mathematics with empiricism, Quine's philosophy distorts mathematical content. The Gödelian views I have discussed in this paper provide a better account of the meaning of mathematics.

A version of this paper was presented in the Berkeley Logic Colloquium, April 1996, and in the Stanford Philosophy Colloquium, May 1996. I thank members of both audiences for comments, and especially Charles Chihara, Sol Feferman, Dagfinn Føllesdal, Thomas Hofweber, David Stump, and Ed Zalta. I also thank Michael Resnik for comments and for a spirited defense of some of Quine's views. I doubt that he will be fully satisfied with my responses. In preparing this paper, I have especially had in mind Charles Parsons' writings on Gödel and Quine, and some of his remarks on Husserl and Kant (Parsons 1979–80, 1983a, b, 1990, 1995a, b). Indeed, my paper can be read as a response to the comments that Charles makes at the end of "Quine and Gödel on Analyticity" (1995b).

REFERENCES

Boolos, G. (1987). "A Curious Inference," *Journal of Philosophical Logic*, 16: 1–12.

Føllesdal, D. (1988). "Husserl on Evidence and Justification," in R. Sokolowski (ed.), *Edmund Husserl and the Phenomenological Tradition* (Washington DC: Catholic University of America Press), 107–29.

Gödel, K. (1944). "Russell's Mathematical Logic," reprinted in S. Feferman *et al.* (eds.), *Kurt Gödel: Collected Works*, Vol. II (Oxford: Oxford University Press, 1990), pp. 119–41.

Gödel, K. (*1951). "Some Basic Theorems on the Foundations of Mathematics and Their Implications," reprinted in S. Feferman *et al.* (eds.), *Kurt Gödel: Collected Works*, Vol. III (Oxford: Oxford University Press, 1995), pp. 304–23.

Gödel, K. (*1953/1959, III and V). "Is Mathematics Syntax of Language?" in S. Feferman *et al.* (eds.), *Kurt Gödel: Collected Works*, Vol. III (Oxford: Oxford University Press, 1995), pp. 334–63.

Gödel, K. (*1961/?). "The Modern Development of the Foundations of Mathematics in the Light of Philosophy," in S. Feferman *et al.* (eds.), *Kurt Gödel: Collected Works*, Vol. III (Oxford: Oxford University Press, 1995), pp. 374–87.

Gödel, K. (1947/1964). "What is Cantor's Continuum Problem?" in S. Feferman *et al.* (eds.), *Kurt Gödel: Collected Works*, Vol. II (Oxford: Oxford University Press, 1990), pp. 176–87 and 254–70.

Gödel, K. (1972a). "Some Remarks on the Undecidability Results," in S. Feferman *et al.* (eds.), *Kurt Gödel: Collected Works*, Vol. II (Oxford: Oxford University Press, 1990), pp. 305–6.

Husserl, E. (1911). "Philosophie als strenge Wissenschaft," *Logos*, 1, 289–341; English translation by Q. Lauer in Husserl, *Phenomenology and the Crisis of Philosophy* (New York: Harper, 1965), pp. 71–147.

Husserl, E. (1913). *Ideen zu einer reinen Phänomenologie und phänomenologishen Philosophie*, Erstes Buch. (Halle: Niemeyer); English translation by F. Kersten as Husserl, *Ideas Pertaining to a Pure Phenomenology and to a Phenomenological Philosophy* (The Hague: Nijhoff, 1982).

Kreisel, G. (1965). "Informal Rigour and Completeness Proofs," in I. Lakatos (ed.), *Problems in the Philosophy of Mathematics* (Amsterdam: North–Holland), pp. 138–86.

Kreisel, G. (1971). "Observations on Popular Discussions of Foundations," in D. Scott (ed.), *Axiomatic Set Theory*, Proceedings of Symposia in Pure Mathematics, Vol. 13, Pt. 1, (Providence, RI: American Mathematical Society), pp. 189–198.

Leonardi, P., and Santambrogio, M. (eds.) (1995). *On Quine* (Cambridge, UK: Cambridge University Press).

Paris, J., and Harrington, L. (1977). "A Mathematical Incompleteness in Peano Arithmetic," in J. Barwise (ed.), *Handbook of Mathematical Logic* (Amsterdam: North–Holland), pp. 1133–42.

Parsons, C. (1979–80). "Mathematical Intuition," *Proceedings of the Aristotelian Society*, 80: 145–68.

Parsons, C. (1983a). *Mathematics in Philosophy* (Ithaca, NY: Cornell University Press).

Parsons, C. (1983b). "Quine on the Philosophy of Mathematics," in *Mathematics in Philosophy* (Ithaca, NY: Cornell University Press), pp. 176–205.

Parsons, C. (1990). "Introductory Note to 1944," reprinted in S. Feferman *et al.* (eds.), *Kurt Gödel: Collected Works*, Vol. II (Oxford: Oxford University Press, 1995), pp. 102–18.

Parsons, C. (1995a). "Platonism and Mathematical Intuition in Kurt Gödel's Thought," *Bulletin of Symbolic Logic*, 1: 44–74.

Parsons, C. (1995b). "Quine and Gödel on Analyticity," in P. Leonardi, and M. Santambrogio (eds.), *On Quine* (Cambridge, UK: Cambridge University Press), pp. 297–313.

Quine, W. V. (1937). "New Foundations for Mathematical Logic," reprinted with revisions in W. V. Quine *From a Logical Point of View* (Cambridge, MA: MIT Press, 1953), pp. 80–101.

Quine, W. V. (1940). *Mathematical Logic* (New York: Norton); Rev. 2nd Ed. (Cambridge, MA: Harvard University Press, 1951).

Quine, W. V. (1951). "Two Dogmas of Empiricism," reprinted in Quine *From a Logical Point of View* (Cambridge, MA: Harvard University Press, 1953), pp. 20–46.

Quine, W. V. (1960). *Word and Object* (Cambridge, MA: MIT Press).

Quine, W. V. (1966). "Necessary Truth," in *The Ways of Paradox and Other Essays* (New York: Random House), pp. 68–76.

Quine, W. V. (1969). *Set Theory and Its Logic*, 2nd Ed (Cambridge, MA: Harvard University Press).

Quine, W. V. (1970). *Philosophy of Logic* (Englewood Cliffs, NJ: Prentice–Hall).

Quine, W. V. (1974). *The Roots of Reference* (La Salle, II: Open Court).

Quine, W. V. (1984). "Review of Charles Parsons' *Mathematics in Philosophy*," *Journal of Philosophy*, 81: 783–94.

Quine, W. V. (1986). "Reply to Charles Parsons," in L. Hahn and P. A. Schilpp (eds.), *The Philosophy of W. V. Quine* (La Salle, II: Open Court), pp. 396–403.

Quine, W. V. (1992). *Pursuit of Truth*, 2nd Ed. (Cambridge, MA: Harvard University Press).

Quine, W. V. (1995). *From Stimulus to Science* (Cambridge, MA: Harvard University Press).

Tieszen, R. (1992). "Kurt Gödel and Phenomenology," *Philosophy of Science*, 59: 176–94.

Tieszen, R. (1994). "Mathematical Realism and Gödel's Incompleteness Theorems," *Philosophia Mathematica* (Series III), 2: 177–201.

Tieszen, R. (1996). "Science Within Reason: Is There a Crisis of the Modern Sciences?" in M. Otte, and M. Panza (eds.), *Analysis and Synthesis in Mathematics: History and Philosophy*, Boston Studies in the Philosophy of Science (Dordrecht, The Netherlands: Kluwer), pp. 243–59.

Tieszen, R. (1997). "Review of *Kurt Gödel: Unpublished Philosophical Essays*, edited by Francisco Rodriquez-Consuegra," *Annals of Science*, 54: 99–101.

Tieszen, R. (1998a). "Gödel's Philosophical Remarks on Logic and Mathematics: Critical Notice of *Kurt Gödel: Collected Works*, Vols. I, II, III," *Mind*, 107: 219–32.

Tieszen, R. (1998b). "Gödel's Path from the Incompleteness Theorems (1931) to Phenomenology (1961)," *Bulletin of Symbolic Logic*, 4: 181–203.

Wang, H. (1986). *Beyond Analytic Philosophy: Doing Justice to What We Know* (Cambridge, MA: MIT Press).

Wang, H. (1987). *Reflections on Kurt Gödel* (Cambridge, MA: MIT Press).

Wang, H. (1996). *A Logical Journey: From Gödel to Philosophy* (Cambridge, MA: MIT Press).

III. NUMBERS, SETS, AND CLASSES

Must We Believe in Set Theory?

GEORGE BOOLOS

According to set theory, by which I mean, as usual, Zermelo–Fraenkel set theory with the axioms of choice and foundation (ZFC), there is a cardinal λ that is equal to $\aleph_\lambda$. Call the least such cardinal κ. The cardinal κ is the limit of $\{\aleph_0, \aleph_{\aleph_0}, \aleph_{\aleph_{\aleph_0}}, \ldots\}$, that is, the least ordinal greater than all $f(i)$, where $f(0) = \aleph_0$ and $f(i+1) = \aleph_{f(i)}$ for all natural numbers i.

Is there such a cardinal? I assume that cardinals are ordinals and ordinals are von Neumann ordinals. Thus if κ exists, there are at least as many as κ sets. *Are* there so many sets?

Much very important and interesting work in set theory these days is concerned with cardinals far, far greater than κ, cardinals whose existence cannot be proved in set theory, and with the consequences of assuming that such large cardinals do exist, particularly those concerning objects at comparatively low levels of the set-theoretic universe. Since κ is the limit of an ω-sequence of cardinals smaller than κ, it is not (even) an inaccessible cardinal, the smallest common sort of cardinal whose existence cannot be proved in set theory, let alone measurable, huge, ineffable, or supercompact.

No, κ is quite small, indeed *teensy*, by the standards of those who study large cardinals.

But it's a *pretty big* number, by the lights of those with no previous exposure to set theory, so big, it seems to me, that it calls into question the truth of any theory, one of whose assertions is the claim that there are at least κ objects.

Zermelo–Fraenkel set theory is an *interpreted* first-order formal theory. Its language has one non-logical constant, the two-place predicate letter $\in$, its variables range over all the (pure) sets there are, and a formula $x \in y$ is true when sets a, b are assigned to the variables x, y if and only if a is a member of b. The logic of the theory is classical first-order predicate logic with identity, and the notion of a theorem of the theory is perfectly standard. Thus one of the theorems of ZFC is the formal sentence σ of the language of ZFC expressing the existence of κ; σ is understood to express the existence of κ because the language of ZFC is understood to be interpreted in the manner just described. σ is a consequence of ZFC; σ asserts that κ exists. If there are not as many as κ objects in existence, κ does not exist, σ is false, and set theory is not true.

Of course κ might exist and set theory be false for some other reason. But κ seems sufficiently large that the claim that it exists might plausibly be regarded as dubious. κ is no gnat; it is a lot to swallow.

Let me try to be as accurate, explicit, and forthright about my beliefs about the existence of κ as I can: It is not the case that I believe that κ exists and it is not the case that I firmly believe that κ does not exist. Without very many or very good reasons, and without strong views on the matter, I tend somewhat to think it probably doesn't exist, but I am really quite uncertain. I am also doubtful that anything could be provided that should be called a *reason* and that would settle the question.

I don't, I say, have what I regard as very good reasons for failing to assent to the existence of κ. But I guess I really don't believe in it, and so of course I think you shouldn't either. Imagine being confronted by a precocious trusting child T. who tells you that three days ago Teacher taught the class about infinity, day before yesterday taught the class that there were not only infinitely many things, numbers and so on, but also infinitely many infinite numbers, and so super-infinitely many things, yesterday taught the class that there were not only super-infinitely many things, but also super-infinitely many infinite numbers, and hence super-super infinitely many things, and today taught the class that you could iterate "super-" as many times as you like, and then asks, "Is it really true what Teacher said? Are there really infinitely and super-infinitely and super-super-infinitely and so on many things . . . ?" What would *you* say to T.? Not, I hope, "Certainly, of *course* there are, Teacher was absolutely right."

But I hope you would also say, "Teacher was completely right," on being told that Teacher had told the class that if $n > 2$, then $x^n + y^n \neq z^n$ (x, y, z, n positive integers), or on being told that Teacher had said that set theory *says that* there are as many things as that.

I said that my own reasons for thinking that κ doesn't exist weren't very good ones. Perhaps they amount only to the sense that there couldn't be *that* many things, that κ is, by ordinary lights, a (literally) unbelievably big number, and that any story according to which there are so many things around ought to be received rather skeptically.

Russell once quipped that there are fewer things in heaven and earth than are dreamt of in *our* philosophy. Dreams are rarely accurate. Why suppose this one correct?

Furthermore, to the best of my knowledge nothing in the rest of mathematics or science requires the existence of such high orders of infinity. The burden of proof should be, I think, on one who would adopt a theory so removed from experience and the requirements of the rest of science (including the rest of mathematics) as to claim that there are κ objects. κ is such an exorbitantly big number (by ordinary standards) that we would seem to need more reason than we now have to think a theory true that tells us that there are κ things in existence. And the apparent absence of reasons to believe in κ itself here seems like some sort of reason for believing in its non-existence.

But perhaps company can make up for the absence of reasons. So let me ask you what *you* think. Do you really think that there are as many sets as that? Really?

I am, of course, quite well aware that certain annoying questions can be put to one who is skeptical about the existence of κ. It is a theorem of ZFC that κ exists, and not a theorem that it is particularly hard to prove either. The proof of any theorem of ZFC appeals to only finitely many axioms of ZFC; and one who would question a theorem must be dubious about the conjunction of the axioms from which it logically follows. But perhaps it is not much more uncomfortable to refuse to accept the conjunction of those axioms of set theory (among them infinity, power set, union, and certain instances of replacement) needed to prove the existence of κ than to refuse to accept the existence of κ.

But what about cardinals less than κ? As it happens, I myself believe in the existence of $\aleph_0$, just as it is not the case that I believe in the existence of κ. Well then, one might ask me, what about $\aleph_1$? What about $\aleph_2$? What about $\aleph_\omega$? Is there an i such that you believe in the existence of $f(i)$, defined above, but fail to believe in the existence of $f(i+1)$? Or do you believe in the existence of all of them, but just not in κ? Or do you believe that there is a set of all the $f(i)$, but that that set has no union?

We all knew, however, that this sort of trouble could always be made. Like almost every other non-mathematical notion, *acceptance* or *belief* is vague and therefore the usual sorts of difficulties connected with the application of vague predicates to indiscriminable objects can be expected to confront one who maintains that we should not (or not yet) accept the existence of some large infinite cardinal but that belief in $\aleph_0$ is warranted. I chose κ only because it is easy to define but sufficiently large for there to be a serious doubt about its existence even for one who believes in the existence of $\aleph_0$.

Along with the more familiar transfinite cardinals $\aleph_\alpha$, the cardinals $\beth_\alpha$ are defined in set theory. Let us recall their definition: $\beth_0 = \aleph_0$, $\beth_{\alpha+1} = 2^{\beth_\alpha}$ (i.e., the cardinality of the set of all subsets of any set whose cardinality is $\beth_\alpha$) and $\beth_\lambda = \lim\{\beth_\beta : \beta < \lambda\}$ (i.e., the least cardinal $\geq$ all $\beth_\beta$, $\beta < \lambda$). Thus $\beth_1$ is the cardinal number of the continuum, and $\aleph_\alpha \leq \beth_\alpha$ for all α. (So the continuum hypothesis states that $\aleph_1 = \beth_1$ and the generalized continuum hypothesis that $\aleph_\alpha = \beth_\alpha$ for all α.) As with the $\aleph$s, there is, similarly, a cardinal λ equal to $\beth_\lambda$. Let us call the least one ρ. (Since $\beth_\rho \leq \rho \leq \aleph_\rho \leq \beth_\rho$, $\beth_\rho = \aleph_\rho$.) I want a number that is problematically big even for one who believes in the existence of the set of real numbers. But since it is consistent that the set of reals has cardinality $\aleph_{(\kappa+1)}$, that is, the next cardinal after κ, κ ought really to be redefined as ρ. But let us stick with κ.

How then should we respond to T. when he asks whether to believe Teacher? There is an obvious response to T.'s question "are there really so many things as that?" which seems to me to be strong evidence that *we* fail to believe that there are κ objects.

We say, "Well, according to set theory, there are."

It is a theorem of set theory that there are cardinals $\kappa = \aleph_\kappa$. According to set theory, there is such a cardinal as κ. Set theory proves, tells us, has it that there are such cardinals.

Are there dragons?

Legend has it that there are.

Legend usually has it wrong, but legend does have it that there are dragons. According to legend, there are dragons. Legend tells us that there are. If it could prove things, legend would prove that there are dragons.

But when we say that set theory proves that there is a least cardinal κ such that $\kappa = \aleph_\kappa$, we mean more than just that a certain proposition is entailed by the axioms or is part of the content or meaning of those axioms: we mean that there is a proof from the axioms of set theory of a certain sentence expressing the existence of κ, a finite sequence of formulas, the last of which expresses the existence of κ, and each of which is either an axiom or follows from earlier formulas by one of the standard logical rules of inference.

And whether or not κ exists, it *is* literally true that there is such a finite sequence of formulas.

Later on I shall have something to say about the response, "It is not literally true that there is such a finite sequence of formulas because formulas are sets or other abstract objects and abstract objects do not exist." Now, though, I want to compare the different responses to the questions "Does κ exist?" and "Does set theory prove that κ exists?"

It's fairly widely believed that one ought not to lie, one ought not to say what one believes to be false. (I take it that one lies who unwittingly speaks the truth if he asserts something contrary to what he believes; if not, substitute "knows" for "believes.") Now it is a somewhat nice question what it means to say that one ought not to lie, or that lying is impermissible, unacceptable, out (for of course there are circumstances in which it is obligatory and praiseworthy to lie), but I don't want to enter into a discussion of the meaning of "ought" or "impermissible."

I merely want to suggest that in whatever sense lying is out, so is saying what one does not know to be true. If one merely believes a thing, *p*, but does not know it, then one ought not to say *p*, it's out, not permissible, to do so. If you don't know the way to Waltham, then you may not answer "yes" to the question "Is this the way to Waltham?" any more than to the question whether you know the way there. If you only think that this is the way to Waltham, you may of course say, "I think this is the way" or "I believe so"; but of course then you are saying something that you know to be true, namely that you think, believe that this is the way. That you believe that it is is indeed something you know.

Thus there seems to me to be – I don't know that I'm right about this, but I think I am (but I know that I think so and that it does seem right to me) – some kind of ban, and roughly the same kind of ban as there is on lying, on saying what one does not know to be the case. You may not do it, it's out, impermissible, wrong.

Perhaps the ban is not quite as strong as that on lying, but a speaker who violates it may justifiably be accused of irresponsibility even if what he says turns out to be true.

So I take it that when responsible speakers like ourselves say that set theory proves that κ exists, we take ourselves to know that set theory proves that it does. I also take it that when we refrain from saying that κ exists, we are simply observing the ban on saying what we do not know to be the case. A point follows that ought to be stated plainly: there are theorems of mathematics to the effect that certain statements are theorems of set theory that are far more certain than those very theorems.

There is a response to the question of whether κ exists that one may expect to hear nowadays. It is that one who asks the question does not realize that he is asking an "external" question, where only an "internal" question is appropriate, or is trying to adopt an "external" standpoint from which to assess his own (or perhaps our) conceptual scheme, or is afflicted with "metaphysical realism." The thought is that since set theory implies that κ exists, to ask whether κ does exist is to attempt to call into question from some external vantage point the one and only theory we have that treats of such matters; to do so is fall into the metaphysical error of thinking we might somehow acquire information concerning the way sets are that is unmediated by any theory at all and then use this information to assess our own current theory of sets. Perhaps, the suggestion continues, we think we have some sort of direct insight into the nature of sets, possibly analogous to perception of physical objects, with the deliverances of which we can see our own best theory of sets to be at variance.

Rubbish, for a number of reasons.

Whatever their strength or source may be, the plausibility considerations about how many things there are that conflict with various theorems of ZFC have as much right to be considered a part of "our conceptual scheme" as does ZFC. ZFC conflicts with certain intuitions about cardinality that we happen to have; those intuitions form part of a fragmentary, inchoate, rival theory. If we think that there may well be fewer things in existence than ZFC tells us there are, we are no more assessing ZFC from some external point of view than we are assessing our intuitions about cardinality from some external vantage point when we say they are contradicted by ZFC.

Furthermore, whose theory is ZFC anyway? The difficulty we are confronted with is that ZFC makes a claim we find implausible. To say we cannot criticize ZFC since ZFC is *our* theory of sets is obviously to beg the question whether we ought to adopt it despite claims about cardinality that we might regard as exorbitant.

Finally, just exactly what is the matter with saying ZFC isn't correct because it tells us that there are κ objects and there aren't that many objects? (As I remarked above, this is not my view; I am agnostic on the question whether κ exists.) To be sure, one who says this may be asked how he knows there aren't. But the reply, "Get serious. Of course there aren't that many things in existence. I can't *prove* that there aren't, of course, any more than I can *prove* that there aren't any spirits shyly but eagerly waiting to make themselves apparent when the *Zeitgeist* is finally ready to acknowledge the possibility of their existence. But there aren't any such spirits and there aren't as many things around as κ. You know that perfectly well, and you also know that any theory that tells you otherwise is at best goofball." – that reply, although it does not *offer reasons* for thinking that there are fewer than κ objects in existence, would not seem to manifest any illusions that could be called metaphysical realist.

The question, in short, seems like a perfectly reasonable one to ask. The difficulty it presents is that it seems that there is very little to be said on either side of the matter, except that standard set theory says that the answer is yes, while common sense or whatever might be inclined to disagree. It is a frequent enough strategy in philosophy to try to dismiss or belittle a philosophical question which cannot be answered as indicating some philosophical confusion, but that there is any confusion at all in asking whether set theory is correct in asserting that κ exists seems to me just not to have been made out.

Part of the problem, of course, is that in the present case, we are considering whether to reject a *part* of a framework. Abandonment of the whole of set theory is not under consideration. It is not to be expected that philosophical theories about the nonsensicality of asking external questions where only internal ones are appropriate will help us in such a situation.

It may be of use here to examine a story often told in connection with set theory and commonly called the iterative conception of set theory.

According to the iterative conception,[1] every set is formed at some stage. There is a relation among stages, *earlier than*, which is transitive. For any two stages, there is stage later than both. There is a stage that is later than some stage, but not *immediately* later than any stage. A set is formed at a stage if and only if its members are all formed before that stage, i.e., formed at a stage earlier than that one. (Thus, on the present version of the iterative conception, each set is formed at every stage later than any one at which it is ever formed.)

Any sets all of which are formed before some one stage are the members of some set.

It is a surprising discovery of Dana Scott that the principles about sets, stages, formation, membership, and *earlier than* just stated, are sufficient to imply (in second-order logic) all the axioms of set theory given by Zermelo, as well as the axiom of foundation.[2] (It is most surprising that they imply that *earlier than* is well founded.) At least one further principle beyond those just recounted is needed to yield the axioms of replacement.

The iterative conception of set is sometimes supposed to "justify" Zermelo set theory (or, with the aid of additional principles about sets and stages, to justify ZFC). It is important to see exactly what sort of justification it does provide.

A less condensed version of the iterative conception will tell of a first stage at which the null set is formed, a second stage at which the null set is (re-) formed and the unit set of the null set is formed, a third stage, at which four sets, two of them new, are formed, . . . an omegath stage, at which, for any sets all of which are formed at finite stages, a set is formed whose members are just those sets, an omegaplusoneth stage, . . . an omegaplusomegath stage, an omegatimestwoth stage, . . . , an omegasquaredth stage, etc.

The account, when presented in full, is a *picturesque* account of the universe of sets. It has another literary virtue. There is a salient partial ordering of sets, *having lower rank than*. The account respects a natural narrative convention ("neatness") by mentioning things that come earlier in some salient order earlier than those that come later in that order. But notice (a point made to me by Dan Leary) that the talk of stages and formation is the most easily dispensable of tropes. One could present the iterative conception just as well by saying, "First there is the null set, then there is the unit set of the null set, then come two more sets, . . . , then there are twelve more, which are the ones not already mentioned whose members . . . . Then after all *those*, there are all the sets whose members are the ones just indicated but not mentioned, etc."

Now, as I remarked earlier, the axioms of replacement, which are needed to guarantee the existence of κ, cannot be derived *just* from the principles about sets and membership naturally inferable from the story just given, nor from the principles about sets, stages, membership, formation, and earlier than inferable from certain common versions of the iterative conception.

And (the far more important point): Even if the iterative conception is supplemented so that replacement follows, what reason have we to think that any such story is *correct*? Certainly, *if* the story from which those principles about sets and stages can be read off is true, then set theory is true, but why should we believe that the story is in fact true? Perhaps after a while, the story turns false and there aren't those sets.

The interest of the iterative conception is that it shows that the axioms of Zermelo(–Fraenkel) set theory are not just a collection of principles chosen for their apparent consistency and ability to deliver desired theorems concerning arithmetic, analysis, and Cantorian transfinite numbers, but not otherwise distinguished from other equally powerful consistent theories. The conception is *natural* in the simple sense that people can and do easily understand, and readily regard as at least plausible, the view of sets that it embodies. The naive conception (any zero or more things whatsoever form a set) is also natural in this sense, but it is of course inconsistent: the things that do not belong to themselves are some things that do not form a set. So the iterative conception is the only view of sets we have that is natural and, apparently, consistent.

In "The Iterative Conception of Set"[3], I expressed the view that neither the axiom schema of replacement nor even the existence of a stage indexed by the first nonrecursive ordinal seemed to me to be implied by the version of the iterative conception described there. Nor would the existence of κ be evident on that account of the conception. But I incline – for whatever accident of psychology – to find principles of set theory acceptable if they can be "read off" *that* presentation. Perhaps there are other theorems that seem evident that cannot be so read off. But I do not regard it as evident that there must be a stage corresponding to every ordinal that is the order-type of some well-ordering formed at some stage. My credulity has limits.

Russell's quip was that there are fewer things in heaven and earth that are dreamt of in *our* philosophy. But might it be that there are fewer things period than are dreamt of in our philosophy? To vary the allusion, if we are sailors rebuilding our ship plank by plank on the open sea, then I know of some cargo we might want to jettison.

I emphasized earlier that I believed in the existence of *proofs*, e.g., of the sentence stating the existence of κ from the axioms of ZFC, and of $\aleph_0$. It will therefore come as no surprise that I am rather a fan of abstract objects, and confident of their existence. It behooves me, I think, to say why smaller numbers, sets, and functions do not offend my sense of reality the way κ does.

Five pages into the first of the *Dialogues between Hylas and Philonous*, Berkeley has Hylas reply to Philonous that we do not immediately perceive by sight any thing beside light and colors, and figures. Of course Hylas is wrong: we immediately perceive by sight some things other than light, colors, and figures, for example, sticks and stones, and baseballs. But we immediately perceive other things too, for example letters and parts of the sky, as Philonous correctly asserts for once. And still other sorts of things too, such as *The Globe* (a Boston newspaper, whose slogan used to be "Have you seen *The Globe* today?"). It would be a rather demented philosopher who would think, "Strictly speaking, you can't *see The Globe*. You can't even see an issue of *The Globe*. All you can really see, really immediately perceive, is a copy of some issue of some morning's *Globe*." To say this, however, reflects a misunderstanding of our word "see": more than a misunderstanding, really, it's a kind of lunacy to think that sound scientific philosophy demands that we think that we see ink-tracks but not words, i.e., word-types.

An observation due to Helen Cartwright is helpful here. It should be called the Helen Cartwright Theorem Theorem. As Richard Cartwright tells the story,[4] he once said that propositions can't be written down. "[S]he correctly replied that she had seen Gödel's Theorem, for instance, written on the board. I replied that to write Gödel's Theorem on the board is just to write on the board a sentence that formulates the theorem. It took me an inordinately long time to see that if it really takes no more than that, then the theorem is easily enough written down." And for that matter, you *can* write a *number*, and not just a numeral on the blackboard. Please consider your view about the meaning of the word "on" before jumping to the conclusion that you can't write a number on the board.

Numbers do not twinkle. We do not engage in physical interactions with them, in which energy is transmitted, or whatever. But we twentieth-century city dwellers deal with abstract objects *all the time*. We note with horror our *bank balances*. We listen to *radio programs*: *All Things Considered* is an abstract object. We read or write *reviews* of *books* and are depressed by *newspaper articles*. Some of us write *pieces of software*. Some of us compose *poems* or *palindromes*. We correct *mistakes*. And we draw *triangles* in the sand or on the board. Moreover bank balances, reviews, palindromes, and triangles are "given" to us "in experience," whatever it may mean to say that. To put the matter somewhat more carefully, no sense of "sensible" or "experience" has been shown to exist under which it is not correct to say that we can have sensible experience of such objects, such things as the zither melody in *Tales from the Vienna Woods*, the front page of the sports section of this morning's *Globe*, a broad grin, or a proof in set theory of the existence of κ.

It is thus no surprise that we should be able to reason mathematically about many of the things we experience, for they are already "abstract." It is very much a philosopher's view that the only objects there are are physical or material objects, or regions of space-time, or whatever it is that philosophers tell us the latest version of physical theory proclaims to be the ultimate constituents of matter. To maintain that there aren't any numbers at all because numbers are abstract and not physical objects seems like a demented way to show respect for physics, which everyone of course admires. But it is nuts to think Wiles could have spared himself all those years of toil if only he had realized that since there are no numbers at all, there are no natural numbers $x, y, z, n > 2\ldots$

The existence of infinitely many natural numbers seems to me no more troubling than that of infinitely many computer programs or the existence of infinitely many sentences of English. Of course, there is no longest (Basic) program: any program is shorter than the one that results when a suitable sentence of the form: [n] PRINT "Hello, world" is subjoined after its last line. Nor is there a longest sentence; any number of "very"s can be inserted into "This is tiresome" as nearly every speaker of English knows.

Irrealism about numbers seems no more tenable than irrealism about programs or sentences. It is an odd view, to say the least, that there are infinitely many programs but no, or only finitely many, natural numbers. What the most effective rebuttal to the view might be could well depend on how the position was articulated. At any rate, I find the existence of natural numbers as unproblematic as that of physical theories or of irrealist tracts in the philosophy of science. (When may we expect *Computer Science Without Programs* to appear?)

In one of the funniest passages in all of contemporary philosophy, David Lewis asks us to imagine what the reaction would be if we were to walk into a mathematician's office and announce that philosophy has discovered that classes do not exist.[5] Lewis concludes – he readily acknowledges that he has not provided an argument – that we have to believe in the existence of singletons, mysterious though he takes them to be, for the theory of classes pervades all of modern mathematics and we certainly do not wish to reject *that*.

Science-worship again. Can we not raise doubts about higher transfinite cardinals? Cannot a philosopher ask a set theorist whether in fact there really is such a cardinal as κ without seeming to be a crank? It does in fact take a small amount of nerve to ask a practicing set theorist whether there really is such a number as κ. I know it from having done so. The response I got from J., as I shall refer to him, was "I have no problem with that, George," even when I asked him what he would say to a child like T. I did have the impression that J was perhaps not entirely speaking *in propria persona*, but rather was making an announcement, as from the standpoint of a set-theorist. I also had the sense that nothing I could think of to say could dislodge J. from that standpoint. (But maybe J. really does believe that there are κ and many many more sets in existence.)

What we are contemplating here, however, is nothing so radical as the rejection of singletons, but only the claim to be a body of *knowledge* on the part of a portion of set theory that treats of objects far removed from ordinary experience, the rest of physical science, the rest of mathematics, and the rest of a certain more "concrete" part of set theory.

In the supplement to the second version of his article "What is Cantor's Continuum problem?" Gödel stated, famously, that "... despite their remoteness from sense experience, we do have something like a perception of the objects of set theory, as is seen from the fact that the axioms force themselves upon us as being true."

I do not believe that all of the axioms of set theory *force* themselves upon us as being true, and, as far as I can tell, Gödel does not argue for this remarkable claim. The axiom of extensionality may do so: if it is false, there would have to be two sets with the same members, and that, we think, just could not be, not two *sets* with the same *members*. Perhaps also the pair set axiom forces itself upon us as being true: how could there not always be a set $\{x, y\}$ for any objects x, y? But I am by no means convinced that any of the axioms of infinity, union, or power so force themselves upon us or that all the axioms of replacement that we can comprehend do.

A pattern of argument in set theory, by which the existence of large sets is often proved, is to define a mapping of the members of ω (customarily proved by appeal to the axiom of infinity) onto certain objects, which then, by an axiom of replacement, form a set, and then to appeal to the axiom of union to infer the existence of a certain other set. It is in this manner that κ can be proved to exist. Even apart from worries about large sets, it seems to me that the axioms utilized in this sort of argument are significantly less evidently true than are extensionality and pair set.

That there are doubts about the power set axiom is of course well known. These doubts have to do not so much with the truth of the axiom but with the clarity or intelligibility of the notion "all subsets of." The diagnosis "unclear" or "unintelligible" has to be the wrong one, however. (It is to be noted that those who hold this view are likely to maintain the extremely suspicious view that "$\exists y \forall x(x \in y \rightarrow \forall w(w \in x \rightarrow w \in z))$" is, but "$\exists y \forall x(x \in y \leftarrow \forall w(w \in x \rightarrow w \in z))$" is not, clear, on the ground that in the latter but not the former the quantifier "$\forall x$" is unbounded. "($\exists y \forall x \in y \forall w \in x\, w \in z$ vs. $\exists y \forall x(\forall w \in x\, w \in z \rightarrow w \in x)$.))" No, there is nothing *unclear* about the power set axiom: not every epistemological drawback is a case of unclarity. There is, to be sure, a lot that we do not know about the set of all subsets of ω, and a lot that we know we cannot, in our present state of knowledge and understanding, find out about it. "Clear," as a number of philosophers have remarked, is an overworked word. What could possibly be *unclear* in, or *unintelligible* about, "$\forall z \exists y \forall x(x \in y \leftrightarrow \forall w(w \in x \rightarrow w \in z))$"?

But it does not seem to me unreasonable to think that perhaps it is not the case that for every set, there is a set of all its subsets. The axiom doesn't, I believe, force itself upon us as true, as extensionality and pair set do, and as, say, $0 \neq sn$, $m \neq n \rightarrow sm \neq sn$, and perhaps also mathematical induction do.

In his 1951 address to the American Mathematical Society (known as the "Gibbs lecture"),[6] Gödel describes both a process of arriving at axioms for set theory and a picture of the set theoretic universe. (It is of course a version of the iterative conception.) Let me quote a bit of it.

> ... evidently this procedure can be iterated beyond ω, in fact up to any transfinite ordinal. So it may be required as the next axiom that the iteration is possible for *any* ordinal, that is, for any order type belonging to some well-ordered set. But are we at an end now? By no means. For we have now a new operation of forming sets, namely forming a set out of some initial set A and some well-ordered set B by applying the operation "set of" to A as many times as the well-ordered set B indicates. And, setting B equal to some

well-ordering of A, now we can iterate this new operation, and again iterate it into the transfinite. This will give rise to a new operation again, which we can treat in the same way, and so on. So the next step will be to require that *any* operation producing sets out of sets can be iterated up to any ordinal number (that is, order type of a well-ordered set). But are we at an end now? No, . . .

Does this view of how matters are with regard to sets really *force itself upon us as true*? Do we find that, on reflection, we are unable to deny in our heart of hearts that matters must be as Gödel has described them? Or do we suspect that, however it may have been at the beginning of the story, by the time we have come thus far the wheels are spinning and we are no longer listening to a description of anything that is the case?[7]

NOTES

Although this article was written for the present volume, it was first published in George Boolos, *Logic, Logic, and Logic*, Cambridge, Mass: Harvard University Press, 1998, pp. 120–132. (The Editors)

1. This version of the iterative conception is given in detail in my "Iteration Again," in G. Boolos, *Logic, Logic and Logic*, Cambridge, Mass.: Harvard University Press, 1998, pp. 88–104.
2. A proof is given in "Iteration Again."
3. In G. Boolos, *Logic, Logic and Logic*, Cambridge, Mass.: Harvard University Press, 1998, pp. 13–29.
4. In the introduction to R. Cartwright, *Philosophical Essays*. Cambridge, Mass.: MIT Press, 1987, p. x.
5. In D. Lewis, *Parts of Classes*, Oxford: Basil Blackwell, 1991, p. 59.
6. Feferman, S. et al. (eds.) *Kurt Gödel: Collected Works*, Vol. III, Oxford: Oxford University Press, 1995, p. 306.
7. In addition to Charles Parsons, two authors whose writings I particularly regret not having discussed in this paper, despite their evident pertinence to its topic and my admiration for them, are Penelope Maddy and Solomon Feferman.

Cantor's *Grundlagen* and the Paradoxes of Set Theory

W. W. TAIT

Foundations of a General Theory of Manifolds (Cantor 1883), which I will refer to as the *Grundlagen*, is Cantor's first work on the *general* theory of sets. It was a separate printing, with a preface and some footnotes added, of the fifth in a series of six papers under the title of "On infinite linear point manifolds." I want to describe briefly some of the achievements of this great work, but I also want to discuss its connection with the so-called paradoxes in set theory. There seems to be some agreement now that Cantor's own understanding of the theory of transfinite numbers in that monograph did not contain an implicit contradiction, but there is less agreement about exactly why this is so and about the content of the theory itself. For various reasons, both historical and internal, the *Grundlagen* seems not to have been widely read compared to later works of Cantor, and to have been even less well understood. But even some of the more recent discussions of the work, while recognizing to some degree its unique character, misunderstand it on crucial points and fail to convey its true worth.

I. Cantor's Pre-*Grundlagen* Achievements in Set Theory

Cantor's earlier work in set theory contained

1. a proof that the set of real numbers is not denumerable, that is, is not in one-to-one correspondence with or, as we shall say, is not *equipollent to* the set of natural numbers (Cantor 1874);
2. a definition of what it means for two sets M and N to have the same *power* or *cardinal number*, namely that they be equipollent (Cantor 1878);
3. a proof that the set of real numbers and the set of points in n-dimensional Euclidean space have the same power (Cantor 1878).

So, on the basis of this earlier work, one could conclude that there are at least two infinite cardinal numbers, that of the set of natural numbers and that of the set of real numbers, but could not prove that there are more than two infinite powers. (It was in 1878 that Cantor stated his Continuum Hypothesis, from which followed that, in mathematics prior to 1883, there were precisely two infinite powers.)

Cantor's clarification of the notion of set prior to 1883 should also be mentioned, especially in connection with his definition of cardinal number. Bolzano, in his *Paradoxes of the Infinite* (1851), seems to have already clearly distinguished a set *simpliciter* from the set armed with some structure: in §4, he writes "An aggregate so conceived that it is indifferent to the arrangement of its members I call a *set*." But when he came to discuss cardinal numbers he seems to have forgotten his definition of set and failed to distinguish between, say, the cardinal number of the set of points on the line and the magnitude of the line as a geometric object. For this reason, he was prevented from resolving one of the traditional paradoxes of the infinite, namely, that, for example, the interval (0, 1) of real numbers is equipollent to the 'larger' interval (0, 2). Of course, even without the confusion of the set with the geometric object, there is still a conflict with Euclid's Common Notion 5, that the whole is greater than the part. But this principle does indeed apply to geometric magnitude, and it is likely that, without the confusion, it would have sooner been accepted that infinite sets are simply a counterexample to it.

In fact, Bolzano's understanding of the notion of set was in general less than perfect. For example, the word I translated as "members" in the above quote is actually the word "Teile" for *parts*, which he generally used to refer to the elements of a set. Throughout his discussion, there are signs that he had not sufficiently distinguished the element/set relation from the part/whole relation. For example, the last sentence in §3 asserts that it would be absurd to speak of an aggregate with just one element, and the null set is not even contemplated. But despite this lack of clarity, it is to him that we owe the identification of sets as the carriers of the property *finite* or *infinite* in mathematics (§11). In particular, his analysis of the 'infinity' of variable quantities in §12 and the observation that this kind of infinity presupposes the infinity of the set of possible values of the corresponding variable preceded (as Cantor acknowledges in §7) Cantor's own discussion in §1 of the *Grundlagen*, where he refers to what he calls the 'variable finite' as the 'improper infinite'.[1]

Cantor himself, in using the notation $\overline{\overline{M}}$ for the cardinal number of the set M, where the second bar indicates abstraction from the order of the elements of M, betrays some confusion between the abstract set M and M armed with some structure. Moreover, both Cantor and Dedekind avoided the null set. (After all, no whole has zero parts.) It is food for thought that as late as 1930, Zermelo chose in his important paper (1930) on the foundations of set theory to axiomatize set theory without the null set (using some distinguished urelement in its stead). The concept of set is no Athena: schoolchildren understand it now, but its development was long drawn out, beginning with the earliest counting and reckoning and extending into the late nineteenth century.

Nevertheless, it was Cantor who understood it sufficiently to dissolve the traditional paradoxes and to simply confront Common Notion 5 and define the relation of having the same cardinal number in terms of equipollence. The equivalence of these notions had long been accepted for finite sets, but it was rejected, even by Bolzano, in the case of infinite sets. Prior to Cantor, these paradoxes had led people to believe that there was *no* coherent account of cardinal number in the case of infinite multiplicities.[2] Note, too, that, in the face of the long tradition, from Aristotle through Gauss, of opposition to the infinite in mathematics, it was not only a better understanding of the notion of set that Cantor needed to bring to his definition of cardinal number; it required, also, some intellectual courage.

In the light of these remarks, it is unfortunate that some contemporary writers on philosophy of mathematics and its history insist on referring to Cantor's definition of equality of power as *Hume's principle*, for the philosopher David Hume, who explicitly rejected the infinite in mathematics.

It is instructive to compare Cantor's conception of a set prior to his *Grundlagen* with what he writes about it thereafter. As far as I know, his earliest explanation of what he meant by a set is in the third paper (Cantor 1882) in the series on infinite linear sets of points. He writes

> I call a manifold (an aggregate [Inbegriff], a set) of elements, which belong to any conceptual sphere, well-defined, if on the basis of its definition and in consequence of the logical principle of excluded middle, it must be recognized that it is internally determined whether an arbitrary object of this conceptual sphere belongs to the manifold or not, and also, whether two objects in the set, in spite of formal differences in the manner in which they are given, are equal or not. In general the relevant distinctions cannot in practice be made with certainty and exactness by the capabilities or methods presently available. But that is not of any concern. The only concern is the internal determination from which in concrete cases, where it is required, an actual (external) determination is to be developed by means of a perfection of resources. ((1932), p. 150)

The latter part of this passage is interesting because it reflects the growing tension within mathematics (and one whose history has yet to be written) over the role of properties that are 'undecidable', that is, for which we have no algorithm for deciding for any object in the conceptual sphere, whether or not it has the property. Cantor is saying that the existence of such an algorithm is unnecessary in order for the property to define a set. He gives the example of determining whether or not a particular real number is algebraic or not, which may or may not be possible at a given time with the available techniques. He contends that, nevertheless, the set of algebraic numbers is well defined. Very likely, Cantor took up this issue here because of his proof in (1874) that the set of algebraic numbers is countable and that, therefore, in any interval on the real line, there are uncountably many transcendental numbers. This shows that interesting results about a set may be obtainable even when no algorithm exists for determining membership in the set.[3] On the other end of the ideological scale was Kronecker, who took the view, later associated with Hilbert's *finitism*, not merely that the law of excluded middle should not be assumed, but even more: only those objects that can be finitely represented and only those concepts for which we have an algorithm for deciding whether or not they hold for a given object should be introduced into mathematics. (See Kronecker [1886, p. 156 n.*].) We shall see that Cantor more fully takes up the defense of classical mathematics against the strictures of Kronecker in §4 of *Grundlagen*.

But to return to the main topic, what did Cantor mean by a 'conceptual sphere'? The answer seems to be clearly indicated by a later passage in the same work, where he writes "The theory of manifolds, according to the interpretation given it here, includes the domains of arithmetic, function theory and geometry, if we leave aside for the time being other conceptual spheres and consider only the mathematical." (1882, p. 152) So I think that Cantor was simply using the expression 'conceptual sphere' to refer to different domains of discourse. Presumably, nonmathematical spheres would include that of physical phenomena and of mental phenomena.[4]

Thus I think that in this work and, in fact, up to the discovery of the theory of transfinite numbers, Cantor's notion of a set was the *logical* notion of set, the notion studied in second-order logic, namely, of a collection of elements of some *given* domain. This notion of a set would fairly be described by saying that a set is, as Frege suggested, the extension of a concept, but, as opposed to Frege's notion of concept, which is defined for all objects, the relevant concept here is one defined only for objects of some given domain.[5] Finally, note that this is the conception of set as it is most often applied in mathematics and that the so-called paradoxes of set theory have nothing to do with it.[6] So much for a summary of Cantor's relevant achievements prior to the *Grundlagen*. I turn now to the *Grundlagen* itself.

II. Summary of the Content of the *Grundlagen*

(1) The *raison d'être* of the *Grundlagen* was the theory of transfinite numbers, which Cantor seems to have mentioned for the first time in a letter to Dedekind in November 1882. He defines the numbers to be what can be obtained, starting with the initial number and applying the two operations of taking successors (the *first principle of generation*) and taking limits of increasing sequences (the *second principle of generation*). (See §III, below.) He in fact took the initial number to be 1, but it will make no difference if we adopt the now more common practice of starting with 0.[7]

(2) The transfinite numbers – and henceforth I will often just speak of the *numbers*, both finite and transfinite – were stratified into the number classes. (See §IV, below.) Whereas only two infinite powers were known to exist prior to *Grundlagen*, the number classes represent an increasing sequence of powers or

cardinal numbers, in one-to-one correspondence with the numbers themselves. Cantor believed the number classes to represent all powers, and it was this application of his theory of transfinite numbers to the theory of powers that he mentioned in his letter to Dedekind and that he mentions first in the *Grundlagen*. In (1890–1), Cantor introduced his diagonal argument, proving that the set of two-valued functions on a set is of higher power than the set itself (1891), and thus providing another, cofinal, sequence of powers; but, in 1883, the number classes were the *only* examples that existed of higher powers.

(3) In §2, Cantor analyzed the notion of a counting number, for which he used the term *Anzahl*, and extended it to the transfinite. He saw that counting a set determines a (total) ordering of it and, indeed, a well-ordering. This notion of well-ordering was introduced into mathematics here for the first time. As he noted, the requirement of well-ordering had been obscured by the fact that, in the case of finite sets, *all* total orderings of the set are well-orderings and are isomorphic to one another.

He noted, too, that the set of predecessors of any transfinite number forms a well-ordered set and that every well-ordered set is isomorphic to such a proper segment of the numbers. For this reason, his transfinite numbers have come to be called *ordinal numbers* in the literature on the *Grundlagen*, although Cantor himself refers to them simply as *numbers* (*Zahlen*) or as *real whole* (*reale ganze*) numbers. He seems to have wanted to distinguish the numbers themselves from their application as measures of well-ordered sets, just as we may consider the finite whole numbers as they are in themselves, independently of applications as counting numbers or as measures of finite sets.

Failure to see this has led to a minor mystification about Cantor's use of the term *Anzahl* and how it is to be distinguished from his use of the term *Zahl* when speaking about his transfinite numbers. Hallett (1984) translates '*Anzahl*', as 'enumeral', which seems a reasonable alternative to 'counting number,' but he interprets the term *Zahl* as referring to ordinal numbers – and then worries about the distinction. An ordinal number for Cantor, when he later introduced the term, is the order type of a well-ordered set, just as a cardinal number is for him the equipollence type of an abstract set. It is true that the numbers (*Zahlen*) represent ordinal numbers, in the sense that every well-ordered set is measured by some number, but the numbers must first be regarded as given before the proper initial segments of them can be taken to be measure sticks of the well-ordered sets. Among philosophers, especially, the confusion has been exacerbated by the influence of Frege, who used the term *Anzahl* for the cardinal numbers. Contrary to the general tendency in the late nineteenth century on foundations of arithmetic, Frege considered the natural numbers primarily in their role as cardinals.

The view that Cantor regarded the transfinite numbers essentially as ordinals stands out rather strongly in Hallett's book, where Cantor's actual definition of the numbers seems to be counted as a mistake, and it is only their role as measures of well-ordered sets that gives the numbers substance. One difficulty with this view, aside from the obvious ones that there is nothing wrong with Cantor's definition in the *Grundlagen* as it is and that he does *not* identify the numbers as ordinals there, is that if the numbers exist essentially as measures of well-ordered sets, then how is one to understand the numbers in the higher number classes? The problem is that, at that time, as we have already noted, the only infinite well-ordered sets known to be not isomorphic to well-orderings of the natural numbers or the continuum were those represented by proper segments of the system of transfinite numbers themselves. In response to this, Hallett attempts to construe the construction of the sequence of numbers as a procedure whereby, having constructed a segment of the numbers and recognized it as well-ordered, one then may introduce its order type (p. 57). But this baroque construction is not at all the way in which Cantor introduces the numbers.[8]

Certainly Hallett is right that Cantor believed the application of the transfinite numbers as measures of well-ordered sets constituted an argument for admitting them into mathematics – of 'legitimitizing' them, if you like. Cantor himself mentions this at the beginning of §2 of *Grundlagen*.[9] But, note that the application to well-ordered sets was not the only application of the transfinite numbers that Cantor had in mind: as we have already mentioned, it was not even the first. The first application that he mentioned in the *Grundlagen* was to the theory of powers. Moreover, in his earlier letter to Dedekind, he also wrote about the issue of legitimatizing his theory, but it was in terms of the application to the theory of powers.[10] But this matter of legitimitization, which we will discuss below, is distinct from the question of whether the notion of transfinite number depends on that of a well-ordered set – whether numbers need to be *explained*, as Hallett puts it, as enumerals. And, in fact, Cantor's definition of the numbers stands on its own feet and is entirely independent of their application, for example, as measures of well-ordered sets.

(4) Not only did Cantor introduce the notion of a well-ordered set in *Grundlagen*, but at the beginning of §3, he proposed the Well-Ordering Principle, that every set is well-orderable, as a fundamental law of thought. It follows from this principle that every infinite power is represented in the sequence of number classes. (See §4.) Zermelo, in 1904, deduced the Well-Ordering Principle from the axiom of choice, which seems to have been (implicitly) regarded as a part of logic by earlier writers such as Cantor and Dedekind.

(5) The *Grundlagen* also introduces for the first time the distinction between sets and what later came to be called *proper classes*. Every well-defined set has a power (§1), but, as we shall see, Cantor recognized that there are totalities, such as the totality of all whole numbers or of all powers, which have no power.

(See *Grundlagen*, note 2, and §III, below.) Often we use the term 'proper class' in a relative sense, to refer to the subsets of the domain of a model of set theory that are not coextensive with some element of the domain. But we also use it in the absolute sense to refer, for example, to the totality of all ordinals, independently of the (well-founded) models in which they are represented. It is in this latter sense that Cantor discovered the distinction between a set and a proper class – his distinction between the determinate infinites (represented by the number classes) and the absolute infinite (represented by the totality of transfinite numbers or the totality of the number classes or powers).

(6) In §8, in defending the introduction of the transfinite numbers, Cantor gives what may be the first statement and defense of the autonomy of what we would call pure mathematics and which he prefers to call 'free' mathematics. I am not referring to the thesis that reasoning in mathematics should proceed purely deductively, without reference to empirical phenomena, but rather the thesis that pure mathematics may be concerned with systems of objects that have no known relation to empirical phenomena at all. Especially when one remembers how long it took for the various extensions of the number system, 0 and the negative numbers and the complex numbers, to be accepted, it is not remarkable that Cantor felt the need to discuss this matter. Let me quote Cantor on free mathematics:

> Mathematics is in its development entirely free and is only bound in the self-evident respect that its concepts must both be consistent with each other and also stand in exact relationships, established by definitions, to those concepts which have previously been introduced and are already at hand and established. In particular, in the introduction of new numbers it is only obligated to give definitions of them which will bestow such a determinancy and, in certain circumstances, such a relationship to the older numbers that they can in any given instance be precisely distinguished. As soon as a number satisfies all these conditions it can and must be regarded in mathematics as existent and real.

We are justified in regarding the numbers as real insofar as the system of transfinite numbers has been consistently defined and integrated with the finite numbers.

Cantor's argument for the 'freedom' of mathematics and the reality of the transfinite numbers is based on his distinction between 'immanent' or 'intrasubjective' reality and 'transient' or 'transsubjective' reality. Of the former, he writes

> First, we may regard the whole numbers as real insofar as, on the basis of definitions, they occupy an entirely determinate place in our understanding, are well distinguished from all other parts of our thought and stand to them in determinate relationships, and thus modify the substance of our minds in a determinate way.

So, the reality that he has claimed for the numbers is immanent reality. Concerning transient reality he wrote:

> But then, reality can also be ascribed to numbers to the extent that they must be taken as an expression or copy of the events and relationships in the external world which confronts the intellect, or to the extent that, for instance, the various number classes are representatives of powers that actually occur in physical or mental nature.[11]

But he further argues in §8 that mathematics – that is, 'free' mathematics – is constrained *only* by the requirements of immanent reality.

Note that, up to this point in his argument, the application of the transfinite numbers either as measures of well-ordered sets or to the theory of powers plays no role at all. But I think that we may see that role in the 'legitimitization' of the transfinite numbers in the following passage in §8:

> It is not necessary, I believe, to fear, as many do, that these principles [admitting into mathematics objects satisfying the criteria for immanent existence] present any danger to science. For in the first place the designated conditions, under which alone the freedom to form numbers can be practiced, are of such a kind as to allow only the narrowest scope for discretion (*Willkür*). Moreover, every mathematical concept carries within itself the necessary corrective: if it is fruitless or unsuited to its purpose, then that appears very soon through its uselessness and it will be abandoned for lack of success.

Quite simply, the two applications of the transfinite numbers were important to Cantor in establishing his theory of transfinite numbers as a legitimate part of mathematics because it was part of his argument that the theory is fruitful. *This has nothing to do with the internal logic of the theory, only with the question of whether it is worth pursuing*. In particular, it has nothing to do with the immanent reality of the numbers.

Note also that the issue of legitimacy is separate from that of transient reality: the applicability of the transfinite numbers to the theory of higher powers or well-ordered sets is no guarantee that they have any empirical application.

In fact, Cantor expresses his faith that whatever has immanent reality also has transient reality:

> there is no doubt in my mind that these two forms of reality always occur together in the sense that a concept said to exist in the first sense also always possesses in certain, even infinitely many, ways a transient reality. To be sure, the determination of this transient reality is often one of the most troublesome and difficult problems in metaphysics, and must frequently be left to the future, when the natural development of one of the other sciences will uncover the transient meaning of the concept in question.

Metaphysics here seems to refer to the inventory of the basic structures of the empirical world – since it may be left to the other sciences to uncover them. I take Cantor to be simply expressing an article of faith here, but not one on which his theory of transfinite numbers in any sense rests. This could be questioned on the grounds that he goes on to write

> The mention of this connection [between the two realities] has here only one purpose: that of enabling one to derive from it a result which seems to me of very great importance for mathematics, namely, that mathematics in the development of its ideas has only to take account of the *immanent* reality of its concepts....

It might then seem that Cantor believes that in some sense the validity of free mathematics depends on this article of faith. But in the ensuing discussion he writes

> If I had not discovered this property of mathematics [i.e. its freedom] by means of the reasoning I have described, then the entire development of the science itself, as we find it in our century, would have led me to exactly the same opinions.

I think that the point of these remarks is not to qualify the autonomy of free mathematics; rather, it is to argue that, *even if what one is interested in is metaphysics, that is, the basic structures of the natural world*, one should allow mathematics to proceed freely, because what it develops freely will in the end turn out to be instantiated in nature.

I especially mention this because Hallett takes a quite different stance with regard to this discussion in §8 and one according to which Cantor's view is identified with an unintelligible doctrine that is often attributed to Plato. Hallett writes (1984, p. 18) "And crucially, immanent and transient reality are intimately connected." After quoting note 6, in which Cantor suggests that his view that what is immanently real will also turn out to be transiently real is in agreement with Plato, he writes

> As Cantor himself says..., what he proposes is a Platonic principle: the 'creation' of a consistent coherent concept in the human mind is actually the uncovering or discovering of a permanently and independently existing real abstract idea.
>
> It is now clear why Cantor considered mathematics as so free. It does concern itself with objective truth and an independent (Platonic) realm of existents in so far as its objects of study are transiently real. But it need not attempt to investigate this transient reality directly, or even worry about the precise transient 'significance' of a concept. All that mathematics need worry itself with is 'intrasubjective' reality, and once this is established it is *guaranteed* that the concepts are also transiently real.

It is evident that, for Hallett, perhaps because he was misled by Cantor's use of the term 'metaphysics,' transient reality refers, not to the instantiation of the concept in nature, but to an 'independent (Platonic) realm of existents'. But that is not the point of Cantor's reference to Plato: rather, he seems to be assuming that Plato also advocated the free development of mathematics and believed that what it created freely would as a matter of fact turn out to be exemplified in the natural world. This reading of Plato, though better than what we usually get, is incorrect; but it *is* the basis of Cantor's note.

Purkert (1989, p. 58) seems to share Hallett's view: he cites as evidence a letter to Everhard Illigens from May 1886 in which Cantor writes

If I have known the internal consistency of a concept that represents a being, then I am forced to believe by the idea of the omnipotence of God that the being which is stated by the concept under discussion must be realized in some way. With regard to this, I call it a *possible* being but this does not mean that it is realized somewhere and some time and somehow in reality.

Purkert concludes from this that "For Cantor's Platonistic ontology of mathematical objects, consistency was a necessary but not a sufficient condition." But, as I understand it, this is not at all what Cantor is saying. He is saying that his theological beliefs lead him to believe that the immanent being must be empirically realized in some way. That it could be so realized is the respect under which he refers to it as 'possible'; but calling it 'possible' does not mean that it has been, is or will be realized, nor does it mean that it lacks immanent being if it fails to be so realized.

It is simply impossible to make sense of the reading of Hallett and Purkert, according to which immanent reality is in some sense wanting and transient reality means 'really exists' in some 'Platonic' sense: Cantor speaks of the further development of the sciences (other than mathematics) uncovering the transient reality; and he refers to the development of function theory as an instance of mathematics proceeding freely, without having first secured the transient reality of its concepts in mechanics, astronomy, and so on. Hallett does not ignore this; rather, he introduces another kind of reality: there is not only immanent and transient reality, but there is also having 'physical applications'. He writes (p. 18) "There may be all kinds of ways in which transient reality is manifested by; in particular, concepts might be represented or instantiated in the physical world." Thus, being transiently real is manifested by, but not identical with, having empirical application (rather like a Calvinist being among the elect is manifested by, but not identical with, having a prosperous farm). But what in the text or in sweet reason prevents Hallett from identifying the latter kind of reality with transient reality?[12]

(7) In §4, Cantor defended what has come to be called classical mathematics, in particular, the methods in function theory associated with Bolzano, Cauchy, Riemann, and Weierstrass, as well as his new theory of transfinite numbers, against the opposition of Kronecker. We noted that he had already in 1882 defended the use of the law of excluded middle in reasoning in arithmetic, geometry, and function theory. But in *Grundlagen* the emphasis is more on a related, but somewhat different, stricture of Kronecker, namely, concerning the kinds of objects that should be admitted into mathematics. For Kronecker, all

genuine mathematical propositions must ultimately be interpretable as statements about the natural numbers. Let me quote Cantor:

> In this manner a definite (if also rather prosaic and obvious) principle is recommended to all as a guideline; it should thereby serve to confine the playing out of the passion for speculation and conceptual invention in mathematics within the true boundaries, within which it runs no danger of falling into the abyss of the "Transcendent," in which, it is said in order to inspire dread and wholesome terror, "everything is possible." It is uncertain (who knows?) whether it was not just from the point of view of expediency alone that the originators of this doctrine decided to recommend it to the soaring powers, which so easily endanger themselves through enthusiasm and extravagance, as an effective regulation for protection against all error; but a *fruitful* principle cannot be found in it. For I cannot accept the assumption that the originators of this view themselves, in the discovery of new truths, started from these principles. And I, no matter how many good things I may cull from these maxims, must strictly speaking regard them as *erroneous*: no real progress has stemmed from them, and if science had proceeded precisely in accordance with them, it would have been retarded or at least confined within the narrowest of boundaries.

This precedes by four years the well-known footnote of Dedekind (1887) in which the author challenges Kronecker to justify his constraints on mathematics.

This concludes my summary of the content of the *Grundlagen*. Given such a rich assortment of original material and given the prominence anyway of the problem of the infinite in the history of philosophy, one would a priori have expected the *Grundlagen* to be regarded as one of the great philosophical classics of all time; but, in fact, until recently, even in discussions of Cantor's work, it has been largely neglected and, when considered at all, has tended to be viewed through the window of his later papers, leading to serious misunderstanding. This is especially so when, as is often the case, Cantor's theory of sets as a whole is interpreted as 'naive set theory', the ground for Frege's later inconsistent *Begriffsschrift* – as a somewhat imprecise formulation of naive intuitions, which Frege, at tragic cost to himself, merely made precise. Of course, the exposition in the *Grundlagen* has also contributed to the lack of appreciation of it. Cantor's exposition of technical arguments is generally quite lucid, but this paper, with its wealth of conceptual, philosophical analysis, does not share that property; and it easily falls prey to those who (following the model of Frege) too quickly attack the choice of words without sufficiently searching for the intended meaning.

Recent scholars such as Michael Hallett (1984), Walter Purkert (1989), and Shaughan Lavine (1994), have recognized the special nature of *Grundlagen*, even within Cantor's *œuvre* on set theory, and to varying degrees have rejected the myth of Cantor's 'naive' set theory[13]; but, interesting and enlightening as they are, they still leave without satisfactory answers a number of questions concerning the text. Indeed, each of the three cited works, while advancing our understanding of the *Grundlagen* on the whole, introduces interpretations on important points which, as in some cases I have already made clear, seem to me entirely wrong.

III. The *Grundlagen* and the Paradoxes of Set Theory

There is a significant change in Cantor's conception of a set in the *Grundlagen*. In note 1, he writes:

> By a 'manifold' or 'set' I understand any multiplicity which can be thought of as one, i.e. any aggregate [inbegriff] of determinate elements which can be united into a whole by some law.

The most notable changes in this from his earlier explanation of the concept of set is the absence of any reference to a prior conceptual sphere or domain from which the elements of the set are drawn and the modification according to which the property or 'law' that determines elementhood in the set "unites them into a whole." But I think that a convincing explanation for this change can be found in his theory of the transfinite number, and it will provide a natural transition to the question of the relation of the *Grundlagen* to the paradoxes of set theory. In introducing the transfinite numbers, Cantor employs the notion of set in an *entirely new way*: numbers are defined in terms of the notion of a set of numbers. Essentially, he introduces the transfinite numbers as follows:

$$X \text{ is a subset of } \Omega \Rightarrow S(X) \in \Omega.$$

Here, Ω denotes the class of all numbers and, let me emphasize: X ranges over *sets*. $S(X)$ is intended to be the least number greater than every number in X. So, assuming the existence of the null set and unit sets of numbers,

$$S(\emptyset) \text{ should be the least number } 0,$$
$$S(\{\alpha\}) \text{ should denote the successor of } \alpha,$$

and, if X is a set of numbers with no greatest element, then $S(X)$ is the limit of the sequence of numbers in X arranged in natural order. We omit the definitions (which Cantor also does not bother to give) of what it means for two numbers to be equal and for one number to be less than another. Cantor also took it to be implicit in his definition of the numbers that there are no infinite descending sequences

$$\alpha_0 > \alpha_1 > \alpha_2 > \cdots$$

so that Ω is well-ordered by '<'.

Cantor's definition of Ω has the familiar look of an inductive definition, but that is deceptive. An inductive definition picks out a subset of some *given* domain of objects by means of some closure condition. The definition of the transfinite numbers, on the contrary, is intended to introduce a whole *new* domain of objects, not a subcollection of a given domain.

But he is not only introducing a new domain of objects, he is introducing it in terms of the notion of a *set* of objects *of that very same domain*. That is, unlike his previous notion of a set, according to which a set is a set of objects from some *given* domain, already well defined, here the notion of an object of the domain and that of a set of objects of the domain are dependent on each other. This is an entirely new context for the notion of a set.[14]

A symptom of the problem that arises from the interdependence of the notions of (transfinite) number and set of numbers is the

Theorem. Ω is not a set.

The proof is simply that, otherwise, $S(\Omega)$ would be a number, and so, $S(\Omega) > S(\Omega) > \cdots$ would be an infinite descending sequence of numbers.

Cantor clearly knew this simple theorem. This, I suggest, accounts for the fact that, in his later explanations of the notion of set, including the one in *Grundlagen*, Cantor does not simply take the notion of set to refer to the extension of a concept in some conceptual sphere. For in the conceptual sphere of transfinite number theory, this leads to a contradiction.

There has been much discussion of whether he knew the Burali–Forti paradox, the paradox of the greatest ordinal, but behind that discussion lies a misconception that we have already noted, namely, of thinking of the transfinite numbers as ordinals. This has led some commentators to forget Cantor's actual definition of the numbers in the *Grundlagen* and to think of them simply as order types of well-ordered sets. Thus, although I entirely agree with Purkert (1989) that it would be difficult to believe that Cantor was not aware that it would be contradictory to assume that Ω is a set, I disagree with him on exactly what contradiction Cantor was likely to have discerned. Purkert cites essentially the Burali–Forti paradox in this connection (p. 57). This paradox involves showing that the totality of all order types of well-ordered sets, if a set O, is a well-ordered set under the natural ordering but one whose order type θ cannot be in O, since otherwise $\theta + 1$ would be in O and so would be $<\theta$. But, given Cantor's actual definition of the numbers, reference to well-orderings is grotesquely prolix. The contradiction we described above is based, *not on the property of ordinals as order types of well-ordered sets*, but directly on Cantor's definition of the numbers, which admits the more direct and immediately evident argument. It is this latter argument, and not the Burali–Forti paradox, that I am reasonably convinced could not have escaped Cantor's eye. Here, I think, is one of many instances in the literature on Cantor where a failure to read *Grundlagen* on its own terms rather than through the window of later works somewhat distorts the picture, though it is far from being the worst such instance.

Whether or not Cantor was aware that it would be contradictory to assume that Ω is a set, we have already noted that, in note 2 in *Grundlagen,* he had certainly excluded Ω and the totality of all powers as sets. Yet, Cantor's comments in this note do not refer to a contradiction as the ground for rejecting the totality of numbers or powers as sets. Rather he writes

The absolute can only be acknowledged but never be known – and not even approximately known. For just as in the [first number class] every finite number, however great, always has the same power of finite numbers greater than it, so every supra-finite number, however great, of any of the higher number classes is followed by an aggregate of numbers and number-classes whose power is not in the slightest reduces compared to the entire absolutely infinite aggregate of numbers, starting with 1.

In fact, this is a strange argument that ultimately makes no sense. For, not only the first number class and the totality of all numbers or alephs, but *any* number class N, have the property that it has the same power as the set M of all of its elements $\geq$ a given one α. For the function $\beta \mapsto \alpha + \beta$ is a bijection of N onto M. Moreover, when γ is a fixed point $\gamma = \omega_\gamma$ of the initial ordinal function $\alpha \mapsto \omega_\alpha$, then for any α in N_γ the function $\beta \mapsto N_{\alpha+\beta}$ is a bijection from N_γ onto the set of number classes N_β for $\alpha \leq \beta < \gamma$. So, when Purkert refers to a letter to Hilbert in 1897 in which Cantor writes that "Totalities that cannot be regarded as sets (an example is the totality of all alephs as is shown above), I have already many years ago called absolute infinite totalities, which I sharply distinguish from infinite sets," he is almost certainly right in taking the reference to be to note 2 in *Grundlagen*, but his claim that this is evidence that Cantor at that time already knew the paradox of the greatest aleph is less well-founded.

IV. What Numbers/Sets of Numbers Are There?

It was Cantor's construction of the system of transfinite numbers employing the concept 'set of numbers' that opened a Pandora's box of foundational problems in mathematics, namely, the question of what cardinal numbers there are. One can, in a way, understand the resistance to Cantor's ideas on the part of the mathematical law-and-order types – in the same way that one can understand the church terrorizing the elderly Galileo: in defense of a closed, tidy universe. In that respect, Hilbert's reference to Cantor's 'Paradise' is ironic: it was the Kroneckers who wanted to stay in Paradise and it was Cantor who lost it for us – bless him. Note, though, that Kronecker went some way beyond the rejection of just Cantor's theory of transfinite numbers. His brand of finitism would have cut back much of the Garden of Eden itself – not just classical analysis, but even constructive analysis in the sense of Brouwer or Bishop.

The latter point is significant because there are many mathematicians who will accept the Garden of Eden, that is, the theory of functions as developed in the nineteenth century, but will, if not reject, at least put aside the theory of transfinite numbers, on the grounds that it is not needed for analysis. Of course, on such grounds, one might also ask what analysis is needed for, and if the answer is basic physics, one might then ask what that is needed for. When it comes down to putting food in one's mouth, the "need" for any real mathematics becomes somewhat tenuous. Cantor started us on an intellectual journey. One can peel off at any point, but no one should make a virtue of doing so.

The question of what numbers $S(X)$ exist is precisely equivalent to the question of what sets X of numbers exist. Before proceeding further, I want to remark on how this fact bears on Lavine's (1994) account of the notion of set in *Grundlagen*. He understands Cantor to have *defined* a set to be a totality in one-to-one correspondence with a proper initial segment of the numbers (see Definition 2.5 on p. 81); but this cannot be right. In the first place, in §3, Cantor describes the Well-Ordering Principle as a "*law of thought*" – "that it is always possible to bring any *well-defined* set into the form of a *well-ordered* set – a law which seems to me fundamental and momentous and quite astonishing by reason of its general validity." For some reason, Lavine quotes this very passage as confirmation of his view, but it is a strange use of language to count a definition as a law of thought. Moreover, we have already quoted a passage from note 1 in which Cantor explained what he meant by a set without any reference to the notion of well-ordering. Lavine again takes this passage as confirmation of his view because "Cantor's typical use of the word 'law' in the [*Grundlagen*] is 'natural succession according to law', which suggests quite a different picture [from the one in which 'law' simply refers to some property]": a "law is, for Cantor, a well-ordering or 'counting.'" But the note in question attaches to the first sentence of §1 and the notion of a well-ordering is introduced *for the first time ever* in §2. Moreover, there are, according to my own count, exactly nineteen uses of the term "Gesetz" or its plural in the *Grundlagen* besides the one in question and the occurrence of "Denkgesetz" mentioned above and counting one occurrence of "gesetzmässig"; and not one of them supports Lavine's contention.[15]

In the second place, and much more decisive: The question of whether an initial segment X of numbers is a proper segment is, as we have noted, *precisely* the question of whether it is a set: it is proper just in case $S(X)$ exists. So, in the particular case of transitive classes X of numbers, Lavine's proposed definition is circular.[16]

Lavine goes on to account for the changes in Cantor's later views by pointing to his discovery in 1891 of the hierarchy of powers arising from iterating the operation of passing from a set M to the set of two-valued functions on M, starting with the set of finite numbers. His point is that, since it is not clear that these function spaces can be well ordered, Cantor had to give up Lavine's Definition 2.5 of the notion of set. But that does not really make sense. Cantor already knew of the totality of real numbers, and any problems that he had with the well-orderability of the set of two-valued functions on a set he would already have faced in the case of the real numbers. But he explicitly speaks of sets of real numbers and subsets of R^n.

Lavine expresses the view that the alternative to his reading of the explanation of the notion of set in Note 1 is to hold that Cantor's set theory in *Grundlagen* is *naive* set theory (1994, p. 85). But, it is only with his introduction of the transfinite numbers in terms of sets of transfinite numbers that the notion of set became problematic; prior to that, naive set theory – viz., the Comprehension Principle, that every property determines a set – was perfectly valid, since property meant property of objects of some conceptual sphere. With the introduction of the transfinite numbers, though, Cantor immediately recognized that the notion of set was problematic, to the extent of understanding that not *every* property of numbers 'unites the objects possessing it into a whole', thereby determining a set. So, even if we read the passage in note 1 in the most natural way, he was not naive. Let me note again that defining sets to be totalities equipollent to some proper initial segment of the numbers in no way eliminates this problem of set theory, not for Cantor and not for us, since the question of whether or not an initial segment of the numbers is proper is precisely the question of whether or not it is cofinal with a set.

But Pandora's box is indeed open: Under what conditions *should* we admit the extension of a property of transfinite numbers to be a set – or, equivalently, what transfinite numbers are there? No answer is final, in the sense that, given any criterion for what counts as a set of numbers, we can relativize the definition of Ω to sets satisfying that criterion and obtain a class Ω' of numbers. But there would be no grounds for denying that Ω' is a set: the preceding argument that Ω is not a set merely transforms in the case of Ω' into a proof that Ω' does not satisfy the criterion in question. So, $S(\Omega')$ is a number, and we can go on. In the foundations of set theory, Plato's dialectician, searching for the first principles, will never go out of business.

Cantor himself offered the first answer to the question of what sets exist. For totalities M and N, let

$$M \prec N$$

mean that there is a function defined on N whose range of values is all of M – that is, M is of power $\leq$ that of N. Cantor repeatedly took as a sufficient condition for the sethood of M that $M \preceq N$ for some set N, which is essentially the axiom of replacement. Relativizing the notion of set in the definition of Ω to totalities of power $\leq$ that of N, we obtain the set Ω', which we denote by $\Omega(N)$.

Cantor proved (assuming the Well-Ordering Principle), that $\Omega(N)$ represents the next highest power after N. By iterating the operation $\Omega(-)$ starting with the finite numbers, we obtain (essentially) Cantor's number classes:

$$N_0 \text{ is the set of finite numbers}$$
$$N_{\alpha+1} = \Omega(N_\alpha)$$

and for γ a limit number

$$N_\gamma = \bigcup_{\alpha<\gamma} N_\alpha.$$

So, the power of N_α is $\aleph_\alpha$.

But, of course, Cantor's answer is only the first in an open sequence of nontrivial answers to the question of what numbers exist. Suitably understood, Cantor's hierarchy yields all the numbers less than the least weakly inaccessible number, that is, the least regular fixed point $\alpha = \omega_\alpha$ of the initial ordinal function. Namely, if C is the least transitive class of numbers that contains 0, is closed under successors, includes N_α for all $\alpha \in C$, and contains $\text{Lim}_{\alpha_i<\gamma}\alpha_i$ when γ and the α_is are in C, then the least upper bound of C is the least weakly inaccessible number.[17]

NOTES

This paper was written in honor of Charles Parsons, from whom I have profited for many years in my study of the philosophy of mathematics and expect to continue profiting for many more years to come. In particular, listening to his lecture on "Sets and classes," (Parsons 1974), motivated my first attempts to understand proper classes and the realm of transfinite numbers. I read a version of the paper at the APA Central Division meeting in Chicago in May 1998. I thank Howard Stein, who provided valuable criticisms of an earlier draft, ranging from the correction of spelling mistakes, through important historical remarks, to the correction of a mathematical mistake, and Patricia Blanchette, who commented on the paper at the APA meeting and raised two challenging points, which have led to improvements in this final version.

1. Bolzano was, in turn, anticipated by Galileo in his *Two New Sciences*, First Day, where Salviati remarks "The very ability to continue forever the division into quantified parts implies the necessity of composition from infinitely many unquantifiables."
2. In the fourteenth century, Henry of Harclay and, perhaps more clearly, Gregory of Rimini seem to have been at least close to Cantor's analysis; but the soil was thin and the idea did not take root.
3. There has been some confusion about the nonconstructive character of Cantor's proof that every interval contains transcendental numbers. His proof that no one-to-one enumeration of real numbers exhausts an interval, while not constructive, is easily reformulated as a constructive proof. His inference from the existence of an enumeration of the algebraic numbers to the existence of one without repetitions is, of course, nonconstructive. However, as Howard Stein has pointed out to me, one can constructively enumerate the algebraic numbers in a one-to-one manner,

although this requires methods, e.g. Sturm's Theorem, that were not available to Cantor.

4. Cantor's explanation of the notion of set does present another apparent difficulty. His words suggest that equality of objects is relative to the set in question: A set is determined when we have determined what objects of the conceptual sphere are in it and, of two objects in it, whether or not they are equal – despite possible formal differences in the manner in which they are given. But here I think that the most reasonable interpretation is that what Cantor actually writes is misleading and that he is thinking of cases such as defining a set of rational numbers by means of a property of pairs of integers or a set of real numbers by means of a property of Cauchy sequences of rational numbers. Thus two Cauchy sequences may have the property in question and so the real numbers they determine are in the set. What more is required is that we have a criterion for when two such sequences, distinct *qua* sequences, determine the same number and that the property in question of the Cauchy sequences in question respect that criterion.
5. Notice that, despite Cantor's avoidance of the null set, it is hard to avoid it on this conception – as Frege pointed out.
6. It is striking how clear both Cantor's conception of set and his treatment of it are compared to Bolzano's some thirty-one years earlier. Despite his double abstraction of the set to obtain the cardinal number, mentioned earlier, he in fact is able to distinguish properties of the set in his definition of power from properties that the set may have by virtue of some particular structure on it. Perhaps the explanation of why Cantor was clearer than earlier writers, and Bolzano in particular, about this lies in his earlier work. The historical confusion was between a geometric object and the set of points constituting it – for example, between a line segment as a geometric object and the set of points on it; and the effect of that confusion was the further one between the measure of the segment (a property of the geometric object) and the number of points on the segment (a property of the set). The explanation I am proposing is that, in his earlier research concerning the uniqueness problem for trigonometric series, he got used to considering sets of points (viz. the domains of convergence of series) to which the notion of measure (certainly as it was understood then) did not obviously extend – in other words, sets devoid of geometric significance. In fact, Cantor did not even bother to explicitly address the confusion between the set and the geometric object as a problem. He was far more concerned with pointing out the distinction between point sets armed with *topological* structure and abstract sets and with explaining why his proof of the equipollence of point sets of different dimension (e.g., the set of points on the line and the set of points in the plane) does not lead to topological paradoxes. (See Cantor [1932, p. 121].) Howard Stein has pointed out to me that, initially, Cantor himself seems to have been not entirely comfortable with his result. In a letter to Dedekind on June 25, 1887 (Nöther and Cavaillès 1937, p. 35), he expresses concern that it undermines the common assumption that no one-to-one map exists between continuous domains of different dimension. It was Dedekind who pointed out to him in a reply on July 2 (Nöther and Cavaillès 1937, p. 37) that what had been assumed was that there be no *continuous* such map.
7. The date I assign, November 1882, may seem to contradict Cantor's own statement in §1 that he had already had the notion of transfinite number in earlier works (Cantor 1880), where he wanted to iterate the operation of taking the derived set (the set of limit points) of a set of real numbers into the transfinite. He introduced what he called the "definite defined infinity symbols," which, in contemporary terms, we

would call a *system of ordinal notations*. In fact, Cantor's symbols constitute a very familiar system of ordinal notations, one that represents precisely all of the ordinals $< \epsilon_0$ – an ordinal in the second number class. The essential difference between this system of notations and Cantor's later definition of the second number class is not, as it is often stated, that the former is a system of *symbols* and the latter of *numbers*. Rather, the important thing is that *any* system of ordinal notations, being countable, will be bounded by some number in the second number class. In fact, in a reasonable sense of the notion of a system of ordinal notations, there is an absolute bound on the ordinal of any such system, namely, the least nonprojectable ordinal, which is still in the second number class. To obtain *all* of the second number class, much less the higher number classes, a new idea was needed, namely, the idea of introducing numbers as limits of *arbitrary* increasing sequences of numbers (ω-sequences in the case of the second number class). As far as I know, at no time before his letter to Dedekind did Cantor explicitly introduce this idea. So, Purkert and Ilgauds, who, in their fine book (1987), claim that the origin of set theory was in Cantor's invention of the the transfinite numbers, a claim with which I entirely agree, may be making a mistake in giving the date of this origin as 1870 (p. 39), when, apparently, Cantor first thought of the ordinal notations. But this may be a matter for more historical investigation. Cantor, himself, in the letter to Dedekind explains that he calls the transfinite numbers *numbers*, in contrast with his earlier name "infinity symbols," because he has introduced the fundamental arithmetic operations on them.

8. Hallett (1984, pp. 50–1) refers to letters to Kronecker and Mittag-Leffler, both in 1884, in which Cantor seems to be supporting the view that the foundation for the transfinite numbers should really be their application as measures of well-ordered sets, that is, that they should be identified with the ordinal numbers. In the letter to Kronecker, he writes "I have for some time had a foundation for these numbers which is somewhat different from that given in my written works, and this will certainly suit you better." He goes on to describe a number as the "symbol or concept" of an order type of a well-ordered set. But notice that it is *Kronecker* that the new foundation would suit better. Cantor was unjustifiably optimistic in thinking that Kronecker, who resisted the introduction of the irrational numbers in terms of ω-sequences of rationals, would be interested in even the second number class, that is, the totality of all well-orderings of the rationals; but surely his optimism did not extend to thinking that he could interest him in the theory of the higher number classes. But it is connection with them, as we have noted, that the conception of number as order type would have been inadequate. The letter to Mittag-Leffler refers to a manuscript that Cantor never published, but is most likely the one published in Grattan-Guinness (1970). But in this manuscript, Cantor is interested in the *general theory of order types* and in this context, certainly, it is reasonable to identifiy the numbers with the ordinal numbers.

9. Section 2 begins

> Since this concept [i.e., of *Anzahl*] is always expressed by a completely determinate number of our extended domain of numbers and since on the other hand the concept of *Anzahl* has an immediate objective representation in our inner intuition (*Anschauung*), then through this connection between *Anzahl* and number, the reality that I stress for the latter, even in the determinate-infinite case, is proven.

The reference to inner intuition is surprising. Later, in the discussion in §10 of the idea of a continuum, Cantor rejects the idea that this can be founded on the concept or intuition of time or on spatial intuition; rather, he argues, following Dedekind in (1872) and, indeed, Bolzano (1817), that just the reverse is true: our adequate conception of space and time depend upon the analysis of the mathematical idea of a continuum. But, he says even more: "even with the help of this latter [i.e., an independent concept of continuity] [time] can be conceived neither objectively as a substance, nor subjectively as a necessary *a priori* form of intuition." Thus he seems in this passage to be rejecting Kant's doctrine of inner intuition, at least insofar as it is identified with time. But, it seems, he is not rejecting it as a basis of the concept of *Anzahl*. On the other hand, note that he is not asserting that it is the basis of the concept of *number*, either finite or infinite. So, in this respect, he is not in conflict with the view of Frege (1884) and Dedekind (1887) that arithmetic can and should be developed purely logically, without reference to inner intuition. But what does he mean when he writes that the concept of *Anzahl* has an objective representation in our inner intuition? My surmise – and it can only be that because I have found no other place in which he discussed this – is that he is referring to the intuitiveness of the idea of iterating an operation, even into the transfinite. Certainly, it was that idea, in the case of iterating the operation of taking the derivative of a set of points on the real line, which led him to the transfinite numbers in the first place.

10. "In this way, by complying with all three elements [i.e., the two principles of generation and the inhibiting principle] one can with the greatest certainty succeed to ever new number classes and powers; and furthermore the new numbers obtained in this way are all entirely of the same concrete determinateness and reality as the old [i.e., the finite numbers]. Thus I truly don't know what holds us back from this process of constructing new whole numbers, so long as it is shown that, for the progress of science, the new introduction of these uncountably many number classes has become desirable or indeed indispensible. And the latter appears to me to be the case in the theory of sets – and perhaps also in a wide range of other cases – at least, without this extension [of the domain of numbers], I can make no further progress and, with it, I obtain much that is entirely unexpected."
11. It is more or less clear what Cantor meant by "*intrasubjektive*" or "*immanente Realität*" and by "*transsubjektive*" or "*transiente Realität*." But where do these *terms* come from?
12. Perhaps some hint of his thinking about this is contained in his remarks on the above quote from *Grundlagen* concerning inner intuition. He writes

> one can make sense of the passage as follows. We invent the concept of (ordinal) number and even postulate that such numbers exist; but the concept obtains legitimacy and significance and we realize that the postulated objects actually *do* exist by recognizing that they correspond to enumerals of well-ordered sets. This is the explanation of the perhaps hazily understood concept of number in terms of, for Cantor, the clear concept of enumeral. (pp. 53–4)

Well, *Cantor* indeed did invent the concept of transfinite whole number; but I doubt that it would have occurred to him to *postulate* that they exist: on the basis of their definition, he would prove that some exist with this or that property. The concept did indeed obtain legitimacy and significance for Cantor by the recognition that the numbers measure well-ordered sets. But this is *not* an *explanation* of the concept of number in terms of that of an enumeral. In whatever sense the concept of number is hazily understood, namely, because (as we will see) it is essentially open-ended, the fact that the numbers measure well-ordered sets does not in any way disperse the haze. And, moreover, as we noted, being enumerals of well-ordered sets does not bestow transient reality on the numbers, anyway, unless the well-ordered sets in question have transient reality.

13. The opening paragraphs of Lavine's book are especially pleasing.
14. That Cantor speaks of *sequences* of numbers rather than sets is inconsequential, since the sequences in question are in their natural order and so are determined by the set of their members.
15. Of the nineteen, fourteen unambiguously refer to theorems or laws of arithmetic. The occurrence of "gesetzmässig" is in a discussion of subsets of R^n defined by any law. Here, Cantor clearly is including the case of sets defined by equations, and so, the usage is not what Lavine suggests. That leaves four occurrences to consider. Two of them, and these are possibly the two that Lavine considered, are contained in the discussion of well-ordered sets and refer to 'laws of counting' or laws of succession, according to which a set can be counted. But there is no implication in these passages that the set is to be *initially* given by such a law. Of the final two occurrences, one refers to the natural numbers in their natural order 'according to law' and the other refers to an arbitrary subset of the second number class, determined by some law.
16. Moreover, the case of classes of numbers is the crucial case. In the usual conception of set theory, every set has a rank, and a class of sets is a set just in case the class of ranks of its elements is a set of numbers.
17. I do not know to what extent Cantor understood the open-ended character of the notion of transfinite number. In particular, did he believe that the class C that we just described contains all numbers? I do not think that the evidence so far is decisive. At the end of note 2, he writes that the smallness of γ compared to ω_γ, when γ is a successor number, "mocks all description; and all the more so, the greater we take γ to be." By excluding the case of limit numbers γ, he excludes the case of fixed points $\gamma = \omega_\gamma$, but let γ be any number and δ a fixed point greater than γ. Then, $\gamma \leq \omega_\gamma < \delta$. So, the smallness of γ compared to ω_γ is hardly increasing the greater we take γ and the difference between γ and ω_γ hardly mocks description. So, despite his exclusion of limit numbers in his statement, it is not clear that he understood the behavior of the initial ordinal function – or of normal functions in general (i.e., order-preserving number-valued functions defined on the numbers that are continuous at limit numbers) – and, in particular, knew that it had fixed points. Much later, in 1897, he discussed the fixed points of the normal exponential function $\alpha \mapsto \omega^\alpha$ (the so-called ϵ-*numbers*), where α ranges over the second number class. But, given the date of this work and the restriction to the second number class, this is not conclusive.

REFERENCES

Bishop, E., and Bridges, D. (1985). *Constructive Mathematics* (Berlin: Springer–Verlag).

Bolzano, B. (1817). "Rein analytischer Beweis des Lehrsatzes, dass zwischen je zwey Werthen, die ein entgegengesetztes Resultat gewähren, wenigstens eine reelle Wurzel der Gleichen liege" (Prague); translation by S. B. Russ, "A translation of Bolzano's Paper on the Intermediate Value Theorem," *Historia Mathematica* 7(1980): 156–85.

Bolzano, B. (1851). *Paradoxien des Unendlichen* (Leipzig); translated by D. A. Steele, *Paradoxes of the Infinite* (London: Routledge and Kegan Paul, 1959).

Cantor, G. (1874). "Über eine Eigenschaft des Inbegriffes aller reelen algebraischen Zahlen," *Journal für die reine und angewandte Mathematik*, 77: 258–62.

Cantor, G. (1878). "Ein Beitrag zur Mannigfaltigkeitslehre," *Journal für die reine und angewandte Mathematik*, 84: 242–58.

Cantor, G. (1880). "Über unendliche, lineare Punktmannigfaltigkeiten, 2," *Mathematische Annalen*, 17: 355–8.

Cantor, G. (1882). "Über unendliche, lineare Punktmannigfaltigkeiten, 3," *Mathematische Annalen*, 20: 113–21.

Cantor, G. (1883). "Über unendliche, lineare Punktmannigfaltigkeiten, 5," *Mathematische Annalen*, 21: 545–86.

Cantor, G. (1891). "Über eine elementare Frage der Mannigfaltigkeitslehre," *Jahresbericht der deutschen Mathematiker-Vereiningung*, 1: 75–8.

Cantor, G. (1897). "Beiträge zur Begründung der transfiniten mengenlehre, 2," *Mathematische Annalen*, 49: 207–46.

Cantor, G. (1932). *Gesammelte Abhandlungen mathematischen und philosophischen Inhalts*, ed. E. Zermelo (Berlin: Springer).

Dedekind, R. (1872). *Stetigkeit und irrationale Zahlen* (Braunschweig: Vieweg). In Dedekind (1932); republished in 1969 by Vieweg and translated in Dedekind (1963).

Dedekind, R. (1887). *Was sind und was sollen die Zahlen*? (Braunschweig: Vieweg). In Dedekind (1932). Republished in 1969 by Vieweg and translated in Dedekind (1963).

Dedekind, R. (1932). *Gesammelte Werke*, ed. R. Fricke, E. Noether, and O. Ore. Vol. 3 (Braunschweig: Vieweg).

Dedekind, R. (1963). *Essays on the Theory of Numbers* (New York: Dover). English translation by W. W. Berman of Dedekind (1872) and Dedekind (1887).

Frege, G. (1884). *Grundlagen der Arithmetik* (Breslau: Verlag von Wilhelm Koebner). German/English edition entitled *The Foundations of Arithmetic*, translation into English by J. L. Austin (Oxford: Basil Blackwell, 1950).

Grattan-Guinness, I. (1970). "An unpublished paper by Georg Cantor: *Principien einer Theorie der Ordnugstypen. Erste Mittheilung*," *Acta Mathematica*, 124: 65–107.

Hallett, M. (1984). *Cantorian Set Theory and Limitation of Size* (Oxford: Oxford University Press).

Kronecker, L. (1886). "Über einige Anwendungen der Modulsysteme auf elementare algebraische Fragen," in K. Hensel (ed.), *Leopold Kronecker's Werke*, Vol. 3 (New York: Chelsea), pp. 147–208.

Lavine, S. (1994). *Understanding the Infinite* (Cambridge, MA: Harvard University Press).

Nöther, E., and Cavaillès, J. (eds) [1937]. *Briefwechsel Cantor-Dedekind*, Paris: Hermann.

Parsons, C. (1974). "Sets and classes," *Noûs* 8: 1–12; reprinted in Parsons (1983).

Parsons, C. (1983). *Mathematics in Philosophy*, Ithaca, NY: Cornell University Press, 1983. pp. 209–20.

Purkert, W. (1989). "Cantor's Views on the Foundations of Mathematics," in D. Rowe and J. McCleary (eds), *The History of Modern Mathematics, Volume I: Ideas and Their Reception* (San Diego, Calif.: Academic Press).

Purkert, W., and Ilgauds, H. (1987). *Georg Cantor: 1845–1918* (Basel: Birkhäuser Verlag).

Zermelo, E. (1930). "Über Grenzzahlen und Mengenbereiche. Neue Untersuchungen über die Grundlagen der Mengenlehre," *Fundamenta Mathematicae* 16: 29–47.

Frege, the Natural Numbers, and Natural Kinds

MARK STEINER

Philosophers identify Frege with the doctrine that arithmetic truths are "analytic." But Frege's main goal, as many now realize,[1] was mathematical: to show that the truths of arithmetic are provable, against many who denied this, each for a different reason. Mathematicians had by Frege's time shown that analysis could be grounded in what we would today call second-order arithmetic: arithmetic plus "logic." Frege's goal was to remove the dependence on arithmetic by beefing up the "logic."

This observation explains what otherwise seems a puzzling digression in the *Grundlagen*. At *Grundlagen* §10, Frege suddenly considers the question of whether the laws of arithmetic might not be "inductive[2] truths," that is, established nondeductively by verifying particular cases. This question is not reducible to the question of whether mathematics is "analytic" in Kant's (or even Frege's) sense. Nor is Frege concerned here, as he is elsewhere, with the more specific claim of John Stuart Mill that arithmetic is empirical. Even if arithmetic is non-empirical, even if it is analytic, it might still be inductive – a possibility that few great philosophers had considered, but that could not escape a mathematician.

Frege's reply is:

> [H]ere there is none of that uniformity, which in other fields can give the method [of induction] a high degree of reliability. Leibniz recognized this already: for to his Philalèthe, who had asserted that "the several modes of number are not capable of any other difference but more or less; which is why they are simple modes, like those of space," he returns the answer: "That can be said of time and of the straight line, but certainly not of the figures and still less of the numbers, which are not merely different in magnitude, but also dissimilar. An even number can be divided into two equal parts, an odd number cannot; three and six are triangular numbers, four and nine are squares, eight is a cube, and so on. And this is even more the case with the numbers than with the figures; for two unequal figures can be perfectly similar to each other, but never two numbers."
>
> We have no doubt grown used to treating the numbers[3] in many contexts as all of the same sort, but that is only because we know a set of general propositions hold for all numbers. For the present purpose, however, we must put ourselves in the position that none of these has yet been discovered. The fact is that it would be difficult to find an

example of an inductive inference to parallel our present case. In ordinary inductions we often make good use of the proposition that every position in space and time is as good in itself as any other. Our results must hold good for any other place and any other time, provided only that the conditions are the same. But in the case of the numbers this does not apply since they are not in space and time. Position in the number series is not a matter of indifference like position in space.

The numbers, moreover, are related to one another quite differently from the way in which the individual specimens of, say, a species of animal are. It is in their nature to be arranged in a fixed, definite order of precedence; and each one is formed in its own special way and has its own unique peculiarities which are specially prominent in the cases of 0, 1, and 2 . . . .

. . . [deleted passage, quoted below, p. 293]

The procedure of induction, we may surmise, can itself be justified only by means of general propositions of arithmetic – unless we understand by induction a mere process of habituation, in which case it has of course absolutely no power whatever of leading to the discovery of truth. The procedure of the sciences, with its objective standards, will at times find a high probability established by a single confirmatory instance, while at others it will dismiss a thousand as almost worthless; whereas our habits are determined by the number and strength of the impressions we receive and by subjective circumstances, which have no sort of right at all to influence our judgment. Induction [then, properly understood,] must base itself on the theory of probability, since it can never render a proposition more than probable. But how probability theory could possibly be developed without presupposing arithmetical laws is beyond comprehension ((Frege 1959, pp. 14–17).

In this admirably clear passage (from which I have dislodged one section, to be discussed later), Frege makes the following points:

1. Inductive knowledge about a set of objects requires that the set must have, and be known to have, common properties.
2. The only way to get this knowledge is by deduction.
3. Therefore, mathematical knowledge is not inductive.
4. Also, inductive knowledge, where objective, depends upon the mathematical theory of probability, which itself obviously depends on, among other theories, arithmetic.
5. Hence, again, mathematical knowledge is not inductive.

These arguments could be resisted, to be sure, by pointing out that, if they were valid, they would make most inductive knowledge impossible: to support any hypothesis about *F*s, one would have to first prove other general properties of *F*s – and for most such predicates, this could not be done deductively, leading to an infinite regress.[4] But suppose Frege's arguments are correct, at least for the cases that he is talking about: What do the arguments prove? What did Frege think they would prove? Consider the following thesis:

6. *Strong Deductivism.* No arithmetic proposition is knowable except by deductive proof.[5]

Now, although Leibniz may have been a strong deductivist, Frege's argument does not support such a sweeping conclusion. Once at least some general propositions about the natural numbers have been proved, they may well be reckoned a "natural kind," "projectible set," upon which induction may now be legitimate. What Frege's argument supports is only

7. *Weak Deductivism. Some* arithmetic propositions must be known by deductive proof.

However, weak deductivism is itself ambiguous. Two possible meanings of doctrine 7 are:

a. It is necessary that *one or another* arithmetic proposition be known by proof.
b. There are *particular* arithmetic propositions that can be known only by proof.

Of these two, the second is more interesting (and logically stronger), so let weak deductivism be the doctrine 7b. Weak deductivism now amounts to the view that there are particular arithmetic propositions, whose proof renders the natural numbers a "natural kind," so that subsequent arithmetic generalizations (but not previous ones) can be projected from special cases.

Weak deductivism is the best Frege (or Leibniz) could have hoped for anyhow, had he known Goedel's theorem about unprovable generalizations in Peano arithmetic, each of whose special cases is provable. Frege believed in a universal deductive system for arithmetic, rather than a hierarchy of systems, but any such universal system, if consistent, would contain Goedel sentences whose establishment would have to be nondeductive.

* * * *

But here is an argument against (even) weak deductivism: the natural numbers *are* a natural kind, since they have an essential common property – following zero by iteration of the "successor" operation! Nothing needs to be proved about them, prior to establishing by induction various generalities about them. This argument is particularly strong, it seems, against Frege, because he himself so defines the natural numbers! (He also reduces the concept of "iteration" to logic, but that does not mitigate the force of the objection to deductivism.)

Interestingly, Frege raises this issue himself. Let me restore the previously excised passage:

. . . Elsewhere when we establish by induction a proposition about a species, we are ordinarily in possession already, merely from the definition of the concept of the species, of a whole series of its common properties. But with the numbers we have difficulty in finding even a single common property which has not actually to be first proved common.

The following is perhaps the case with which our putative induction might most easily be compared. Suppose we have noticed that in a borehole the temperature increases regularly with the depth; and suppose we have so far encountered a wide variety of differing rock strata. Here it is obvious that we cannot, simply on the strength of the observations made at this borehole, infer anything whatever as to the nature of the strata at deeper levels, and that any answer to the question, whether the regular distribution of temperature would continue to hold good lower down, would be premature. There is, it is true, a concept, that of "whatever you come to by going on boring," under which fall both the strata so far observed and those at lower levels alike; but that is of little assistance to us here. And equally, it will be no help to us to learn in the case of the numbers that these all fall together under the concept of "whatever you get by going on increasing by one." It is possible to draw a distinction between the two cases, on the ground that the strata are things we simply encounter, whereas the numbers are literally created, and determined in their whole natures, by the process of continually increasing by one. *Now this can only mean that from the way in which a number, say 8, is generated through increasing by one all its properties can be deduced. But this is in principle to grant that the properties of numbers follow from their definitions, and to open up the possibility that we might prove the general laws of numbers from the method of generation which is common to them all, while deducing the special properties of the individual numbers from the special way in which, through the process of continually increasing by one, each one is formed.* In the same way in the geological case too, we can deduce everything that is determined simply and solely by the depth at which a stratum is encountered, namely its spatial position relative to anything else, from the depth itself, without having any need of induction; but whatever is not so determined, cannot be learned by induction either . . . . [my italics]

That is, it is true that we begin with at least one property common to the natural numbers, namely, their definition (whatever it will turn out to be). But if we have such a definition, we expect to derive from it all the properties of the defined set. Hence induction is unnecessary.

It is easy to see that Frege is far from today's "formalist" or "metamathematical" view of a definition, according to which a definition is not a premise (but rather an abbreviation), and hence nothing "follows" from a definition.[6] But leaving Frege's views of definition aside for a moment, his reply still leaves us with the question: In all other species, except for the natural numbers, Frege allows, the common property or definition of the species permits us to perform inductive reasoning. And, he allows, inductive reasoning is also legitimate, since, he implies, we cannot in general derive all the properties of a species from its definition. Why are the numbers different?

Let us go back to Frege's own[7] words: "the numbers are literally created, and determined in their whole natures, by the process of continually increasing by one." This at least sounds like "constructionist" rhetoric: the natural numbers are *constructed* by their definition, and therefore have no other properties than those given in the definition. When pressed to explain the metaphor of "construction," particularly "construction by definition," some philosophers and mathematicians make the natural numbers *mental* constructions, of which it is presumably more reasonable to say that they could be "constructed" by a mere definition.

But we know Frege was adamantly opposed to all forms of mentalism and psychologism in mathematics. Again and again he subjected the view that the numbers are mental to withering criticism. The numbers, he emphasized without end, are objective; and to make sure that the message was understood, he concluded a diatribe against formalism with the following words: "even the mathematician cannot create things at will, any more than the geographer can; he too can only discover what is there and give it a name." (pp. 107–8, §96). I can only agree with Dauben's remark: "This certainly seems a version of Platonism, and hardly an 'empty' one . . ."[8]

We are forced to the conclusion that there are different kinds of mathematical objects. The fact that each natural number is an iterated successor of zero has something to do with its "generation" – not in the sense of mental construction, but in the objective world of mathematical objects. The definition of the natural numbers does not, of course, create them, but the existence of an inductive definition reflects a *pre-existing metaphysical reality*, that the existence of the numbers is dependent upon their relation to zero. Thus, with respect to the natural numbers and arithmetic, Frege might be called a "constructive Platonist."

The points of Euclidean space, by contrast, have no such generating relation to which their existence is due. For this reason, there is no reason to believe that we can prove all the properties of space *from a definition*, even if we could define the concept "point in space." On the other hand, Euclidean geometry demonstrates that we can prove at least the mathematical properties of space *from axioms*. Since the axioms are not those of logic, Frege agrees with Kant[9] that geometry is a synthetic a priori theory. More specifically, he agrees with Kant that geometry, unlike arithmetic, is derived from a sensory intuition – that of space. But the Kantian standpoint is that a synthetic a priori proposition can also be known – and is often known – a posteriori, that is, by induction.[10] This is because synthetic a priori propositions, at least for Kant, are about the empirical world of space-time. For example, one can check by measurement that the shortest distance between two points is a straight line; one can also note (says Kant) that the proposition holds of any possible "intuition" (sense perception).

Since space is connected with sense perception (even though it is not perceived itself), it makes sense to the Kantian (or Fregean) that regions of space have empirical properties. Some of these properties could be known only by induction, for example, the existence of various kinds of fields. Note, though, that the natural numbers are completely different, as mathematical objects, from spatiotemporal points. For Frege, numbers have no empirical properties whatever. "Being the number of planets" is not, for Frege, an empirical property of 9. What is the case, for him, is that 9 = the number of planets. What is predicated of 9 here is not "the number of planets" but "being identical to the number of planets" – and there is no sense in which this, even if a property, is an empirical property. In other words, unlike spatial points, of which it can be said of one of them, x, that x is X, where X is an empirical property, there is not any true sentence of the form "9 is X" where X is an empirical property. As Frege aficionados know, the real logical form of "9 is the number of planets" is, for Frege, just the opposite: we are not predicating anything of 9, but rather something of the *concept* of being a planet, namely, that *it* has the number 9. Numbers do not relate to the world, but to our conceptualization of the world.[11]

Some readers may now object to this dichotomy between arithmetic and geometry. Doesn't modern mathematics – doesn't modern philosophy, for that matter – arithmetize geometry? Is it not the case that the Euclidean plane is nothing but the set $R \times R$ (i.e., the set of ordered pairs $\langle x, y \rangle$ where x and y are real numbers) with the distance defined between any two points as $\sqrt{(x_2 - x_1)^2 - (y_2 - y_1)^2}$? Didn't Descartes show that a straight line is nothing but the points $\langle x, y \rangle$ that solve an equation of the form $y = mx + b$? Isn't geometry reducible to analysis, and didn't Frege hold that analysis is logic just as arithmetic?

The answer is that Frege did hold that analysis is logic, but he did not hold that geometry is analysis. Viewing geometry as the study of $R \times R$ with the Pythagorean metric is not reducing, but modeling geometry in analysis. Unlike Quine, Frege never confused a model with a reduction though Wittgenstein can be understood as saying that he did.

It was self-evident to Frege that a point in space is simply a different object from an ordered pair $\langle x, y \rangle$ of numbers, and that the Euclidean plane is not the mathematical object $R \times R$ under the Pythagorean metric – even if we can model the axioms of Euclid in this way. Frege would probably have retorted to "Quine" that we can just as well model non-Euclidean geometry as Euclidean, but only one of these can be the geometry of space. (Of course, one can deny the existence of "space" altogether, as did Leibniz, but that is not the issue between Frege and Quine. Both Frege and Quine are willing to "quantify over" spatial points, albeit for different reasons.) Whether Euclidean or non-Euclidean geometry describes space, further, has great implications for the applications of geometry in physics. Logic cannot decide this issue. It is true, of course, that Frege (unfortunately) followed Kant in making the issue one to be decided a priori. But it is a synthetic a priori, which means that the issue could also be decided empirically, that is, by measurement, by those not aware that geometry is a priori.

Recall, now, that for Frege in any case, spatial points (unlike numbers) have other than geometrical properties. Now the application of geometry in physics follows precisely from the fact that the spatial points have other than mathematical properties. We can place physical rulers, for example, between spatial points. These other properties are not preserved by simply modeling geometry in $R \times R$. The applications of arithmetic, on the other hand, do not depend on the numbers themselves having nonarithmetic properties. Numbers, as before, relate only to the world of concepts. Arithmetic is applicable to physical science simply because we can instantiate the concepts of physical science in the universal conceptual truths of arithmetic. It is for this reason that, as Benacerraf first pointed out, the actual identity of the numbers need not be fixed by Frege's logicist program, and a number of "models" are possible. (For example, Frege says that the numbers could have been concepts, rather than the extensions of concepts.) The paradigmatic applications of arithmetic do not depend on the identity of the numbers or even whether they are objects (rather than – higher-order – concepts). What is important to applications of arithmetic is what Frege has called "Hume's Principle," which links equinumerosity with 1-1 correlations.

NOTES

Parsons' groundbreaking article (Parsons 1965) inspired me to begin thinking about Frege's philosophy of mathematics. Other works of Parsons (1979–80, 1982) led me to think about mathematical objects. I had the good fortune of being Charles' colleague at Columbia for seven years so that I could learn from him directly. I am honored to dedicate this essay to my teacher, colleague, and friend.

1. Benacerraf (1981), reprinted in Demopoulos (1995).
2. Whenever the term "induction" is used in this article, I refer not to "mathematical induction," but to standard induction, as in "the problem of induction."
3. I take Frege to be referring here to the natural numbers, although, in principle, Frege regards the natural numbers as only part of the numbers that there are.
4. This point is not new, of course. It is the basis of Goodman's New Riddle of Induction. Goodman's theory of "entrenchment" was supposed to have stopped the infinite regress.
5. If it is asked, what about the axioms of arithmetic – how are they known? Frege would answer that the ultimate axioms of arithmetic proofs are logical axioms, not arithmetic ones.
6. I am not the first to point this out. Cf. Dummett (1991).
7. It's true that these words are put into the mouth of Frege's imaginary interlocutor, but Frege does not disavow them in his subsequent remarks – he says only that this point of view supports logicism rather than inductivism.
8. Dauben (1993, p. 619).
9. This statement is complicated because Kant and Frege had differing notions both of the analytic and of the a priori. Cf. "Frege: the Last Logicist" (Benacerraf 1981) for a discussion of Frege's concept of the a priori.

10. Exegetically, I am not sure that Kant would agree that the "same" proposition can be known empirically or a priori. It may be internal to a proposition that it is empirical, say, or a priori. For Kant sometimes says that a proposition changes when it becomes known stated a priori. But, first, this is a conflation that Frege explicitly avoids: in the *Grundlagen* he explicitly says that "a priori" has to do with the justification, not the content, of a proposition. Second, even Kant would have to explain why philosophers have often been confused over whether a proposition is a priori or empirical, if they are not the same propositions even in principle.
11. Am I saying, then, that "9 is the number of planets" is an a priori truth for Frege? Of course not: Frege's deep point is that empirical truths need not be "about" empirical objects – they can be "about" empirical *concepts*.

REFERENCES

Benacerraf, Paul (1981). "Frege: The Last Logicist," *Midwest Studies in Philosophy*, 6: 17–35.

Dauben, Joseph W. (1993). "Review of Joan Weiner's *Frege in Perspective*," *Isis* 84: 618–19.

Demopoulos, William (1995). *Frege's Philosophy of Mathematics*, ed. William Demopoulos (Cambridge, MA: Harvard University Press).

Dummett, Michael (1991). *Frege: Philosophy of Mathematics* (Cambridge, MA: Harvard University Press).

Frege, Gottlob (1959). *The Foundations of Arithmetic*, translated by J. L. Austin (Oxford: Blackwell); republished by Northwestern University Press.

Parsons, Charles (1965). "Frege's Theory of Number," in M. Black (ed.), *Philosophy in America* (Ithaca, NY: Cornell University Press), pp. 180–203.

Parsons, Charles (1979–80). "Mathematical Intuition," *Proceedings of the Aristotelian Society* 80 New Series: 142–68.

Parsons, Charles (1982). "Objects and Logic," *The Monist*, 65: 491–516.

A Theory of Sets and Classes

PENELOPE MADDY

The nature of classes – particularly proper classes, collections "too large" to be sets – is a perennial problem in the philosophy and foundations of set theory. Logicians worry that the unrestricted quantifiers of set theory must range over the collection of all sets, a collection that cannot itself be a set, and hence, a collection that is ill-understood; philosophers puzzle over the existence of properties (such as $x \notin x$) that seem to have no extensions; set theorists ponder heuristic arguments that involve performing operations on the entire universe, V, of sets as if it were a set. Existing theories of sets and classes seem unsatisfactory because their 'proper classes' are either indistinguishable from extra layers of sets or mysterious entities in some perpetual, atemporal process of becoming. In the spirit of Cantor's bold introduction of the completed infinite, we might hope for a theory of sets and classes that both distinguishes plausibly between the two and treats classes as bona fide entities. In the end, this may be too much to ask, but it seems at least the right place to begin.

Several of Charles Parsons' papers have addressed the difficult problem of sets and classes in insightful and influential ways.[1] One central thrust of his treatment has been to emphasize the strong analogy between paradoxes of truth, such as the liar, and paradoxes of classes, such as Russell's paradox. Another has been his clear distinction between sets as mathematical entities, determined combinatorially in a series of stages, and classes as logical entities, determined as extensions of predicates or properties. As Parsons notes, the goals of set theory as a foundation and a branch of mathematics can be met by restricting attention to the mathematical collections, as understood in ZFC. This makes it seem that the paradoxes of classes have been resolved, while the paradoxes of truth have not. But what has really happened is that the mathematical sets have been distinguished from the logical classes, and the emphasis of study has shifted from the latter to the former. The paradoxes of the logical classes remain as stubborn as the paradoxes of truth.

As is well known, Kripke and his successors present an approach to truth that resolves the paradoxes by allowing truth-value gaps.[2] Here I propose an analogous theory of logical classes that resolves their paradoxes by allowing gaps in the membership relation.[3] As it happens, this approach is something of an anachronism, for it seems that Kripke-like constructions for the case of classes preceded Kripke's own work on truth.[4] The difference between that work and what I propose here is that those theories aimed to supplant typed mathematical sets with untyped logical classes, while my goal is to provide a theory that includes and coordinates both kinds of entities. The earlier constructions could be carried out over any ground model, in particular, over a ground model of set theory, and the resulting structure would have a new class membership relation, distinct from any relation, including set membership, that might be present in the ground model; in contrast, the result of my construction will be a model with one membership relation involving both sets and classes. This seems to me preferable, as, for example, both the set of natural numbers and the class of all infinite collections[5] ought to be members of the class of all infinite collections, members in the same sense that two is a member of the set of natural numbers.

I will be aiming, then, for a theory of sets and classes that characterizes sets as usual, that is, as combinatorially determined in stages, and classes as extensions, capable of such things as self-membership. The construction and accompanying discussion will take up Section I. Section II contains a discussion of the oddities of equality and extensionality, Section III takes up axiomatization and fixed points, and Section IV (given the Fregean inspiration behind much interest in classes) explores the notion of equinumerosity.

I. The Construction

Begin with a first-order language with equality and '$\in$' as its only nonlogical symbol. To this language, add a term-forming operator '$\hat{}$' (called 'hat'), to form terms such as $\hat{x}(x = x)$, $\hat{x}(x \in \emptyset)$, and $\hat{x}(\exists y(x \in \hat{z}(z \notin y)))$.[6] This machinery alone would be enough to handle cases like these three – $\hat{x}(x \in \emptyset)$ could be simulated by assignment of $\emptyset$ to the free variable in $\hat{x}(x \in y)$ – but we make two further additions: for added expressive power, a constant $\overline{V}$ to stand for the class of all sets,[7] and for added simplicity, a constant $\bar{a}$ for every set a. The second addition was apparently standard in the older theories,[8] but it is stronger here, when the intended ground model contains all sets. Though the simplicity gained is great, it could be done without, as will be indicated below.

Define 'formula of $\mathscr{L}$' and 'term of $\mathscr{L}$' simultaneously, as follows[9]:

(i) All constants and variables are terms.
(ii) If t and t' are terms, then $t = t'$ and $t \in t'$ are formulas.
(iii) If ϕ and ψ are formulas, and x is a variable, then $\sim\phi$, $\phi \wedge \psi$, and $\forall x\phi$ are formulas.

(iv) If ϕ is a formula, and x is among the free variables of ϕ, then $\hat{x}\phi$ is a term.

The collection of all terms, T, is the union of S, the collection of all set constants, C, the collection of all $\mathscr{L}$-terms of the form $\hat{x}\phi$, and $\{\overline{V}\}$. C^* is the collection of closed class terms in C; T^* is the collection of all closed terms.[10]

Semantically, we think of $\mathscr{L}$ as a partly interpreted language: the intended domain includes all sets, the constant $\bar{a}$ stands for the set a, the constant $\overline{V}$ stands for a class whose extension includes all sets and whose antiextension includes all classes. The only variable part of the interpretation of $\mathscr{L}$ is the extensions and antiextensions assigned to class terms in C^*.

DEFINITION: $\mathfrak{C} = \{(t, t^+_{\mathfrak{C}}, t^-_{\mathfrak{C}}) \mid t \in C^*\}$ is an $\mathscr{L}$-structure iff for all $t \in C^*$, $t^+_{\mathfrak{C}} \subseteq T^*$ and $t^-_{\mathfrak{C}} \subseteq T^*$, and $t^+_{\mathfrak{C}} \cap t^-_{\mathfrak{C}} = \emptyset$.

The idea is that $t^+_{\mathfrak{C}}$ and $t^-_{\mathfrak{C}}$ represent the extension and the antiextension of the class term t; sets and classes are represented in $t^+_{\mathfrak{C}}$ and $t^-_{\mathfrak{C}}$ by their terms. We leave open the possibility that $t^+_{\mathfrak{C}} \cup t^-_{\mathfrak{C}} \neq T^*$, that is, the possibility that some sets or classes will appear in neither the extension nor the antiextension of t.[11]

Because $\mathfrak{C}$ is a partial interpretation in this sense, we must distinguish three possibilities for a given sentence σ: $\mathfrak{C} \models \sigma$ ($\mathfrak{C}$ thinks σ is true), $\mathfrak{C} \not\models \sigma$ ($\mathfrak{C}$ thinks σ is false), and $\mathfrak{C} \overset{?}{\models} \sigma$ ($\mathfrak{C}$ has no opinion about σ). The 'thinks' relation will be defined only for sentences, beginning with the atomic case:

DEFINITION: If σ is of the form $t \in t'$, for t, t' in T^*, then $\mathfrak{C} \models \sigma$ iff

(i) t is $\bar{a}$, t' is $\bar{b}$, and $a \in b$, or
(ii) $t \in S$ and t' is $\overline{V}$, or
(iii) $t' \in C^*$ and $t \in (t')^+_{\mathfrak{C}}$.

$\mathfrak{C} \not\models \sigma$ iff

(i) t is $\bar{a}$, t' is $\bar{b}$, and $a \notin b$, or
(ii) t is $\overline{V}$ and $t' \in S \cup \{\overline{V}\}$, or
(iii) $t \in C^*$ and $t' \in S \cup \{\overline{V}\}$, or
(iv) $t' \in C^*$ and $t \in (t')^-_{\mathfrak{C}}$.

DEFINITION: If σ is of the form $t = t'$ for t, t' in T^*, then $\mathfrak{C} \models \sigma$ iff t and t' are the same term, and $\mathfrak{C} \not\models \sigma$ iff t and t' are different terms.[12]

Finally, with these interpretations of the atomic sentences, truth and falsity for complex sentences are defined using the strong Kleene rules.

DEFINITION: For $\mathscr{L}$-sentences σ and τ,

$\mathfrak{C} \models \sim\sigma$ iff $\mathfrak{C} \not\models \sigma$; $\mathfrak{C} \not\models \sim\sigma$ iff $\mathfrak{C} \models \sigma$.
$\mathfrak{C} \models \sigma \wedge \tau$ iff $\mathfrak{C} \models \sigma$ and $\mathfrak{C} \models \tau$; $\mathfrak{C} \not\models \sigma \wedge \tau$ iff $\mathfrak{C} \not\models \sigma$ or $\mathfrak{C} \not\models \tau$.
$\mathfrak{C} \models \forall x\phi$ iff for all $t \in T^*$, $\mathfrak{C} \models \phi(t/x)$.
$\mathfrak{C} \not\models \forall x\phi$ iff for some $t \in T^*$, $\mathfrak{C} \not\models \phi(t/x)$.

Then '$\vee$', '$\supset$', '$\equiv$', and '$\exists$' can be defined from these in the usual ways.[13]

It is easy to see that these notions are monotonic; that is, if we define

DEFINITION: If $\mathfrak{C}$ and $\mathfrak{C}'$ are two $\mathscr{L}$-structures, then $\mathfrak{C} \sqsubseteq \mathfrak{C}'$ iff for all $t \in C^*$, $t^+_{\mathfrak{C}} \subseteq t^+_{\mathfrak{C}'}$ and $t^-_{\mathfrak{C}} \subseteq t^+_{\mathfrak{C}'}$.

Then it is an easy induction on the complexity of formulas to show that

PROPOSITION: If $\mathfrak{C} \sqsubseteq \mathfrak{C}'$, then for all $\mathscr{L}$-sentences σ, if $\mathfrak{C} \models \sigma$, then $\mathfrak{C}' \models \sigma$, and if $\mathfrak{C} \not\models \sigma$, then $\mathfrak{C}' \not\models \sigma$.

In other words, once a sentence is decided, adding more elements to the extensions and antiextensions of classes does not disturb that decision.

This simple semantic theory is bought at the expense of a fairly extravagant syntax, that is, at the expense of 'proper class many' constants. As mentioned earlier, this could be avoided by sticking to a language without set constants and allowing assignments to free variables in the usual Tarskian fashion. The semantics for the language would then include interpretations (i.e., extensions and antiextensions) for terms coupled with assignments of constants to free variables, that is, for items of the form $(t, \overline{p})$, where $\overline{p}$ assigns members of $S \cup \{\overline{V}\}$ to the free variables of t. The drawback is the complexity of the definition of satisfaction. When an assignment s to the free variables of ϕ assigns one of these $(t, \overline{p})$s to a free variable that itself occurs within a class term, the scheme becomes notationally complex. In general, to determine whether or not $\mathfrak{C}$ thinks ϕ at s, we must first adjust the free variables of any terms that s assigns so that they do not clash with free variables of ϕ or each other; second, substitute the adjusted t-part of any assigned $(t, \overline{p})$s into the relevant places in ϕ; and third, adjust s to make the assignments dictated by the appropriate $\overline{p}$s.[14] To avoid all this, I prefer to complicate the syntax.

With this machinery in place, we construct a sequence of $\mathscr{L}$-structures, $\mathfrak{C}_0$, $\mathfrak{C}_1$, $\mathfrak{C}_2$, as follows:

$$\mathfrak{C}_0 = \{(\hat{x}\phi, \hat{x}\phi^+_0, \hat{x}\phi^-_0) \mid \hat{x}\phi \in C^*\}, \text{ where } \hat{x}\phi^+_0 = \hat{x}\phi^-_0 = \emptyset.$$
$$\mathfrak{C}_{\alpha+1} = \{(\hat{x}\phi, \hat{x}\phi^+_{\alpha+1}, \hat{x}\phi^-_{\alpha+1}) \mid \hat{x}\phi \in C^*\}, \text{ where}$$
$$\hat{x}\phi^+_{\alpha+1} = \{t \in T^* \mid \mathfrak{C}_\alpha \models \phi(t/x)\} \text{ and } \hat{x}\phi^-_{\alpha+1} = \{t \in T^* \mid \mathfrak{C}_\alpha \not\models \phi(t/x)\}.$$

For λ a limit ordinal,

$$\mathfrak{C}_\lambda = \{(\hat{x}\phi, \hat{x}\phi^+_\lambda, \hat{x}\phi^-_\lambda) \mid \hat{x}\phi \in C^*\}, \text{ where}$$
$$\hat{x}\phi^+_\lambda = \bigcup_{\alpha<\lambda} \hat{x}\phi^+_\alpha \text{ and } \hat{x}\phi^-_\lambda = \bigcup_{\alpha<\lambda} \hat{x}\phi^-_\alpha.$$

Finally, we define an $\mathscr{L}$-structure U (for 'universe'):

$$U = \{(\hat{x}\phi, \hat{x}\phi^+, \hat{x}\phi^-) \mid \hat{x}\phi \in C^*),$$

where

$$\hat{x}\phi^+ = \bigcup_{\alpha\in\mathrm{Ord}} \hat{x}\phi^+_\alpha \quad \text{and} \quad \hat{x}\phi^- = \bigcup_{\alpha\in\mathrm{Ord}} \hat{x}\phi^-_\alpha.$$

This U and the theory it generates are what interest us. By monotonicity, whatever becomes true (or false) at one of the $\mathfrak{C}_\alpha$s remains true (or false) in U. To take one easy example, $\mathfrak{C}_0 \models \bar{\emptyset} \in \overline{\{\emptyset\}}$, and so, $\bar{\emptyset} \in \hat{x}(x \in \overline{\{\emptyset\}})^+_1$; thus, $\mathfrak{C}_1 \models \bar{\emptyset} \in \hat{x}(x \in \overline{\{\emptyset\}})$, which in turn implies that $U \models \bar{\emptyset} \in \hat{x}(x \in \overline{\{\emptyset\}})$.

The classes of U are strictly monadic, but 'ordered tuples' can be defined in various ways. The simplest plays on our strict definition of '='.

DEFINITION: For $t, t' \in T^*$, '(t, t')' is $\hat{z}(z = t \vee z = t')$.[15]

Since $\hat{z}\,(z = t \vee z = t')$ is not the same symbol as $\hat{z}\,(z = t' \vee z = t)$, it is easily established that

PROPOSITION: For $t,\ t' \in T^*, U \models ((t, t') = (u, u'))$ iff $U \models (t = t' \wedge u = u')$.

Notice that these 'ordered classes' are total – that is, every element of T^* is in either the extension or the antiextension – so it follows that $U \not\models ((t, t') = (u, u'))$ iff $U \not\models (t = t' \wedge u = u')$. The force of 'ordered n-tuples' can then be recovered as usual – $(t, t', t'') = ((t, t'), t'')$ – and the following useful notation introduced.[16]

DEFINITION: If $x_0, x_1, \ldots$ are among the free variables of ϕ, then $\hat{x}_0 \cdots \hat{x}_n\phi$ abbreviates $\hat{z}(\exists x_0 \cdots \exists x_n(z = (x_0, x_1, \ldots x_n) \wedge \phi))$, where z is the first variable (in some canonical listing) that does not appear in ϕ.

Continuing the earlier example, $\mathfrak{C}_0 \models \bar{\emptyset} \in \overline{\{\emptyset\}}$, so, $\mathfrak{C}_0 \models \exists x \exists y((\bar{\emptyset}, \overline{\{\emptyset\}}) = (x, y) \wedge x \in y)$, which means that $(\bar{\emptyset}, \overline{\{\emptyset\}}) \in \hat{z}(\exists x \exists y(z = (x, y) \wedge x \in y))^+_1$, and thus that $\mathfrak{C}_1 \models ((\bar{\emptyset}, \overline{\{\emptyset\}}) \in \hat{x}\hat{y}(x \in y))$. By monotonicity, U agrees.

More substantively, U underwrites our original intuition that $\hat{x}$ (x is infinite) is self-membered. For a sketch of the proof,[17]call a collection, x, 'relational' iff $\forall y(y \in x \supset \exists u \exists v(y = (u, v)))$. A non-empty relational collection will be a class, not a set, because it has classes as members. We can then use the standard methods to define what it is to be a domain or range of a relational class[18] and what it is for a relational class to be functional or one-to-one. For any set function f, let f^* be $\hat{x}(\exists y \exists z(\langle y, z\rangle \in \overline{f} \wedge x = (y, z)))$, where $\langle y, z\rangle \in \overline{f}$ is the usual statement that the Kuratowski ordered pair of y and z is in $\overline{f}$, but with y, z, and all quantifiers relativized to $\overline{V}$. Then it is not hard (though somewhat tedious) to see that

PROPOSITION: For any sets a and b,

(1) if $\langle a, b\rangle \in f$, then $\mathfrak{C}_1 \models (\overline{a}, \overline{b}) \in f^*$;
(2) if $\langle a, b\rangle \notin f$, then $\mathfrak{C}_1 \not\models (\overline{a}, \overline{b}) \in f^*$;
(3) $\mathfrak{C}_1 \models \forall x(x \in f^* \vee x \notin f^*)$ (i.e., 'f^* is total').

In general, f^* will behave as a class surrogate for f in $\mathfrak{C}_1$.

We can then define 'infinite' in a familiar way.

DEFINITION: 'x is infinite' abbreviates $\exists f$('f is functional' $\wedge$ 'f is one-to-one' $\wedge$ '$\overline{\omega}$ is a domain of f' $\wedge$ 'x contains a range of f').

For any $n \in \omega$, let $n^* = \{n+1, n+2, \ldots\}$. This set is obviously infinite; using f^*, where f is a set function that maps ω one-to-one into n^*, we can easily show that

PROPOSITION: For all $n \in \omega$, $\mathfrak{C}_2 \models \overline{n^*} \in \hat{x}$ (x is infinite).

Let $\phi(y, z)$ be the formula $y \in \overline{\omega} \wedge z \in \overline{\wp(\omega)} \wedge \forall u(u \in \overline{V} \supset (u \in z \equiv u \in \overline{\omega} \wedge y \in u))$. Then,

PROPOSITION: For all $n \in \omega$, $\mathfrak{C}_1 \models (\overline{n}, \overline{n^*}) \in \hat{y}\hat{z}\phi$.

$\mathfrak{C}_1$ also thinks that $\hat{y}\hat{z}\phi$ is functional and one-to-one, and that ω is a domain; by the preceding proposition, $\mathfrak{C}_2$ thinks that $\hat{x}$ (x is infinite) contains a range. It follows that

Theorem. $\mathfrak{C}_3 \models \hat{x}$ (x is infinite) $\in \hat{x}$ (x is infinite).

Finally, as expected, there are gaps in the membership relation of U. The famous example behaves as it should.

PROPOSITION: $U \overset{?}{\models} \hat{x}(x \notin x) \in \hat{x}(x \notin x)$.

Proof: If $\hat{x}(x \notin x)$ were in $[\hat{x}(x \notin x)]^+$, then it would have to enter at some $\hat{x}(x \notin x)^+_\alpha$. Because α cannot be a limit, it must be of the form $\beta + 1$. But then $\mathfrak{C}_\beta \models \hat{x}(x \notin x) \notin \hat{x}(x \notin x)$, and so, $\mathfrak{C}_\alpha \models \hat{x}(x \notin x) \in \hat{x}(x \notin x)$ (because $\hat{x}(x \notin x) \in \hat{x}(x \notin x)^+_\alpha$) and $\mathfrak{C}_\alpha \models \hat{x}(x \notin x) \notin \hat{x}(x \notin x)$ (because $\mathfrak{C}_\beta \sqsubseteq \mathfrak{C}_\alpha$). Contradiction. Similarly, $\hat{x}(x \notin x) \notin \hat{x}(x \notin x)^-$. ■

II. Equality and Extensionality

In a context with membership gaps, extensionality can be formulated as follows: If collections A and B have the same extension and the same antiextension, then $A = B$. (Essentially, we have taken the extension of a set, a, to consist of its members, and its antiextension to consist of everything else (i.e., if we represent it by $T^* - \{\bar{b} \mid b \in a\}$).] Clearly, our definition of '$=$' is not extensional: For example, the terms $\bar{a}$ and $\hat{x}(x \in \bar{a})$ have the same extension and antiextension, but they are not identical. This outcome is perhaps not unwelcome; after all, one of our motivational intuitions was the conviction that sets and classes are entities of two quite different kinds. As a practical matter, identifying classes with coextensional and antiextensional sets would have the result that sets are not total – for example, it would be indeterminate whether $\hat{z}(z \in \bar{a} \wedge (\hat{x}(x \notin x) \in \hat{x}(x \notin x)))$ is a member of $\overline{\{a\}}$ – a decidedly unwelcome result.

Another form of identity relation that might seem attractive is the following natural notion.

DEFINITION: For $t, t' \in T$, '$t \simeq t'$' abbreviates $\forall z(z \in t \equiv z \in t')$, where z is the first variable (in some canonical listing) not in t or t'.

This relation does hold (in U) between coextensive sets and classes like $\bar{a}$ and $\hat{x}(x \in \bar{a})$, and between mildly varying classes like $\hat{x}(x \in \bar{a})$ and $\hat{x}(x \in \bar{a} \wedge x \in \bar{a})$, as well. But the oddities of the connective '$\equiv$' in this context produce some surprises. For U to think, for example, that $u \in t \supset u \in t'$, it must think $u \in t$ is false or $u \in t'$ is true. If U should happen to be undecided about $u \in t$, then it must think $u \in t'$, which in turn means that it must think $u \in t$. This means that for U to think $u \in t \equiv u \in t'$, it cannot be undecided about $u \in t$ or $u \in t'$. Indeed '$t \simeq t$' is a way of expressing the fact that t is total; it is easy to check that $U \models t \simeq t$ if and only if $U \models \forall x(x \in t \vee x \notin t)$. So, $U \overset{?}{\models} (\hat{x}(x \notin x) \simeq \hat{x}(x \notin x \wedge x \notin x))$. For that matter, $U \overset{?}{\models} (\hat{x}(x \notin x) \simeq \hat{x}(x \notin x))$!

Of course, if we confine our attention to sets, the trivial definition of '$=$' adopted here will still guarantee that coextensive sets are equal. (If a and b are coextensive, they are in fact the same set, and so, $\bar{a} = \bar{b}$.) But attempted identifications across the set/class boundary are not the only failures of extensionality produced by our definition; classes can be coextensive without coinciding if they are picked out by different terms. To a certain extent, this seems appropriate, because classes are understood as closely tied to the properties that determine them, and coextensive properties are not identified. Still, it seems a bit much to distinguish $\hat{x}(x \in \bar{a})$ from $\hat{y}(y \in \bar{a})$, and some might balk at distinguishing $\hat{z}(z \in x \vee z \in y)$ from $\hat{z}(z \in y \vee z \in x)$.[19] It might be possible to modify my 'trivial' definition of '$=$' to suit various notions of how properties should be individuated, but I will not take up this topic here.

III. Axiomatics and Fixed Points

Of course, we cannot expect to axiomatize all the truths of a structure as rich as U, but we might hope for an axiomatization that provides as much information about sets as ZFC as well as a usefully large body of truths about classes. Unfortunately, the nonclassical context created by the membership gaps is a serious obstacle; for an authoritative discussion, see Feferman (1984). A telling symptom is that we cannot hope for an axiom of the form $x \in \hat{x}\phi \equiv \phi$, because the biconditional is indeterminate if either side is. The general problem is well illustrated by this version of Curry's paradox, drawn from Flagg and Myhill (1987):

PROPOSITION: Any system with the following properties:

(i) $\Gamma \vdash x \in \hat{x}\phi$ iff $\Gamma \vdash \phi$ (Frege's principle)
(ii) $\Gamma \cup \{\phi\} \vdash \psi$ iff $\Gamma \vdash \phi \supset \psi$ (deduction theorem)
(iii) if $\Gamma \vdash \phi$ and $\Gamma \vdash \phi \supset \psi$, then $\Gamma \vdash \psi$ (Modus ponens)

is inconsistent.

Proof: Let σ be any sentence. Then,

$$\hat{x}(x \in x \supset \sigma) \in \hat{x}(x \in x \supset \sigma) \vdash \hat{x}(x \in x \supset \sigma) \in \hat{x}(x \in x \supset \sigma).$$

So, by Frege's principle,

$$\hat{x}(x \in x \supset \sigma) \in \hat{x}(x \in x \supset \sigma) \vdash (\hat{x}(x \in x \supset \sigma) \in \hat{x}(x \in x \supset \sigma)) \supset \sigma.$$

By modus ponens,

$$\hat{x}(x \in x \supset \sigma) \in \hat{x}(x \in x \supset \sigma) \vdash \sigma.$$

By the deduction theorem,

$$\vdash [\hat{x}(x \in x \supset \sigma) \in \hat{x}(x \in x \supset \sigma)] \supset \sigma.$$

By Frege's principle again,

$$\vdash \hat{x}(x \in x \supset \sigma) \in \hat{x}(x \in x \supset \sigma).$$

So, by modus ponens, $\vdash \sigma$. ■

It is hard to know how to proceed without modus ponens and the deduction theorem. Nevertheless, for the case at hand, Myhill has suggested a straightforward infinitary axiomatization. Though the deduction theorem fails, the system is worth considering for the further questions it raises.

Because the comprehension axiom cannot be effectively expressed using the Kleene '$\equiv$', Myhill uses a battery of rules in addition to axioms. Call the following system M.

AXIOMS: For any $a, b \in V$, and any $t \in C^*$,

(i) $\bar{a} \in \bar{b}$ when $a \in b$,
(ii) $\bar{a} \notin \bar{b}$ when $a \notin b$,
(iii) $\bar{a} \in \overline{V}$,
(iv) $t \notin \bar{a}$,
(v) $t \notin \overline{V}$,

for any distinct $t, t' \in T^*$,

(vi) $t = t$,
(vii) $t \neq t'$.

RULES:

(1) $\dfrac{\phi}{\sim\sim\phi}$

(2) $\dfrac{\phi, \psi}{\phi \wedge \psi} \qquad \dfrac{\sim\phi}{\sim(\phi \wedge \psi)} \qquad \dfrac{\sim\psi}{\sim(\phi \wedge \psi)}$

(3) $\dfrac{\phi(t/x) \text{ for all } t \in T^*}{\forall x\phi} \qquad \dfrac{\sim\psi(t/x) \text{ for some } t \in T^*}{\sim\forall x\phi}$

(4) $\dfrac{\phi(t/x)}{t \in \hat{x}\phi} \qquad \dfrac{\sim\phi(t/x)}{t \notin \hat{x}\phi}$

We write $\vdash_M \sigma$ when σ can be derived in this system. Notice that the set constants play an ineliminable role here, for the first time.

M is a success if the same sentences are provable as are true, that is, if for all σ, $\vdash_M \sigma$ if and only if $U \models \sigma$ (and $\vdash_M \sim \sigma$ iff $U \not\models \sigma$). To begin, it is fairly easy to see (by a double induction, first on α, then on the complexity of σ) that if $\mathfrak{C}_\alpha \models \sigma$, for some α, then $\vdash_M \sigma$ (and if $\mathfrak{C}_\alpha \not\models \sigma$, for some α, then $\vdash_M \sim \sigma$). To get from here to the completeness of M requires that $U \models \sigma$ imply the existence of an α such that $\mathfrak{C}_\alpha \models \sigma$ (and that $U \not\models \sigma$ imply the existence of an α such that $\mathfrak{C}_\alpha \not\models \sigma$). This would obviously be true if the construction reached a fixed point, that is, if there were an $\mathfrak{C}_\alpha$ such that $\mathfrak{C}_\alpha = \mathfrak{C}_{\alpha+1}$. As for soundness, we know that $\mathfrak{C}_\alpha \models \sigma$, for some α, implies that $U \models \sigma$ (and $\mathfrak{C}_\alpha \not\models \sigma$, for some α, implies $U \not\models \sigma$). To get from here to soundness, we need to know that $\vdash_M \sigma$ implies the existence of an α such that $\mathfrak{C}_\alpha \models \sigma$ (and $\vdash_M \sim \sigma$ implies the existence of an α such that $\mathfrak{C}_\alpha \not\models \sigma$). Because all the axioms are true in all $\mathfrak{C}_\alpha$s, we could begin by showing that if each antecedent of a rule is true at some $\mathfrak{C}_\alpha$ or other, then there is an α' such that the consequent is true at $\mathfrak{C}_{\alpha'}$. This is obvious for rules 1, 2, and 4, but the quantifiers present a problem: assuming that for every $t \in T^*$, there is an α_t such that $\mathfrak{C}_{\alpha_t} \models \phi(t/x)$, the usual replacement argument[20] does not guarantee the existence of an α larger than all the $\alpha_t s$ (where $\mathfrak{C}_\alpha$ would think $\phi(t/x)$ for all $t \in T^*$, and thus think $\forall x\phi$), because T^* is not a set. In this case, too, the existence of a fixed point would save the day.

So, the existence of a fixed point would imply that the infinitary axiom system M successfully codifies the truths of the structure U. In fact, however, the construction does not reach a fixed point, as can be seen from the following theorem of Tait:

Tait's Theorem. For any formula $\phi(y, x)$, there is a formula $\psi(z)$ such that $\hat{z}\psi^+ = \hat{x}\phi(\hat{z}\psi, x)^+$ and $\hat{z}\psi^- = \hat{x}\phi(\hat{z}\psi, x)^-$. In fact, for all α, $\hat{z}\psi^+_{\alpha+1} = \hat{x}\phi(\hat{z}\psi, x)^+_\alpha$ and $\hat{z}\psi^-_{\alpha+1} = \hat{x}\phi(\hat{z}\psi, x)^-_\alpha$.

Proof: Let A be $\hat{y}\hat{x}\phi(\hat{z}((y, z) \in y), x)$. Let $\psi(z)$ be $(A, z) \in A$, so that $\hat{z}\psi$ is $\hat{z}((A, z) \in A)$. Then, for any $t \in T^*$ and any α, if $t \in \hat{z}\psi^+_{\alpha+1}$, then there is a $\beta < \alpha + 1$ such that $\mathfrak{C}_\beta \models \psi(t)$, that is, such that $\mathfrak{C}_\beta \models (A, t) \in A$. By definition of A, it follows that there is a $\gamma < \beta$ such that $\mathfrak{C}_\gamma \models \phi(\hat{z}((A, z) \in A), t)$, which is just to say that $\mathfrak{C}_\gamma \models \phi(\hat{z}\psi, t)$. Thus, $t \in \hat{x}\phi(\hat{z}\psi, x)^+_{\gamma+1}$, and $\gamma < \beta < \alpha + 1$, so $t \in \hat{x}\phi(\hat{z}\psi, x)^+_\alpha$. Conversely, for any $t \in T^*$ and any α, if $t \in \hat{x}\phi(\hat{z}\psi, x)^+_\alpha$, then there is a $\beta < \alpha$ such that $\mathfrak{C}_\beta \models \phi(\hat{z}\psi, t)$, which is to say that $\mathfrak{C}_\beta \models \phi(\hat{z}((A, z) \in A), t)$. By definition of A, it follows that $\mathfrak{C}_{\beta+1} \models (A, t) \in A$, that is, that $\mathfrak{C}_{\beta+1} \models \psi(t)$. So, $t \in \hat{z}\psi^+_{\beta+2}$, which implies that $t \in \hat{z}\psi^+_{\alpha+1}$. From these two facts, it follows that $\hat{z}\psi^+ = \hat{x}\phi(\hat{z}\psi, x)^+$. Similarly for $\hat{z}\psi^-$. ■

This theorem yields a number of interesting examples, beginning with the one that undermines the possibility of a fixed point.

To see this, suppose that $\phi(y, x)$ is $\forall w(w \in x \supset w \in y)$. If ψ is formed as in the proof of Tait's theorem, then we have $\hat{z}\psi^+ = \hat{x}(\forall w(w \in x \supset w \in \hat{z}\psi)^+$ and $\hat{z}\psi^- = \hat{x}(\forall w(w \in x \supset w \in \hat{z}\psi))^-$. Roughly, then, a collection is in $\hat{z}\psi$ when all its elements are.[21] We see that $\mathfrak{C}_0 \models \forall w(w \in \overline{\emptyset} \supset w \in \hat{z}\psi)$, so $\overline{\emptyset} \in \hat{x}(\forall w(w \in x \supset w \in \hat{z}\psi))^+_1$; applying Tait's theorem, we get $\overline{\emptyset} \in \hat{z}\psi^+_2$. Then, $\mathfrak{C}_2 \models \forall w(w \in \overline{\{\emptyset\}} \supset w \in \hat{z}\psi)$, so $\overline{\{\emptyset\}} \in \hat{x}(\forall w(w \in x \supset w \in \hat{z}\psi))^+_3$, and by Tait's theorem, $\overline{\{\emptyset\}} \in \hat{z}\psi^+_4$. The pattern is clear. By induction on ordinals, it is straightforward to check that:

COROLLARY: For all ordinals α, $\alpha \notin \hat{z}\psi^+_\alpha$.

COROLLARY: For any ordinal α, there is an natural number n such that $\alpha \subseteq \hat{z}\psi^+_{\alpha+n}$ and $\overline{\alpha} \in \hat{z}\psi^+_{(\alpha+n)+2}$.

New ordinals will be entering $\hat{z}\psi$ arbitrarily high up, and so, the construction cannot become constant, and there is no fixed point.[22]

For the record, we can say more about the class $\hat{z}\psi$. We have seen that every ordinal enters $\hat{z}\psi^+$ at some stage not long after its rank, and it is easy to see that every set also will. On the other hand,

PROPOSITION: $\hat{z}\psi^-$ is empty.

Proof: Suppose not, that is, that $\hat{z}\psi^-$ is non-empty; suppose that $t \in T^*$, $t \in \hat{z}\psi^-$, and t enters $\hat{z}\psi^-$ at the first stage at which it is non-empty, say at $\hat{z}\psi^-_\alpha$. Then

α cannot be a limit; let $\alpha = \beta + 1$. By Tait's theorem, we have $t \in \hat{x}(\forall w(w \in x \supset w \in \hat{z}\psi))^{-}_{\beta}$, and thus, $\mathfrak{C}_{\beta} \not\models \forall w(w \in t \supset w \in \hat{z}\psi)$. For some $t' \in T^*$, $\mathfrak{C}_{\beta} \not\models t' \in t \supset t' \in \hat{z}\psi$, and so, $\mathfrak{C}_{\beta} \models t' \in t$ and $\mathfrak{C}_{\beta} \not\models t' \in \hat{z}\psi$. So, $t' \in \hat{z}\psi^{-}_{\beta}$, which contradicts the choice of α. ■

Another interesting example is generated by taking $\phi(y, x)$ to be $x \notin y$. Then the class – call it E – generated by Tait's construction includes all collections that are not members of E!

PROPOSITION: $E^{+} = E^{-} = \emptyset$. (So, for all $t \in T^*$, $U \overset{?}{\models} t \in E$.)

Proof: If $t \in E^{+}_{\alpha}$, then $t \in \hat{x}(x \notin E)^{+}_{\alpha}$, so there is a $\beta < \alpha$ such that $\mathfrak{C}_{\beta} \not\models t \in E$. But then $\mathfrak{C}_{\alpha} \models t \in E \wedge t \notin E$. Contradiction. Similarly, for E^{-}. ■

All class terms (except $\overline{V}$) begin with empty extensions and antiextensions, and E also ends up that way.

In any case, we see that we cannot hope to settle our questions about the soundness and completeness of M by appeal to a fixed point. Still, as it happens, we are now in a position to resolve these questions directly. We have noted that

PROPOSITION: For all sentences σ,

(1) if for some α, $\mathfrak{C}_{\alpha} \models \sigma$, then $\vdash_{M} \sigma$; and
(2) if for some α, $\mathfrak{C}_{\alpha} \not\models \sigma$, then $\vdash_{M} \sim \sigma$.

We attempted to prove the converse by showing that if each antecedent to a rule is true at some $\mathfrak{C}_{\alpha}$ or other, then there is a $\mathfrak{C}_{\alpha'}$ where the consequent is also true, and this led to the question of the existence of a fixed point. But the converse can be proved in a slightly subtler way, if we assume (as seems reasonable) that any proof in M will employ α-many applications of M-rules, for some ordinal α.

PROPOSITION: For all sentences σ and all ordinals α,

(1) if σ is M-provable in α steps, then $\mathfrak{C}_{\alpha} \models \sigma$;
(2) if $\sim\alpha$ is M-provable in α steps, then $\mathfrak{C}_{\alpha} \not\models \sigma$.

This can be proved by an easy induction on α. So, proofs in M and the construction of U match up as they go along.

But we are interested in what happens at U itself, and here the information gained from Tait's theorem is relevant. So far, we have an example of a class whose extension gains members at arbitrarily late stages. Can we convert this into an example of a ϕ such that (1) for every $t \in T^*$, there is an α_t such that $\mathfrak{C}_{\alpha_t} \models \phi(t/x)$; and (2) for all α, $\mathfrak{C}_{\alpha} \overset{?}{\models} \forall x\phi$? If so, it follows both that M is incomplete and that one of its rules is unsound. For incompleteness, note that for such a ϕ: not $\vdash_{M} \forall x\phi$ (because this would imply that $\forall x\phi$ is true at some $\mathfrak{C}_{\alpha}$, which it is not), but $U \models \forall x\phi$. For unsoundness of an M-rule, given such a ϕ, consider the formula $\forall x\phi \vee y \in E$ and let t be any closed term. Then, $U \models (\forall x\phi \vee y \in E)(t/y)$ (because $U \models \forall x\phi$), but $U \overset{?}{\models} t \in \hat{y}(\forall x\phi \vee y \in E)$ (because t never enters a $\hat{y}(\forall x\phi \vee y \in E)^+_\alpha$ or a $\hat{y}(\forall x\phi \vee y \in E)^-_\alpha$]. ($M$ would still be sound in the sense that $\vdash_M \sigma$ implies $U \models \sigma$, but it could not be used as a dependable way to move from truths of U to truths of U.) So, the existence of such a ϕ would be of considerable interest.

Knowing that ordinals enter $\hat{z}\psi$ at arbitrarily late stages, we might begin our search for such a ϕ with something like [Ord$(x) \supset x \in \hat{z}\psi$), where 'Ord$(x)$' is '$x$ is transitive and $\in$ is connected on x'. The trouble with this thought is that the Kleene interpretation of '$\supset$' would require that Ord(t) be determinately false for any t that is not determinately in $\hat{z}\psi$. But the very empty E is a counter-example.

PROPOSITION: $U \overset{?}{\models}$ Ord(E) and $U \overset{?}{\models} E \in \hat{z}\psi$.

Proof: Consider just the transitivity clause of Ord(E): $\forall x \forall y(x \in y \wedge y \in E \supset x \in E)$. If $U \models t \in t'$ for some $t, t' \in T^*$, as it often does, then for this pair, U must think $t' \notin E$ or $t \in E$. But as we have just seen, U thinks neither of these things. So, not $U \models$ Ord(E). Similarly, for U to think Ord(E) is false, it must think E is not transitive or $\in$ is not connected on E, both of which require that it think something is definitely in E. But it does not. So not $U \models\!\!\!\!/$ Ord(E).

We have already seen that not $U \models\!\!\!\!/ E \in \hat{z}\psi$. For U to think that $E \in \hat{z}\psi$, there would have to be an α such that $\mathfrak{C}_\alpha \models \forall w(w \in E \supset x \in \hat{z}\psi)$. In other words, for all $t \in T^*$, we would need $\mathfrak{C}_\alpha \models t \notin E \vee t \in \hat{z}\psi$. Now, no $\mathfrak{C}_\alpha$ ever thinks $t \notin E$, and so, we would need: For all $t \in T^*$, $\mathfrak{C}_\alpha \models t \in \hat{z}\psi$. It is easy to check that $\hat{x}(x \notin x)$ is a counterexample, indeed that for all α, $\mathfrak{C}_\alpha \overset{?}{\models} \hat{x}(x \notin x) \in \hat{z}\psi$. (We have already seen that $\hat{z}\psi^-$ is empty, so not $\mathfrak{C}_\alpha \models\!\!\!\!/ \hat{x}(x \notin x) \in \hat{z}\psi$. If, on the other hand, there is an α such that $\mathfrak{C}_\alpha \models \hat{x}(x \notin x) \in \hat{z}\psi$, let α^* be the least. This α^* cannot be a limit, so let $\alpha^* = \beta + 1$. Then $\mathfrak{C}_\beta \models \forall w(w \in \hat{x}(x \in x) \supset w \in \hat{z}\psi)$, so that, in particular, $\mathfrak{C}_\beta \models \hat{x}(x \notin x) \notin \hat{x}(x \notin x) \vee \hat{x}(x \notin x) \in \hat{z}\psi$. $\mathfrak{C}_\beta$ cannot think the first disjunct, and so, it must think the second. But this contradicts the choice of α^*.) So, not $U \models E \in \hat{z}\psi$. ■

So, our first try at the desired ϕ is unsuccessful.

As it happens, the key idea can be revived simply by modifying the formula Ord(x) to make it total. In particular, it can be made explicitly false of all non-ordinals simply by relativizing everything to $\overline{V}$. Let Ord$^*(x)$ be $x \in \overline{V} \wedge$ [Ord$(x)]^{\overline{V}}$. Let $\phi(x)$ be Ord$^*(x) \supset x \in \hat{z}\psi$. Then,

PROPOSITION: (1) For any $t \in T^*$, there is an α_t such that $\mathfrak{C}_{\alpha_t} \models \phi(t/x)$. (2) For all α, $\mathfrak{C}_\alpha \overset{?}{\models} \forall x\phi$.

Proof: (1) We have seen that, for any α, there is a natural number n such that $\mathfrak{C}_{\alpha+n} \models \overline{\alpha} \in \hat{z}\psi$. For $t \in C \cup \{\overline{V}\}$ and for t that are set constants $\bar{a}$, where a is not an ordinal, we have $\mathfrak{C}_0 \not\models \text{Ord}^*(t)$.

(2) This is just a restatement of our observation that the ordinals enter $\hat{z}\psi^+$ in a cofinal series of stages. ■

As noted earlier, this example compromises both the soundness and the completeness of M.

IV. Equinumerousity and Number

One of the enduring fascinations of classes is the possibility of reviving some approximation of the Fregean theory of numbers. Something along these lines can be accomplished here, though the irritations of the nonclassical context naturally persist.

Begin with a version of Frege's notion of equinumerosity.

DEFINITION: For $t, t' \in T$, '$t \approx t'$' abbreviates

$$\exists z(\forall u \forall v \forall w(((u, v) \in z \wedge (u, w) \in z \supset v = w) \wedge ((u, v) \in z \wedge (w, v) \in z \supset u = w)) \wedge \forall u((u \in t \supset \exists v(v \in t' \wedge (u, v) \in z)) \wedge (u \in t' \supset \exists v(v \in t \wedge (v, u) \in z)))),$$

where z, u, v, w are the first variables (in some canonical listing) that do not appear in t or t'.

If f is a term that truly instantiates this sentence, we write '$t \approx_f t'$'. If a set f is a one-to-one correspondence between sets a and b, in the ordinary sense, recall that there is a class f^* such that $\bar{a} \approx_{f^*} \bar{b}$.

It is perhaps unsurprising that '$\approx$' only captures the intended notion for total collections. If t is not total, and u is in neither t^+ nor t^-, then the second clause of the definition requires that there be a u' in $(t')^+$ such that u is mapped to u'. But since u' is in $(t')^+$, the second clause also requires that there be a u'' in t^+ such that u'' is mapped to u'. But, then, the first clause requires that $u = u''$, so u is in t^+ and u is not in t^+. So, '$\approx$' can never hold between nontotal classes.[23] But it is fairly well behaved for total classes:

PROPOSITION: For all α, for all $t, t', t'' \in T^*$, if $\mathfrak{C}_\alpha \models t \simeq t \wedge t' \simeq t' \wedge t'' \simeq t''$, then

(1) $\mathfrak{C}_\alpha \models t \approx t$.
(2) if $\mathfrak{C}_\alpha \models t \approx t'$, then $\mathfrak{C}_{\alpha+1} \models t' \approx t$.
(3) if $\mathfrak{C}_\alpha \models t \approx t'$ and $\mathfrak{C}_\alpha \models t' \approx t''$, then $\mathfrak{C}_{\alpha+1} \models t \approx t''$.

The proofs are as usual: For example, for (3), if $t \approx_f t'$, and $t' \approx_g t''$, and h is $\hat{x}\hat{y}(\exists z((x, z) \in f \wedge (z, y) \in g))$, then $t \approx_h t''$.

But if the failure of '$\approx$' in the case of nontotal classes is unsurprising, other of its shortcomings are more troublesome. For example, there are very few explicit failures of equinumerousity.

PROPOSITION: For any $t, t' \in T^*$, if $U \models \exists x(x \in t) \wedge \exists x(x \in t')$ and not $U \models t \approx t'$, then $U \overset{?}{\models} t \not\approx t'$.

Proof: For U to think $t \approx t'$ is false, it must think that every member of T^* falsifies one of the two clauses in the definition. Consider E. For E to falsify either conjunct of the first clause, it would need to have members, which it does not have. For E to falsify the first conjunct of the second clause, we would need $\exists u(u \in t \wedge \forall v(v \notin t' \vee (u, v) \notin E))$. Because the antiextension of E is empty, this can only happen if the extension of t' is empty. Similarly, the second conjunct of the second clause can only be falsified if t^+ is empty. ■

So, for example, U is undecided about whether or not $\overline{\{\emptyset\}}$ is equinumerous with $\overline{\{\emptyset, \{\emptyset\}\}}$.

Furthermore, if we attempt to define the number of a collection as its equivalence class under equinumerousity, we cannot (given our strict interpretation of '$=$') recover the fundamental Hume's principle in its usual form: $[t]_\approx = [t']_\approx$ iff $t \approx t'$. We could replace '$=$' by '$\simeq$', but this runs up against the problem that the equivalence class $[t]_\approx$ is not total, even if t is.

PROPOSITION: For all $t \in T^*$, $U \overset{?}{\models} E \approx t$.

Proof: For $U \models E \approx t$, we would need a $z \in T^*$ which, among other things, satisfies $\forall u(u \notin t \vee \exists v(v \in E \wedge (v, u) \in z))$. The second disjunct is never true, and so, we would need $\forall u(u \notin t)$. On the other hand, we also need $\forall u(u \notin E \vee \exists v(v \in t \wedge (u, v) \in z))$. The first disjunct is never true, so for this sentence to be satisfied, t would have to have members. Contradiction. For $U \not\models E \approx t$, we need one clause in the definition of '$\approx$' to be false for any $z \in T^*$. Consider E itself. As noted in the proof of the preceding proposition, for E to falsify the first clause or the first conjunct of the second clause would require E's extension to have members, which it does not have. The only hope is to falsify the final conjunct, that is, to satisfy $\exists u(u \in t \wedge \forall v(v \notin E \vee (v, u) \notin E))$. To do this, E's antiextension would need to have members, which it does not have. ■

Given these constraints, it seems the closest we can get to Hume's principle is something like this: $\hat{x}(x \approx t)^+ = \hat{x}(x \approx t')^+ \neq \emptyset$ iff there is an α such that $\mathfrak{C}_\alpha \models t (\approx t'$. ($\Rightarrow$: if $u \in \hat{x}(x \approx t)^+$ and $u \in \hat{x}(x \approx t')^+$, then there are α and β such that $u \in \hat{x}(x \approx t)^+_\alpha$ and $u \in \hat{x}(x \approx t')^+_\beta$. If $\gamma = \max(\alpha, \beta)$, then $\mathfrak{C}_\gamma \models u \approx t \wedge u \approx t'$. So, by an earlier proposition, $\mathfrak{C}_{\gamma+1} \models t \approx t'$. $\Leftarrow$: if $\mathfrak{C}_\alpha \models t \approx t'$, then $\mathfrak{C}_\alpha \models t \approx t$, so $\hat{x}(x \approx t)^+ \neq \emptyset$. So, if $u \in \hat{x}(x \approx t)^+$, then there is a β such that $u \in \hat{x}(x \approx t)^{+}_{\beta}$. If $\gamma = \max(\alpha, \beta)$, then $\mathfrak{C}_\gamma \models t \approx t' \wedge u \approx t$. So, by an earlier proposition, $\mathfrak{C}_{\gamma+1} \models u \approx t'$, and $u \in \hat{x}(x \approx t')^{+}$. Similarly, if $u \in \hat{x}(x \approx t')^{+}$, then $u \in \hat{x}(x \approx t)^{+}$.)

There is another, more concrete, but still quasi-Fregean,[24] approach to number that works a bit better. The general idea, as in contemporary set theory, is to use the von Neumann ordinals as a standard measure of cardinality.[25] Limiting our attention first to the natural numbers, consider the following proposal:

DEFINITION: For $t \in T^*$, let '$[t]_\approx$' be $\hat{x}(x \approx t)$. Let $\mathbf{N}$ be $\hat{x}(\exists\alpha(\alpha \in \overline{\omega} \wedge x = [\alpha]_\approx))$.

Notice that $\mathbb{N}$ is total. (If t is $\overline{V}$ or $\bar{a}$, for some set a, then $t = [\alpha]_\approx$ is false for any $\alpha \in \omega$; if $t \in C^*$, then $t = [\alpha]_\approx$ is determinately true or false for any $\alpha \in \omega$.)

DEFINITION: 'Num(x, y)' abbreviates $x \in \mathbf{N} \wedge y \in x$.

0 is $[\overline{\emptyset}]_\approx$, 1 is $[\overline{\{\emptyset\}}]_\approx, \ldots$

Sxy abbreviates $x \in \mathbf{N} \wedge y \in \mathbf{N} \wedge \exists u \exists v(x = [u]_\approx \wedge y = [v]_\approx$
$\wedge \forall w(w \in u \equiv w \in v \vee w = v))$.

In these terms, our approximation to the Fregean equivalence is a bit better.

PROPOSITION: For all α, and all $t, t', u \in T^*$,

(1) if $\mathfrak{C}_\alpha \models t \approx t' \wedge \text{Num}(u, t)$, then $\mathfrak{C}_{\alpha+2} \models \text{Num}(u, t')$;
(2) if $\mathfrak{C}_\alpha \models \exists x[\text{Num}(x, t) \wedge \text{Num}(x, t')]$, then $\mathfrak{C}_{\alpha+1} \models t \approx t'$.

Proof: (1) if $\mathfrak{C}_\alpha \models t \approx t' \wedge \text{Num}(u, t)$, then there is a $\beta \in \omega$ such that $\mathfrak{C}_\alpha \models u = [\beta]_\approx \wedge t \in u$. So, $\mathfrak{C}_\alpha \models t \approx t' \wedge t \approx \beta$. By an earlier proposition, $\mathfrak{C}_{\alpha+1} \models t' \approx \beta$, so $\mathfrak{C}_{\alpha+2} \models \text{Num}(u, t')$. (2) If $\mathfrak{C}_\alpha \models \text{Num}(u, t) \wedge \text{Num}(u, t')$, then there is a $\beta \in \omega$ such that $\mathfrak{C}_\alpha \models t \approx \beta \wedge t' \approx \beta$. By the same earlier proposition, $\mathfrak{C}_{\alpha+1} \models t \approx t'$. ■

Notice that it is possible to count classes as well as sets using these numbers. (For example, $\hat{x}(x \in x)$ and $\hat{x}(x \approx \overline{\{\emptyset, \{\emptyset\}\}})$ (alias, '2') can be collected into $\hat{y}(y = \hat{x}(x \in x) \vee y = 2)$, and it is then easy to see that $\hat{y}(y = \hat{x}(x \in x) \vee y = 2) \approx \overline{\{\emptyset, \{\emptyset\}\}}$. At the next stage, we get $\hat{y}(y = \hat{x}(x \in x) \vee y = 2) \in 2$, and thus Num$(2, (y = \hat{x}(x \in x) \vee y = 2))$.)

It is straightforward, though occasionally tedious, to check that the first four Peano axioms become true in the early stages of the construction of U.

PA(1) $0 \in \mathbf{N}$.
PA(2) $\forall x(x \in \mathbf{N} \supset \exists y(y \in \mathbf{N} \wedge S(y, x))$.
PA(3) $\forall x(x \in \mathbf{N} \supset {\sim}S(0, x))$
PA(4) $\forall x \forall y(x \in \mathbf{N} \wedge y \in \mathbf{N} \wedge \exists z(S(z, x) \wedge S(z, y) \supset x = y)$.

The induction axiom is more delicate. It is false in its usual form – $\forall x((x \subseteq \mathbf{N} \wedge 0 \in x \wedge \forall y(y \in \mathbf{N} \wedge y \in x \supset \exists z(z \in \mathbf{N} \wedge S(z, y) \wedge z \in x))) \supset x = \mathbf{N})$ – because of the strict interpretation of '='. This can be avoided by replacing '=' with '$\simeq$', but the result still fails for nontotal classes like E. (With 'E' in for 'x', none of the antecedents is definitely falsified, nor is the consequent definitely satisfied.) The solution is to trade the axiom for a rule:

$$\frac{t \simeq t, t \subseteq \mathbf{N}, 0 \in \mathbf{N}, \forall y(y \in \mathbf{N} \wedge y \in t \supset \exists z(z \in \mathbf{N} \wedge S(z, y) \wedge z \in t))}{t \simeq \mathbf{N}}$$

More arithmetic can be developed from here, within the limitations of the nonclassical context.

This account of number can obviously be carried into the infinite, using the transfinite ordinals. Let me close this discussion with an few rather fanciful observations. To begin, consider $\hat{x}(\mathrm{Ord}^*(x))$.

PROPOSITION:
(1) $\mathfrak{C}_1 \models \hat{x}(\mathrm{Ord}^*(x)) \simeq \hat{x}(\mathrm{Ord}^*(x))$.
(2) $\mathfrak{C}_2 \models \hat{x}(\mathrm{Ord}^*(x)) \in \hat{x}(\mathrm{Ord}(x))$.

Proof: (1) $\mathfrak{C}_0$ knows which sets are ordinals and which are not, and that only sets are members of $\overline{V}$, to which the quantifiers of Ord* are relativized. So, $\hat{x}(\mathrm{Ord}^*(x))$ is total at $\mathfrak{C}_1$.

(2) For all $t, t' \in T^*$, $\mathfrak{C}_1 \models t \notin t' \vee t' \notin \hat{x}(\mathrm{Ord}^*(x)) \vee t \in \hat{x}(\mathrm{Ord}^*(x))$ (transitivity) and $\mathfrak{C}_1 \models t \notin \hat{x}(\mathrm{Ord}^*(x)) \vee t' \notin \hat{x}(\mathrm{Ord}^*(x)) \vee t \in t' \vee t = t' \vee t' \in t$ (connectedness of $\in$), and so, $\mathfrak{C}_1 \models \mathrm{Ord}(\hat{x}(\mathrm{Ord}^*(x)))$. Thus, $\mathfrak{C}_2 \models \hat{x}(\mathrm{Ord}^*(x)) \in \hat{x}(\mathrm{Ord}(x))$. ■

Given that $\hat{x}(\mathrm{Ord}^*(x))$ is so ordinal-like, it seems natural to regard $[\hat{x}(\mathrm{Ord}^*(x))]_{\approx}$ as number-like. To my knowledge, the only previous discussion of such numbers is due to Himmel,[26] who called them 'supernaturals'. There would also be additional 'superordinals', for example, $\hat{y}(y \in \hat{x}(\mathrm{Ord}^*(x)) \vee y = \hat{x}(\mathrm{Ord}^*(x)))$. Some of our earlier problems might be helped by iterating the construction through these superordinal stages, though new superordinals would still be entering Tait's class at each superordinal stage.

This is only the bare beginning of an exploration of the proposed theory of sets and classes, and it is not clear to what extent it will repay further study. Still, we must be grateful to Parsons for the analogy that suggested this possibly illuminating approach.

NOTES

I am indebted to Anil Gupta, John Myhill, and Bill Tait for their suggestions and observations concerning my 1983 article, and to Gupta and Tait again for numerous helpful suggestions on earlier drafts of the present paper (as will become abundantly clear in footnotes to follow). My thanks also to the National Science Foundation, whose grant #SBR-9320220 supported this research.

1. See, for example, Parsons (1974a,b, 1977).
2. Kripke (1975). See also Martin and Woodruff (1975).
3. This is an extension and improvement of my earlier theory (Maddy 1983) (which also contains more detail on the history of treatments of proper classes).
4. See Feferman (1984) for discussion of these earlier theories.
5. I use 'collection' as a term neutral between sets and classes.
6. The theory in Maddy (1983) did not permit free variables within class terms, and so, it was impossible to quantify into class contexts (as in the third example).
7. Though $\overline{V}$ may be redundant, the question of whether or not there is a $t \in T^*$, not involving $\overline{V}$, with $t^+ = \{\bar{a} \mid a \in V\}$ and $t^- = T^* - \{\bar{a} \mid a \in V\}$ remains open. (This addition to the earlier theory was suggested by both Gupta and Myhill.)
8. See Feferman (1984, p. 88). This improvement of the earlier theory was suggested by Myhill.
9. This simultaneous definition, suggested by Gupta, replaces the cumbersome definitions of Maddy (1983).
10. Obviously, I am speaking here of collections too big to be sets; in other words, my account of sets and classes presupposes some rudimentary understanding of classes, much as the iterative picture of sets presupposes some rudimentary notion of set.
11. The symbols '=' and '∈' (and others defined in terms of these) are ambiguous here: they appear both in the formulas of the language $\mathscr{L}$, and in meta-linguistic uses (as in this paragraph). The context will disambiguate.
12. The earlier theory lacked '=' as a primitive, because I had not realized that this trivial definition would work. For the shortcomings of less trivial definitions, see Section II.
13. Alas, '⊃' behaves rather strangely: $\mathfrak{C} \models p \supset q$ iff $\mathfrak{C} \neq p$ or $\mathfrak{C} \models q$, so if $\mathfrak{C} \overset{?}{\models} p$, then $\mathfrak{C} \models q$. As a result, if $\mathfrak{C} \models p \equiv q$, then $\mathfrak{C}$ cannot be undecided about either p or q. This fact will come back to haunt.
14. For example, to determine whether or not $\mathfrak{C} \models x \in \hat{y}(y \in z)[s]$, where $s(x)$ is $\emptyset$ and $s(z)$ is $(\hat{x}(x \in y), y \rightarrow \{\emptyset\})$, we must first remove the clash of variables by replacing $(\hat{x}(x \in y), y \rightarrow \{\emptyset\})$ with $(\hat{u}(u \in v), v \rightarrow \{\emptyset\})$; second, substitute $\hat{u}(u \in v)$ for z; third, adjust s to s' where $s'(x)$ is $\emptyset$ and $s'(v)$ is $\{\emptyset\}$; and fourth (to see whether or not $\mathfrak{C} \models (x \in \hat{y}(y \in \hat{u}(u \in v))[s'])$, determine whether or not $\emptyset$ is in $(\hat{y}(y \in \hat{u}(u \in v)), v \rightarrow \emptyset)^+_{\mathfrak{C}}$.
15. Any of the usual set-theoretic definitions of ordered pairs can be imitated here, but Gupta, again, pointed out that this extremely simple idea also works.
16. Tait bemoaned the absence of such class relations in my (1983) and suggested that they be added outright, but the presence of the trivial '=' relation makes this unnecessary. Tait's fixed-point theorem (in Section III, below) depends on class relations; the validity of the theorem for the earlier theory remains open.
17. The corresponding proof in Maddy (1983) could be carried over, but the added flexibility of the extended system makes a more direct proof possible.
18. 'A' domain or range because of the failures of extensionality discussed in Section II.
19. If we identified the last two, we would need to complicate the definition of (x, y).
20. Cf. Kripke (1975, p. 704).
21. The oddities of ⊃ come into play here, but they will not affect what follows, because we concentrate on ordinals, which are total.
22. Tait conjectured and Flagg later proved that if the construction is carried out over a ground model of ZF with real ordinals, then it will reach a fixed point at the first admissible ordinal greater that the least upper bound of the ordinals in the ground model.

23. And thus, '$t \approx t$' is another way of expressing the claim that t is total, that is, $U \models t \approx t$ iff $U \models \forall x(x \in t \vee x \notin t)$.
24. 'Fregean' because numbers are identified with equivalence classes; 'quasi-Fregean' because of the element of arbitrariness that enters with the conspicuous use of the von Neumann ordinals. This last would be less bothersome if, for example, $\hat{x}(x \approx \{\emptyset, \{\emptyset\}, \{\emptyset, \{\emptyset\}\}\}) = \hat{x}(x \approx \{\emptyset, \{\emptyset\}, \{\{\emptyset\}\}\})$, but our definition of '=' is too fine-grained for this.
25. See Maddy (1990, Ch. 3), for a philosophical discussion of this approach. The particular approach taken here is suggested at the end of Maddy (1983).
26. See Goldstein (1983).

REFERENCES

Feferman, S. (1984). "Toward Useful Type-Free Theories, I," *Journal of Symbolic Logic*, 49: 75–111.

Flagg, R., and Myhill, J. (1987). "Implication and Analysis in Classical Frege Structures," *Annals of Pure and Applied Mathematics*, 34: 33–85.

Goldstein, R. (1983). *The Mind-Body Problem* (New York: Random House).

Kripke, S. (1975). "Outline of a Theory of Truth," *Journal of Philosophy*, 72: 690–716.

Maddy, P. (1983). "Proper classes," *Journal of Symbolic Logic*, 48: 113–39.

Maddy, P. (1990). *Realism in Mathematics* (Oxford: Oxford University Press).

Martin, R., and Woodruff, P. (1975). "On Representing 'True in *L*' in *L*," *Philosophia*, 5: 113–17.

Parsons, C. (1974a). "Sets and Classes," reprinted in Parsons (1983), pp. 209–20.

Parsons, C. (1974b). "The Liar Paradox," reprinted in Parsons (1983), pp. 221–67.

Parsons, C. (1977). "What Is the Iterative Conception of Set?" reprinted in Parsons (1983), pp. 268–97.

Parsons, C. (1983). *Mathematics in Philosophy* (Ithaca, NY: Cornell University Press).

Challenges to Predicative Foundations of Arithmetic

SOLOMON FEFERMAN* AND
GEOFFREY HELLMAN

> The White Rabbit put on his spectacles. "Where shall I begin, please your Majesty?" he asked. "Begin at the beginning," the King said gravely, "and go on till you come to the end: then stop."
>
> Lewis Carroll, *Alice in Wonderland*

This is a sequel to our article "Predicative foundations of arithmetic" (Feferman and Hellman, 1995), referred to in the following as PFA; here we review and clarify what was accomplished in PFA, present some improvements and extensions, and respond to several challenges. The classic challenge to a program of the sort exemplified by PFA was issued by Charles Parsons in a 1983 paper, subsequently revised and expanded as Parsons (1992). Another critique is due to Daniel Isaacson (1987). Most recently, Alexander George and Daniel Velleman (1996) have examined PFA closely in the context of a general discussion of different philosophical approaches to the foundations of arithmetic.

The plan of the present paper is as follows: Section I reviews the notions and results of PFA, in a bit less formal terms than there and without the supporting proofs, and presents an improvement communicated to us by Peter Aczel. Then, Section II elaborates on the structuralist perspective that guided PFA. It is in Section III that we take up the challenge of Parsons. Finally, Section IV deals with the challenges of George and Velleman, and thereby, that of Isaacson as well. The paper concludes with an Appendix by Geoffrey Hellman, which verifies the predicativity, in the sense of PFA, of a suggestion credited to Michael Dummett for another definition of the natural number concept.

I. Review

In essence, what PFA accomplished was to provide a formal context based on the notions of finite set and predicative class and on prima facie evident principles for such, in which could be established the existence and categoricity of

*This paper was written while the first author was a Fellow at the Center for Advanced Study in the Behavioral Sciences (Stanford, CA) whose facilities and support, under grants from the Andrew W. Mellon Foundation and the National Science Foundation, have been greatly appreciated.

a natural number structure. The following reviews, in looser formal terms than PFA, the notions and results therein prior to any discussion of their philosophical significance. Three formal systems were introduced in PFA, denoted EFS, EFSC, and EFSC*. All are formulated within classical logic. The language L(EFS), has two kinds of variables:

Individual variables: $a, b, c, u, v, w, x, y, z, \ldots$, and
Finite set variables: $A, B, C, F, G, H, \ldots$

The intended interpretation is that the latter range over *finite sets of individuals*. There is one binary operation symbol (,) for a *pairing function* on individuals, and *individual terms s, t,...* are generated from the individual variables by means of this operation. We have two relation symbols, '$=$' and '$\in$', by means of which *atomic formulas* of the form $s = t$ and $s \in A$ are obtained. *Formulas* $\varphi, \psi, \ldots$ are generated from these by the propositional operations '$\neg$', '&', '$\vee$', '$\rightarrow$', and by the quantifiers '$\forall$' and '$\exists$' applied to either kind of variable. The language L(EFSC), which is the same as that of EFSC*, adds a third kind of variable:

Class variables: $\mathbf{X}, \mathbf{Y}, \mathbf{Z}, \ldots$[1]

In this extended language, we also have a membership relation between individuals and classes, giving further atomic formulas of the form $s \in \mathbf{X}$. Then formulas in L(EFSC) are generated as before, allowing, in addition, quantification over classes. A formula of this extended language is said to be *weak second-order* if it contains no bound class variables. The intended range of the class variables is the collection of weak second-order definable classes of individuals. We could consider finite sets to be among the classes, but did not make that identification in PFA. Instead, we write $A = \mathbf{X}$ if A and $\mathbf{X}$ have the same extension. Similarly, we explain when a class is a subclass of a set, and so on. A class $\mathbf{X}$ is said to be *finite* and we write Fin($\mathbf{X}$) if $\exists A(A = \mathbf{X})$.

The *Axioms of EFS* are denoted (Sep), (FS-I), (FS-II), (P-I), and (P-II), and are explained as follows: The separation scheme (Sep) asserts that any definable subset of a finite set is finite; that is, for each formula φ of EFS, $\{x \in A \mid \varphi(x)\}$ is a finite set B when A is a given finite set. The axiom (FS-I) asserts the existence of an empty (finite) set, and (FS-II) tells us that if A is a finite set and a is any individual then $A \cup \{a\}$ is a finite set. The pairing axioms (P-I) and (P-II), respectively, say that pairing is one-one and that there is an urelement under pairing; it is convenient to introduce the symbol 0 for an individual that is not a pair.

The *Axioms of EFSC* augment those of EFS by the scheme (WS-CA) for weak second-order comprehension axiom, which tells us that $\{x \mid \varphi(x)\}$ is a class $\mathbf{X}$ for any weak second-order φ. In this language, we allow the formula φ in (Sep) to contain free class variables; then it can be replaced by the assertion that any subclass of a finite set is finite. The following theorem (numbered 1 in PFA) is easily proved by a model-theoretic argument, but can also be given a finitary proof-theoretic argument.

Metatheorem. EFSC is a conservative extension of EFS.

In the language of EFSC, (binary) relations are identified with classes of ordered pairs, and functions, for which we use the letters $\mathbf{f}, \mathbf{g}, \ldots,$[2] are identified with many-one relations; n-ary functions reduce to unary functions of n-tuples. Then we can formulate the notion of *Dedekind finite class* as being an $\mathbf{X}$ such that there is no one-one map from $\mathbf{X}$ to a proper subclass of $\mathbf{X}$. By the axiom (Card) is meant the statement that every (truly) finite class is Dedekind finite. The *Axioms of EFSC** are then the same as those of EFSC, with the additional axiom (Card).

Now, working in EFSC, we defined a triple $\langle \mathbf{M}, a, \mathbf{g} \rangle$ to be a *pre-N-structure* if it satisfies the following two conditions:

(N-I) $\quad \forall x \in \mathbf{M}[\mathbf{g}(x) \neq a]$, and

(N-II) $\quad \forall x, y \in \mathbf{M}[\mathbf{g}(x) = \mathbf{g}(y) \rightarrow x = y]$.

These are the usual first two Peano axioms when a is 0 and $\mathbf{g}$ is the successor operation. By an *N-structure* is meant a pre-N-structure that satisfies the *axiom of induction* in the form

(N-III) $\quad \forall \mathbf{X} \subseteq \mathbf{M}[a \in \mathbf{X} \;\&\; \forall x(x \in \mathbf{X} \rightarrow \mathbf{g}(x) \in \mathbf{X}) \rightarrow \mathbf{X} = \mathbf{M}]$.

It is proved in EFSC that we can define functions by primitive recursion on any N-structure; the idea is simply to obtain such as the union of finite approximations. This union is thus definable in a weak second-order way. From that, we readily obtain the following theorem (numbered 5 in PFA):

Theorem. (Categoricity, in EFSC). Any two N-structures are isomorphic.

Now, to obtain the existence of N-structures, in PFA we began with a specific pre-N-structure $\langle \mathbf{V}, 0, \mathbf{s} \rangle$, where $\mathbf{V} = \{x \mid x = x\}$ and $\mathbf{s}(x) = x' = (x, 0)$; that this satisfies (N-I) and (N-II) is readily seen from the axioms (P-II) and (P-I), respectively. Next, define

$$\mathrm{Clos}^{-}(A) \leftrightarrow \forall x[x' \in A \rightarrow x \in A], \tag{1}$$

and

$$y \leq x \leftrightarrow \forall A[x \in A \;\&\; \mathrm{Clos}^{-}(A) \rightarrow y \in A]. \tag{2}$$

In words, $\mathrm{Clos}^{-}(A)$ is read as saying that A is closed under the predecessor operation (when applicable), and so, $y \leq x$ holds if y belongs to every finite set that contains x and is closed under the predecessor operation. Let

$$\mathrm{Pd}(x) = \{y \mid y \leq x\}. \tag{3}$$

The next step in PFA was to cut down the structure $\langle \mathbf{V}, 0, \mathbf{s}\rangle$ to a special pre-N-structure:

$$\mathbf{M} = \{x \mid \mathrm{Fin}(\mathrm{Pd}(x)) \;\&\; \forall y[y \leq x \rightarrow y = 0 \vee \exists z(y = z')]\}. \tag{4}$$

This led to the following theorem (numbered 8 in PFA):

Theorem. (Existence, in EFSC*). $\langle \mathbf{M}, 0, \mathbf{s}\rangle$ is an N-structure.

To summarize: In PFA, categoricity of N-structures was established in EFSC and existence in EFSC*. Following publication of this work, we learned from Peter Aczel of a simple improvement of the latter result obtained by taking in place of **M** the following class:

$$\mathbf{N} = \{x \mid \mathrm{Fin}(\mathrm{Pd}(x)) \;\&\; 0 \leq x\}. \tag{5}$$

Theorem. (Aczel). EFSC proves that $\langle \mathbf{N}, 0, \mathbf{s}\rangle$ is an N-structure.

We provide the proof of this here, using facts established in Theorem 2 of PFA.

(i) $0 \in \mathbf{N}$, because $\mathrm{Pd}(0) = \{0\}$ and $0 \leq 0$.
(ii) $x \in \mathbf{N} \rightarrow x' \in \mathbf{N}$, because $\mathrm{Pd}(x') = \mathrm{Pd}(x) \cup \{x'\}$, and $0 \leq x \rightarrow 0 \leq x'$.
(iii) If $\mathbf{X}$ is any subclass of $\mathbf{N}$ and $0 \in \mathbf{X} \wedge \forall y[y \in \mathbf{X} \rightarrow y' \in \mathbf{X}]$, then $\mathbf{X} = \mathbf{N}$. For, suppose that there is some $x \in \mathbf{N}$ with $x \notin \mathbf{X}$. Let $A = \{y \mid y \leq x \;\&\; y \notin \mathbf{X}\}$; A is finite since it is a subclass of the finite set $\mathrm{Pd}(x)$. Moreover, A is closed under the predecessor operation, and so, A contains every $y \leq x$; in particular, $0 \in A$, which contradicts $0 \in \mathbf{X}$.

The theorem follows from (i)–(iii), since the axioms (N-I) and (N-II) hold on **V** and hence on **N**.

It was proved in PFA that EFSC* is of the same (proof-theoretic) strength as the system PA of Peano axioms and is a conservative extension of the latter under a suitable interpretation. The argument was that EFSC* is interpretable in the system ACA_0, which is a well-known second-order conservative extension of PA based on the arithmetical comprehension axiom scheme together with induction axiom in the form (N-III). Conversely, we can develop PA in EFSC* using closure under primitive recursion on any N-structure. Since any first-order formula of arithmetic so interpreted then defines a class, we obtain the full induction scheme for PA in EFSC*. Now, using the preceding result, the whole argument applies *mutatis mutandis* to obtain the following:

Metatheorem. (Aczel). EFSC is of the same (proof-theoretic)strength as PA and is a conservative extension of PA under the interpretation of the latter in EFSC.

This result also served to answer Question 1 on p. 13 of PFA.

Incidentally, it may be seen that the definition of **N** in (5) above is equivalent to the following:

$$x \in \mathbf{N} \leftrightarrow \forall A[x \in A \ \&\ \mathrm{Clos}^{-}(A) \rightarrow 0 \in A] \ \&\ \exists A[x \in A \ \&\ \mathrm{Clos}^{-}(A)]. \quad (6)$$

For, the first conjunct here is equivalent to the statement that $0 \leq x$, and the second to Fin(Pd(x)). In this form, Aczel's definition is simply the same as the one proposed by George (1987, p. 515).[3] Part of the progress that is achieved by this work in our framework is to bring out clearly the assumptions about finite sets that are needed for it and that are prima-facie evident for that notion.

There is one further improvement in our work to mention. It emerged from correspondence with Alexander George and Daniel Velleman that the remark in footnote 5 on p. 16 of PFA asserting a relationship of our work with a definition of the natural numbers credited to Dummett was obscure. The exact situation has now been clarified by Geoffrey Hellman in the Appendix to this paper, where it is shown that Dummet's definition also yields an N-structure, provably in EFSC.

II. The Structuralist Standpoint and "Constructing the Natural Numbers"

In developing predicative foundations of arithmetic, we have been proceeding from a structuralist standpoint, one that each of us has pursued independently in other contexts. In general terms, structuralism has been described by one of us as the view that "mathematics is the free exploration of structural possibilities, pursued by (more or less) rigorous deductive means" (Hellman 1989, p. 6), along with the claim that,

> In mathematics, it is not particular objects which matter but rather certain 'structural' properties and relations, both within and among relevant totalities. (Hellman 1996, p. 101)

Such general formulations raise questions of scope, for it seems that there must be exceptional mathematical concepts requiring a nonstructural or prestructural understanding so that prior sense can be made of "items *in* a structure," *substructure*, and other concepts required for structuralism to get started.[4] For present purposes, however, this question need not be taken up in a general way, as we may work within a more specialized form of structuralism, one explicitly concerned with number systems. As the other of us has put it:

> The first task of any general foundational scheme for mathematics is to establish the number systems. In both the extensional and intensional approach this is done from the modern *structuralist* point of view. The structuralist viewpoint as regards the basic number systems is that it is not the specific nature of the individual objects which is of the essence, but rather the isomorphism type of the structure of which they form a part. Each structure $\mathcal{A}$ is to be characterized up to isomorphism by a structural property P which, logically, may be of first order or of higher order. (Feferman 1985, p. 48)

So long as this is understood, we may work with a system such as EFSC, leaving open whether this itself is to be embedded in a more general structuralist framework or whether it is thought of as standing on its own.

The central point here is that what we are seeking to define in a predicatively acceptable way is not, strictly speaking, the predicate '*natural number*' *simpliciter*, but rather the predicate '*natural-number-type structure*'. That is, we seek to characterize what it is to be a *structure* of this particular type – what Dedekind (1888) called "*simply infinite systems*" and what set-theorists call "*ω-sequences*" – and also to prove that, mathematically, such structures exist. Once this has been accomplished, we may then, as a *façon de parler*, identify the elements of a particular such structure as "the natural numbers," employing standard numerals and designations of functions and relations, but this is essentially for mathematical convenience. Officially, we *eliminate* the predicate 'is a natural number' in its absolute sense and speak instead of what holds in any natural-number-type structure. And thanks to our (limited) second-order logical machinery, we can render arithmetical statements directly, relativized to structures, as illustrated by the conditions (N-I)–(N-III) (Sec. 1, above); there is no need to introduce a relation of *satisfaction* between structures and sentences.

This standpoint has some implications worth noting. First, since no absolute meaning is being assigned to 'natural number', the same goes for 'nonnumber'. While of course a good definition of 'natural-number-type structure' must rule out anything that does not qualify as such a structure, there is simply no problem of "excluding nonnumbers" such as Julius Caesar (on standard platonist conceptions). This notorious Fregean problem simply does not arise in the structuralist setting. Rather than having to answer the question, "Is Julius Caesar a number?" (and presumably get the right answer), we sidestep it entirely. We even regard it as misleading to ask, "Might Julius Caesar be or have been a number?" for this still employs 'number' in an absolute sense. Of course, Julius Caesar might have been – and presumably is, in a mathematical sense – a member of many natural-number-type structures. On the other hand, we can make sense of standard, mathematically sensible statements such as "3/5 is not a natural number" by writing out "In any structure for the rationals with a substructure for the natural numbers (identified in the usual way), the object denoted '3/5' does not belong to the domain of the latter." And, of course, many elliptical references to "the natural numbers" are harmless.

More significantly, the whole question of circularity in "constructions of the natural numbers" must be looked at afresh. In contrast to 'natural number', 'natural-number-type structure' is an infinitistic concept in the straightforward sense that any instance of such a structure has an infinite domain with (at least) a successor-type operation defined on it. While it might well appear circular to define 'natural number' in terms of a predicate applying to just finite objects – for example, finite sets or sequences from some chosen domain – since it might seem obvious that such objects can do the duty of natural numbers, nevertheless if one succeeds in building up an *infinite structure* of just the right sort from finite objects, using acceptable methods of construction, and then proves by acceptable means that one has succeeded, prima facie one has done as much as could reasonably be demanded.

In predicative foundations, it is quite natural to take the notion 'finite set' as given, governed by elementary closure conditions as in EFSC. The cogency of this can be seen as follows: Within the definitionist framework, a predicatively acceptable domain is one in which each item is specified by a designator, say in a mathematical language. Hence any finite subset of the domain is specificable outright by a disjunction of the form $x = d_1 \vee x = d_2 \vee \cdots \vee x = d_k$, where each d_i is a designator of an object in the domain. Thus, the finite sets correspond to finite lists of designators, and it is reasonable for the definitionist to take *this* notion – "finite list of quasi-concrete objects" – as understood. The claim is, along Hilbertian lines, that this does not depend on a grasp of the *infinite structure of natural numbers*, nor does it depend on an explicit understanding of the even more complex infinite structure of finite subsets ordered, say, by inclusion. Once given such a starting point, the closure conditions of EFSC are then evident.

There is a further related point of comparison between the concepts 'finite set' and 'natural number' that is relevant to our project. Given an infinite domain X of objects, we think of a finite set A of Xs as fully determined by its members. Although certain relations to other finite sets of Xs are also evident for us – for example, adjoining any new element to A yields a finite set – the identity of A *as a finite set* is not conceived as depending on its position in an infinite structure of finite sets of Xs. Yet this "self-standing" character of finite sets is not shared by natural numbers, even on platonist views. To identify a natural number is to identify its position in an infinite structure. Even on a set-theoretic construction, while the sets taken as numbers are of course determined *as sets* by their members, they are not determined *as numbers* until their position in a sequence is determined. Such considerations lead naturally to the structuralist project of PFA.

The significance of these points has perhaps not been sufficiently appreciated because, historically, structuralism has not been articulated independently of platonism. If one succeeds in defining 'natural number' platonistically, say as Frege or Russell did, or as Zermelo or von Neumann did, so that the natural numbers are identified uniquely with particular abstract objects, then, since the whole sequence of natural numbers thus defined together with arithmetic functions and relations are unproblematic as objects in such frameworks, it is a trivial matter to pass to an explicit definition of 'natural-number-type structure': one simply specifies as such a structure any that is isomorphic to the original, privileged one. Then, clearly all the work has gone into the original definition of 'natural number', and questions of circularity are directed there. However, the approach of PFA is different, sharing more with Hilbert's conception of mathematical axioms and reference than with Frege's.[5] For we bypass construction of 'the natural numbers' as particular objects and proceed directly to the infinitistic concept, 'natural-number-type structure' (much as Dedekind [1888] proceeded directly to define 'simply infinite system'). Then, in proving the existence of such structures, we introduce a certain sequence of finite objects available within our framework. Collecting these is predicatively unproblematic, for they are specified as having finitely many earlier elements (including an initial one), not as fulfilling mathematical induction. That they satisfy induction is then proved as a theorem.[6]

Despite this result and the related ones established in PFA – especially the categoricity of our characterization and the proof-theoretic conservativeness of our system over PA – questions have been raised, implicitly and explicitly, concerning circularity and possible hidden impredicativity in the constructions. In the remaining two sections, we will address these specifically.

III. Parsons' Challenge

In his stimulating paper "The Impredicativity of Induction" (I of I in the following), Charles Parsons takes up a number of issues in his typically thoughtful and thorough manner. Our main purpose here is to address the points most directly related as a challenge to what PFA was intended to accomplish, namely, a predicative foundation of the structure of natural numbers, given the notion of finite set of individuals.[7] But it is necessary, first, to make some distinctions in regard to the idea of predicativity. To begin, a putative definition of an object c is said to be *impredicative* if it makes use of bound variables whose range includes c as one of its possible values.[8] Such bound variables may appear attached to quantifiers, or as the variable of abstraction in definitions of sets or functions, or as the variable in a unique description operator, and so on. We do not agree with the position ascribed to Poincaré and Weyl,[9] that impredicative definitions are prima facie viciously circular and to be avoided. For example, we regard the number associated with the Waring problem for cubes – defined as the least positive integer n such that every sufficiently large integer is a sum of, at most, n positive cubes – as a perfectly meaningful and noncircular description of a specific integer; it is known that $n \leq 7$, but beyond that, the exact value of n is not known. While this definition would generally be considered nonconstructive, and is impredicative according to the general idea given above, from a classical predicative point of view it is not viciously circular, since we are convinced by predicative arguments that such a number exists and must have an alternative predicative definition, be it 7 or a smaller integer. So, for us, the issue is to determine when there is a predicative warrant for accepting a prima facie impredicative definition. That cannot be answered without saying what constitutes a predicative proof of existence of objects of one kind or another. Moreover, the above explanation of what it is about the form of a putative definition that makes it impredicative does not tell us what constitutes a *predicative definition*, because it only tells us what should *not* appear in it, and nothing about what (notions, names, etc.) *may* appear in it. Since the latter have to be, in some sense, prior to the object being defined, and since it is not asserted in explaining what is to be avoided just what that is, an answer to this necessarily makes of predicativity a *relative* rather than an *absolute* notion.

Considerations such as this led Kreisel to propose a formal notion of predicative provability *given the natural numbers*, and that was characterized in precise proof-theoretical terms independently (and in agreement with each other) by Feferman (1964) and Schütte (1965). Speaking informally, that characterization takes for granted the notions and laws of classical logic as applied to definitions and statements involving, to begin with, only the natural numbers as the range of bound variables in definitions of sets of natural numbers, and then admits, successively, definitions employing variables for sets ranging over collections of sets that have been comprehended predicatively.[10] The details need not concern us; suffice it to say that Parsons, among others, has found this analysis of predicativity given the natural numbers to be persuasive (I of I, p. 150). However, as he suggests in the latter part of I of I, he also finds it reasonable to ascribe the term 'predicative' to the use of certain generalized inductive definitions that breach the bounds of the Feferman–Schütte characterization. There is no contradiction here from our point of view; the latter simply shifts what the notion of predicativity is taken relative to. One might go further and consider a notion of predicativity relative to the structure of real numbers, if one regarded that structure as well determined, and so on to higher levels of set theory. Though the idea is clear enough, none of these has been studied and characterized in precise proof-theoretical terms.[11]

Now, finally, we return to the program of PFA. There, the aim is to consider what can be done predicatively in the foundations of arithmetic relative to the notion of a finite set of individuals, where the individuals themselves may have some structure as built up by ordered pairs.[12] Philosophically, the significance of this is that we have a prior conception of finite set that does not require the understanding of the natural-number system, and for this notion we have some evident closure principles, which are simply expressed by the axioms (Sep), (FS-I), and (FS-II) of PFA. We do not regard the success of the program PFA to be necessary for the acceptance of the natural-number system, but believe that its success, if granted, is of philosophical interest.

The challenge raised by Parsons in I of I begins with the evident impredicativity of Frege's definition of the natural numbers, in the form

$$\text{(Frege-N)} \qquad Na \leftrightarrow \forall P\{P0 \,\&\, \forall x(Px \rightarrow P(Sx)) \rightarrow Pa\},$$

where the variable P is supposed to range over "arbitrary" second-order entities in some sense or other (Fregean concepts, predicates, propositional functions, sets, classes, attributes, etc.), including, among others, the entity N supposedly being defined. But Parsons enlarges on what constitutes the impredicativity of Frege's definitions in that he says that, to use it to derive induction in the form (say) of a rule,

$$\text{(Ind-Rule)} \qquad \frac{\varphi(0),\, Na \rightarrow [\varphi(a) \rightarrow \varphi(a')],\, Nt}{\varphi(t)}$$

we must allow instantiation of the variable 'P' in (Frege-N) by formulas $\varphi(x)$ which may contain the predicate "N". In this sense, the focus of Parsons' discussion is on the *impredicativity of induction*, rather than the prima facie impredicativity of the putative definition (Frege-N). He expands the implications of this still further as follows:

> The thesis of the present note is that the impredicativity that arises from Frege's attempt to reduce induction to a definition is not a mere artifact of Frege's strategy of reduction. As Michael Dummett observed some years ago, the impredicativity – though not necessarily impredicative second-order logic – remains if we regard induction in a looser way as part of the explanation of the term 'natural number'. If one explains the notion of natural number in such a way that induction falls out of the explanation, then one will be left with a similar impredicativity. (I of I, p. 141; the reference is to Dummett [1978 p. 199].)

Perhaps what we were up to in PFA is orthogonal to the issue as posed in this way by Parsons, but let us see what we can do to relate the two. First, as explained in the preceding section, what we are *not* after is a definition of the notion of natural number in the traditional sense in which this is conceived, but rather it is to establish the existence (and uniqueness, up to isomorphism) of a *natural-number structure*, or N-structure (as it was abbreviated, PFA (i.e., Feferman and Hellman 1995)). Second, induction in the form of the principle (N-III) of Section I, above, is taken to be *part* of what constitutes an N-structure. We agree with Parsons (I of I, p. 145) that "[s]tated as a general principle, induction is about 'all predicates'," but we do not agree with the conclusion that he draws (ibid.) that "[i]nduction is thus inherently impredicative, because . . . we cannot apply it without taking predicates involving quantification over [the domain of natural numbers] as instances." Rather, our position is that our – or, perhaps better, Aczel's – proof of the existence (and categoricity) of an N-structure is predicative, given the notion of arbitrary finite set of individuals, and thence in any such structure we may apply induction to any formula that is recognized to define a class in our framework, including formulas that refer to the particular definition of our N-structure. Specifically, within EFSC, these are the weak second-order formulas, in which only quantification over individuals and finite sets is permitted. Of course, if we want to apply induction to more general classes of formulas in our system, or to formulas in more extensive systems, the question of predicativity has to be re-examined on a case-by-case basis. For example, if we expand the system EFSC by a principle that says that in any N-structure we may apply induction to *arbitrary* formulas of L(EFSC), the resulting system EFSC + FI is no longer evidently predicative, given the notion of finite set, but it is so nonetheless. The reason is that EFSC + FI can be interpreted in the system ACA with full second-order induction – which is predicative given the natural numbers according to the Feferman–Schütte characterization. And since, on our analysis, the natural numbers are predicative, given the finite sets, this also justifies EFSC + FI on that same basis. Naturally, one may expect that if the language is expanded by introducing terms for impredicatively defined sets (specified by suitable instances of the comprehension axiom), or if one adds impredicative higher-type or set-theoretical concepts, then the expanded instances of induction that become available will take us beyond the predicative, whether considered relative to the natural numbers or to finite sets.[13] But this cannot be counted as an objection to what is accomplished in PFA. It is not the general principle of induction that is impredicative, but only various of its instances; and those instances that Parsons argues to be impredicative, in the above quotation, are not examples of such, granted the notion of finite set.

Now, finally, and relatedly, we take up the objection that Parsons raises in I of I, pp. 146–68, to the predicativity of Alexander George's (1987) revision of Quine's definition of the natural numbers using quantification over finite sets, which is equivalent to Aczel's definition of an N-structure as we pointed out in Section I, above. Of this he says: "To the claim that the Quinean definition of the natural numbers is predicative, one can also reply that it is so only because the notion of finite set is assumed." Indeed, as the above discussion affirms, we could not agree more. But the reason for his objection then is that "[o]nce one allows oneself the notion of finite set, it seems one should be allowed to use some basic forms of reasoning concerning finite sets," and in particular (according to Parsons) of induction and recursion on finite sets, which would then allow one to define the natural numbers as the cardinal numbers of finite sets. But it is just this that we do *not* assume in EFSC (or EFSC*); no assumptions are made on finite sets besides the closure principles (Sep), (FS-I), and FS-II) (and [Card] in the case of EFSC*). Of course, within our system, once we have an N-structure, we can formally define what it means to be a finite set by saying that it is in one-one correspondence with an initial segment of that structure, and then derive principles of induction and recursion for *that* notion. But we cannot prove that these exhaust the range of the finite-set variables.

IV. The Challenge of George and Velleman

In their 1996 paper, "Two conceptions of natural number" (TC in the following), Alexander George and Daniel Velleman take up the PFA constructions in connection with two main conceptions of natural number, which they describe as "pare down" (PD) and "build up" (BU) corresponding to two ways of characterizing the minimal closure of a set A under an operation f. On the PD approach, this is defined explicitly as the intersection of all sets including A and closed under f. In the case of the natural numbers, this corresponds to the definitions given by Dedekind, Frege, and Russell, essentially as the intersection of all classes containing zero and closed under successor. In contrast, the BU approach provides an inductive definition, illustrated in the case of the natural numbers by clauses such as

(1) 0 is a natural number, and
(2) If n is a natural number, then so is $S(n)$,

together with an *extremal clause*, which says that natural numbers are only those objects generated by these rules. As their discussion brings out, the PD approach comports with a platonist view, according to which impredicative definitions are legitimate means of picking out independently existing sets, whereas the BU approach comports with a constructivist view that rejects the platonist stance and impredicative definitions in favor of rules for generating the intended set of objects. Not surprisingly, neither camp is satisfied with the other's approach, the constructivist rejecting the PD approach as just indicated, but the platonist also rejecting the BU approach as failing properly to define the intended class by failing explicitly to capture the required notion of "finite iteration" of the rules of construction. Furthermore, neither camp is impressed with the other's critique. And so the impasse persists.

The question arises for George and Velleman: To which type of definition should that of PFA be assimilated? As they recognize, it seeks to avoid impredicativity and so surely should not be thought of as a PD definition. On the other hand, in PFA, "the completed infinite" is recognized; moreover (although George and Velleman do not highlight this), an *explicit* definition of "*natural-number-type structure*" is provided, not merely an inductive or recursive description of "natural numbers," and so, assimilation of PFA to the BU approach is misleading. Here we would suggest that a new, third category of definition be recognized, one that combines the explicitness demanded by PD with the predicative methods demanded by BU; it might be called "predicative structuralist" (PS), if one wants a two-letter label. But before recognizing a qualitatively new product, we want to be sure that at least the labeling is honest and accurate.

In notes, George and Velleman raise questions on this score. The essential worry seems to be that the construction in PFA (or its simplification by Aczel) succeeds only if the range of the finite-set quantifiers is restricted to truly finite sets; otherwise, "nonstandard numbers" will not be excluded. But, for some reason, any effort to impose this restriction must appear circular or involve some hidden impredicativity. They put it this way:

> As Daniel Isaacson (1987) suggests, the predicativist definition will be successful only if (i) the second-order quantifier in the definition ranges over a domain that includes all finite initial segments of N and (ii) the domain contains no infinite sets. He concludes that the definition therefore "does not fare significantly better on the score of avoiding impredicativity than the one based on full second-order logic" (p. 156). Feferman and Hellman argue in response (1995, note 5, p. 16) that the existence of the required finite initial segments can be justified predicatively, but it seems to us that they have failed to answer part (ii) of Isaacson's objection, namely that infinite sets must be excluded from the domain of quantification. As we saw earlier, it is this exclusion of infinite sets from the second-order domain that guarantees that Feferman and Hellman's definition will capture *only* natural numbers. In fact, the difficulty here is in effect the same as the difficulty that the platonist finds with the BU definition; it is not the inclusion of desired elements in the domain that causes problems, but rather the exclusion of unwanted elements. (TC, n.9)

Now an adequate response to this requires distinguishing what may be called "external" and "internal" viewpoints concerning formalization of mathematics. From an external standpoint, one views a formalization from the outside and asks whether and how nonstandard models of axioms or defining conditions can be ruled out. Here the metamathematical facts are clear. So long as one works with a consistent formal system based on a (possibly many-sorted) first-order logic, or indeed any logic that is compact, nonstandard models of arithmetic are inevitable. *But this is true even if an impredicative definition of "N-structure" is given.* Even a PD definition in ZFC is subject to this limitation and will have realizations in which "numbers" with infinitely many predecessors appear. No extent of analysis of 'finite' or 'standard number', and so on, can overcome this limitation. What this shows is that the problem of "excluding nonstandard models" in this sense is "orthogonal," so to speak, to the problem of predicativity. All formal definitions are in the same boat, and the only recourse, from the external vantage point, is somehow to transcend the framework of first-order logic. Let us return to this momentarily.

Alternatively, one can look at matters from an *internal* point of view. One accepts the inevitability of nonstandard models of theories built on formal logic, but then one attempts to lay down axioms that are intuitively evident of the informal notions one is trying to capture, and then one seeks to prove the strongest theorems that one can, which, on their ordinary informal interpretation, express interesting and desirable results. Thus, one can lay down closure conditions, as in PFA, that are evident of finite sets, and, although they can hold of other collections as well, the theorems that one proves, such as mathematical induction in specified pre-N-structures, establish desired results even if they can be nonstandardly interpreted. (Bear in mind that every mathematical result about the continuum, say, recovered in ZFC has nonstandard interpretations.) Indeed, on this score, a good case can be made that the predicativist can prove results on the existence and uniqueness of natural-number-type structures that are just as decisive as those the classicist can prove. Let us return to this after elaborating a bit further about what can be said on behalf of PFA and the improvements described in Section I from the *external* viewpoint.

To effect the desired "exclusion of infinite sets" that can lead to "nonstandard numbers," that is, elements of N-structures with infinitely many predecessors, one takes the bull by the horns, so to speak: the exclusion is imposed by fiat in the meta-language by stipulating that we are only concerned with interpretations in which the range of the finite-set quantifiers contains only finite sets. 'Finite' is taken as absolute. This is the framework of "weak second-order logic" in its semantical sense. As is well known, it is noncompact and not recursively axiomatizable, but this is offset by gains in expressive power, exploited in PFA. For now one can collect items of a pre-N-structure that correspond to genuinely finite initial segments of a linear ordering, and this suffices to characterize N-structures.

There is a limited analogy with the classicist's approach via PD definitions, for example, those of Dedekind, Frege, and Russell, formalized say in second-order notation; for these characterize N-structures only if nonstandard, less-than-full ranges of the second-order quantifiers are excluded (so that second-order monadic quantifiers must range over *all* subsets of the domain, precluding Henkin models). The problem of nonstandard models is overcome by moving to noncompact, nonaxiomatizable "full second-order logic." But the analogy is only partial. For, whereas the classical logicist excludes nonfull interpretations on the basis of a claim to understand "*all subsets* of an infinite set," the predicative logicist merely excludes infinite sets from the range of finite-set quantifiers on the basis of a claim to understand '*all finite subsets*'. If the objection is that this is illegitimate because 'finite' "is as much in need of analysis as the concept 'natural number' " (TC, note 9), then it is appropriate to refer back to Section II, above, and the whole case for grounding the infinitistic notion of "natural-number-type structure" on elementary assumptions on finite objects, together with the point made earlier (Sec. III) that nowhere do we have to invoke finite-set induction in order to prove any of our theorems, including the theorem that says that mathematical induction holds in any special pre-N-structure. (*Mutatis mutandis* for the Aczel theorem.) Indeed, since induction is essential to the natural-number concept and to reasoning "about the natural numbers," the very fact that finite-set induction is *not* needed to recover this much counts in favor of the view that 'finite set' is actually *less* in need of analysis than 'natural number'.

Moreover, on the question of existence, there is a fundamental disanalogy between the PD and the PS approaches. For, as George and Velleman bring out, the impredicative definitions of the logicists still must presuppose existence of the minimal closure, and this is an additional assumption, not guaranteed merely by the restriction to full interpretations. There still must *be* some full interpretation of the right sort, that is, containing the real minimal closure. In contrast, the predicative constructions of PFA, Aczel, and the Appendix below yield the desired classes by a restricted comprehension principle, WS–CA. Given finite sets as objects, such a principle is justified much as arithmetical comprehension is; one can even eliminate talk of classes of individuals in favor of satisfaction of formulas, since these contain only bound individual and finite-set variables but no bound class variables.

Thus, the predicative logicist accompanies the platonist classicist only a relatively small step beyond first-order logic; then construction takes over on the new higher ground, while the platonist continues ascending, eventually into the clouds.

Consistently with this external view, one can, however, also pursue the internalist course of proving desirable theorems. Here, perhaps surprisingly, the predicativist is able to recover predicativist analogues of well-known classical results. The proofs of categoricity or unicity of N-structures and of mathematical induction in the pre-N-structures of PFA, Aczel, and the Appendix already illustrate this. But one can go further and prove theorems that, informally understood, say explicitly that N-structures cannot contain any nonstandard elements. The idea is to formalize the following, familiar reasoning. Let $\langle \mathbf{M}, 0, ' \rangle$ be an N-structure. Induction implies than any non-empty class (subclass of **M**) closed downward under predecessor, $p(x)$, contains 0. Consider the class of nonstandard numbers (of **M**); call it **K**. If $z \in \mathbf{K}$, then also $p(z) \in \mathbf{K}$ (contraposing the Adjunction axiom); therefore, if **K** is non-empty, it contains 0, a blatant contradiction. (Put positively, 0 is standard and if z is standard, so is z', and so, all members of **M** are standard.) Elements with infinitely many predecessors are ruled out directly by Induction.

But in what system is the above reasoning carried out? If we attempt to formalize it in EFSC, expressing "x is nonstandard" by "$\forall A(A \neq \{y : y \leq x\}$," we immediately contradict the definition of **M**! On the other hand, we cannot simply plug in "$\{y : y \leq x\}$ is Dedekind-infinite" or any other second-order analysis of "infinite" involving general class or function quantifiers, for then we would not be able predicatively to form the class **K**. However, there is an alternative method that gets around this. For here we may appeal to the metatheorem mentioned in Section III: If we add to EFSC the axiom schema known as "full induction" (FI), that is, induction for *arbitrary* second-order formulas, the resulting system, EFSC + FI, is interpretable in the subsystem of PA^2 known as ACA. This also contains FI and, moreover, is a predicatively acceptable system relative to the natural numbers (on the Feferman–Schütte characterization) as noted in Section III. But, as was also observed there, since the natural numbers or N-structures are predicative given the finite sets, EFSC + FI is also predicatively acceptable relative to the finite sets. Although it cannot prove the existence of subclasses of an N-structure defined by formulas with class quantifiers, it can prove that induction holds directly for any formula that platonistically defines a subclass, as it were. In particular, now one can formalize the above induction ruling out nonstandard numbers, using, in place of $x \in \mathbf{K}$, a second-order formula $\varphi(x)$ to express "x has infinitely many predecessors," for example, "the predecessors of x form a class, a subclass of which is in one-one correspondence with an unbounded subclass of **M**"; or it could just as well be "the predecessors of x form a Dedekind-infinite class." The predicativist, as well as the classicist, regards these as good formalizations of the intended notion. Thus, the reasoning is formalizable in a predicatively acceptable extension of EFSC without appealing to the special finite-set variables and without any circular or impredicative reference to the class **K**.[14]

Looking at the contrapositive, one sees that one has thus *derived* the consequence of the axiom (Card) directly relevant to ruling out nonstandard members of N-structures, viz., the statement that the predecessors of any such element form a Dedekind-finite set,

$$\forall x[x \in \mathbf{M} \rightarrow \text{Ded Fin}(\text{Pd}(x))].$$

This follows straightforwardly by induction on the formula, $\varphi(z)$, expressing Ded Fin(Pd(z)). Again, we need not be able to collect all elements satisfying this formula in order to reason with it by mathematical induction.

Thus, "nonstandard numbers" are ruled out as decisively as they can be. From the external standpoint, they are excluded by the semantics of weak second-order logic, which, as has been argued, is a good framework for elementary predicative mathematics. From an internal perspective, without falling back on special finite-set variables, we can employ standard, logicist *analyses* of 'finite', 'infinite', and so forth, and derive theorems in predicatively acceptable systems that directly express the desired exclusion. This may seem like "having one's cake and eating it at the same time." But really it is more like having two desserts.

NOTES

1. The class variables are given in boldface, to distinguish them from the finite set variables.
2. As a point of difference with PFA, function variables here are given in boldface to indicate that they are treated as special kinds of classes.
3. That, in turn, was a modification of a definition of the natural numbers proposed by Quine (1961) using only the first conjunct in (6), which is adequate when read in strong second-order form, but not when read in weak second-order form; cf. George (1987, p. 515) and George and Velleman (1996, n.10).
4. For a good discussion of this and related issues, see Parsons (1990).
5. For a valuable discussion of Hilbert's structuralist views of axioms and reference in mathematics and the contrast with Frege's views, see Hallett (1990).
6. Our construction thus improves on Dedekind's, for he relied, for a Dedekind-infinite system, on a totality – of "all things which can be objects of my thought" (Dedekind 1888, Theorem 66) – which, even apart from its unmathematical character, is unacceptable to a predicativist on logical grounds, for, presumably, such a totality would contain itself! Furthermore, for a *simply infinite system*, he then relied on a subtotality impredicatively specified as the intersection of all subtotalities containing an initial element and closed under the given function (1888, Theorems 72 and 44). But it is noteworthy that the particular example that Dedekind sought to invoke to ensure nonvacuity of his definitions was not identified as "the numbers." As it happened, Dedekind did go on to speak of such abstract particulars, but that is another story, and, in any case, it is a further move that we have not been tempted to make.
7. Parsons' paper appeared well before PFA, and so, the challenge was not issued to *it* but rather to the kind of program that it exemplifies. That challenge was addressed briefly in the final discussion section of PFA, pp. 14–15, but is expanded on substantially here.
8. The informal explanation of what constitutes an impredicative definition varies from author to author. A representative collection of quotations is given by George (1987); the explanation given in the text here is closest to that taken by George from an article of Hintikka (1956).
9. Cf. I of I, pp. 152–3 and p. 159, n.24.
10. To be more precise, this is spelled out by means of an autonomous transfinite progression of ramified systems, where autonomy is a bootstrap condition that restricts one to those transfinite levels that have a prior predicative justification; cf. Feferman (1964).
11. The relative notion of predicativity is recast by Feferman (1996) in terms of a formal notion of the *unfolding* of a schematic theory, which is supposed to tell us what more should be accepted once we have accepted basic notions and principles.
12. Parsons has an interesting discussion in I of I (pp. 143–5), of what is reasonable to assume about the range of first-order variables in proposed definitions of the natural numbers. We believe that the assumptions (P-I) and (P-II) are innocuous, in the sense that the notion of ordered pair is a prerequisite to an understanding of any abstract mathematics.
13. Addition of higher types or even set-theoretical language does not per se force us into impredicative territory; cf. Feferman (1977).
14. There is some irony in the fact that George and Velleman, after claiming (TC, note 10) that the Aczel construction cannot rule out nonstandard numbers without a circular appeal to "the complement of N," present an argument of their own for

the predicative acceptability of an extension of EFSC* in which the full induction schema is derivable. (See their note 14.) They argue for a direct extension to include the separation schema for finite sets with arbitrary second-order formulas. This is closely related to the fact that, in a weak subsystem of analysis, FI is equivalent to the so-called "bounded comprehension scheme,"

$$\forall n \exists X \forall m (m \in X \leftrightarrow m < n \ \& \ \varphi(m)),$$

where $\varphi(m)$ is any formula of second-order arithmetic (lacking free 'X'). (See Simpson [1985, p. 150].) This corresponds to the separation scheme for finite sets with arbitrary second-order formulas. We prefer the direct route to full induction via ACA and proof theory, since it is predicatively problematic to say that an arbitrary formula "specifies unambiguously which elements of the [given] finite set are to be included in a subset" (TC, note 14). It then turns out that their proposed stronger separation scheme is derivable from full induction, and so inherits a predicative justification after all. In any case, once full induction is available, the reasoning that N-structures are truly standard is predicatively formalizable without appeal to finite-set variables, even while employing a standard logicist analysis of 'finite' or 'infinite' as just indicated.

REFERENCES

Dedekind, R. (1888). Was sind und was sollen die Zahlen? (Brunswick: Vieweg). Translated as "The Nature and Meaning of Numbers" in R. Dedekind, *Essays on the Theory of Numbers*, ed. W. W. Beman (New York: Dover, 1963), pp. 31–115.

Dummett, M. (1978). "The Philosophical Significance of Gödel's Theorem," in *Truth and Other Enigmas* (London: Duckworth), pp. 186–201 (first published in 1963).

Feferman, S. (1964). "Systems of Predicative Analysis," *Journal of Symbolic Logic*, 29: 1–30.

Feferman, S. (1977). "Theories of Finite Type Related to Mathematical Practice," in J. Barwise (ed.), *Handbook of Mathematical Logic* (Amsterdam: North-Holland), pp. 913–71.

Feferman, S. (1985). "Intensionality in Mathematics," *Journal of Philosophical Logic*, 14: 41–55.

Feferman, S. (1996). "Gödel's Program for New Axioms: Why, Where, How and What?" in *Gödel '96*, P. Hájek (ed.), Lecture Notes in Logic 6, Berlin: Springer, 3–22.

Feferman, S., and Hellman, G. (1995). "Predicative Foundations of Arithmetic," *Journal of Philosophical Logic*, 24: 1–17.

George, A. (1987). "The Imprecision of Impredicativity," *Mind* 96: 514–18.

George, A., and Velleman, D. (1998). "Two Conceptions of Natural Number," in G. Dales and G. Olivieri (eds.), *Truth in Mathematics* (Oxford: Oxford University Press), pp. 311–327.

Hallett, M. (1990). "Physicalism, Reductionism, and Hilbert," in A. D. Irvine (ed.), *Physicalism in Mathematics* (Dordrecht, The Netherlands: Kluwer), pp. 183–257.

Hellman, G. (1989). *Mathematics Without Numbers* (Oxford: Oxford University Press).

Hellman, G. (1996). "Structuralism Without Structures," *Philosophia Mathematica*, 4: 100–23.

Hintikka, J. (1956). "Identity, Variables, and Impredicative Definitions," *Journal of Symbolic Logic*, 21: 225–45.

Isaacson, D. (1987). "Arithmetical Truth and Hidden Higher-Order Concepts," in Paris Logic Group (eds.), *Logic Colloquium '85* (Amsterdam: North-Holland), pp. 147–69.

Parsons, C. (1990). "The Structuralist View of Mathematical Objects," *Synthese*, 84: 303–46.

Parsons, C. (1992). "The Impredicativity of Induction," in M. Detlefsen (ed.), *Proof, Logic and Formalization* (London: Routledge), pp. 139–61 (revised and expanded version of a 1983 paper).

Quine, W. V. O. (1961). "A Basis for Number Theory in Finite Classes," *Bulletin of the American Mathematical Society*, 67: 391–2.

Schütte, K. (1965). "Eine Grenze für die Beweisbarkeit der transfiniten Induktion in der verzweigten Typenlogik," *Archiv für Mathematische Logik und Grundlagenforschung*, 7: 45–60.

Simpson, S. (1985). "Friedman's Research on Subsystems of Second-Order Arithmetic," in L. Harrington et al. (eds.), *Harvey Friedman's Research on the Foundations of Mathematics* (Amsterdam: North-Holland), pp. 137–59.

Wang, H. (1963). "Eighty Years of Foundational Studies," reprinted in his *A Survey of Mathematical Logic* (Amsterdam: North-Holland), pp. 34–56.

Appendix: Realizing Dummett's Approach in EFSC

GEOFFREY HELLMAN

Here the notation of Section I will be followed, including the use of F, G, H, as finite-set variables. Let 0 and $'$, respectively, be the initial element and successor-like function of a given pre-N-structure. Our principal aim is to prove the following:

Theorem. (EFSC):
Let $\mathbf{M} =^{df} \{x : \exists F(0 \in F \,\&\, \forall y(y \in F \,\&\, y \neq x \rightarrow y' \in F) \,\&\, x \in F \,\&\, \forall G[[(0 \in G \,\&\, \forall y(y \in G \,\&\, y \neq x \rightarrow y' \in G)\} \rightarrow F \subseteq G])\}$, that is, x belongs to a (the) minimal finite set containing 0 and "closed upward except at x." Then $\langle \mathbf{M}, 0, ' \rangle$ is an N-structure.

Remark: Note that this definition of **M** incorporates both an existential condition and a universal one, corresponding to the conditions Wang attributes to Dummett (Wang 1963; cf. TC, note 10).

The proof is simplified by adopting the following abbreviations, which also bring out the relationship between this theorem and that of Aczel:

$$\mathrm{Clos}_d(F, [z, x]) \equiv z \in F \,\&\, \forall y(y \in F \,\&\, y \neq x \rightarrow y' \in F),$$

read as "F is closed upward from z to x." (The subscript d is for Dummett.) Next define

$$z \leq_d x \text{ by } F \equiv \mathrm{Clos}_d(F, [z, x]) \,\&\, x \in F \,\&\, \forall G(\mathrm{Clos}_d(G, [z, x]) \rightarrow F \subseteq G).$$

This can be read as "F witnesses $z \leq_d x$." Trivially, if both F_1 and F_2 witness $z \leq_d x$, then $F_1 = F_2$ (extensionally). Now, define

$$z \leq_d x \equiv \exists F(z \leq_d x \text{ by } F).$$

Now, **M** in the Theorem can be defined by $\mathbf{M} = \{x : 0 \leq_d x\}$. For purposes of comparison, recall the Aczel construction, $\mathbf{M} = \{x : \mathrm{Fin}(\mathrm{Pd}(x)) \,\&\, 0 \leq x\}$, where '$\leq$' is the ordering introduced in PFA, as in Section I, above. We now proceed to the proof of the theorem.

Proof: Let F_x denote the unique F that witnesses $0 \leq_d x$. Then, to say that $x \in \mathbf{M}$ is to say that F_x exists. We have

(i) $0 \in \mathbf{M}$, to wit $\{0\}$ as F_0.

(ii) $z \in \mathbf{M} \to z' \in \mathbf{M}$.

Given F_z, set $F_{z'} = F_z \cup \{z'\}$. We must show that this works. $F_z \cup \{z'\}$ is finite, by adjunction (FS-II). $0 \in F_z \cup \{z'\}$ and $z' \in F_z \cup \{z'\}$. Now if $y = z$, then trivially $y' \in F_z \cup \{z'\}$; and if $y \neq z$, then if $y \in F_z \cup \{z'\}$ & $y \neq z'$, then $y \in F_z$, and by hypothesis then so is y', whence $y' \in F_z \cup \{z'\}$. Thus, $\mathrm{Clos}_d(F_z \cup \{z'\}, [0, z'])$. It remains to prove minimality.

Suppose $\exists u \in F_z \cup \{z'\}$ such that $u \notin G$, some G such that $\mathrm{Clos}_d(G, [0, z'])$. Consider $G - \{z'\}$. If $y \neq z$ & $y \in G - \{z'\}$, $y \neq z'$ either, so $y' \in G - \{z'\}$ by hypothesis on G. So, $\mathrm{Clos}_d(G - \{z'\}, [0, z])$, and so, $F_z \subseteq G - \{z'\}$, by hypothesis on F_z. Thus, $z \in G - \{z'\}$, and so, $u \neq z'$, by the closure condition on G that forces $z' \in G$. Therefore, $u \in F_z$ but, by hypothesis that $u \notin G$, $u \notin G - \{z'\}$ either, contradicting the minimality of F_z. This completes the proof of minimality of $F_z \cup \{z'\}$ and of step (ii).

(iii) Induction, N-III: Let $\mathbf{X}$ be a class such that $0 \in \mathbf{X}$ and $y \in \mathbf{X} \to y' \in \mathbf{X}$, all $y \in \mathbf{M}$.

To prove: $z \in \mathbf{M} \to z \in \mathbf{X}$.

We will prove $F_z \subseteq \mathbf{X}$, which suffices since $z \in F_z$ and indeed $z \in \cap[G : \mathrm{Clos}_d(G, [0, z])]$. Let $H = G \cap \mathbf{X}$, for some such G (existence by F_z itself). H is finite by WS-Sep. We have $\mathrm{Clos}_d(H, [0, x])$ by the closure conditions on G and $\mathbf{X}$. Therefore, $F_z \subseteq H = G \cap \mathbf{X}$, whence $F_z \subseteq \mathbf{X}$. ■

By virtue of the unicity of N-structures ("categoricity," Theorem 5 of PFA), an N-structure can be represented as of the form $\langle \mathbf{M}, 0, ' \rangle$ of Theorem 1, that is, the domain $\mathbf{N}$ of any N-structure $= \{x \in \mathbf{N} : 0 \leq_d x\}$, where 0 here is the initial element of $\mathbf{N}$ and $\leq_d$ is defined over $\mathbf{N}$ via the successor-type relation on $\mathbf{N}$.

As expected, the ordering $\leq_d$ is closely related to '$\leq$' of PFA. This is spelled out in the following.

Theorem. (EFSC):

(1) In any pre-N-structure,

$$z \leq_d x \to z \leq x.$$

(2) In any N-structure $\langle \mathbf{M}, 0, ' \rangle$ defined as in Theorem 1, and hence in any N-structure,

$$z \leq_d x \leftrightarrow z \leq x.$$

Proof: (1) and $\to$ of (2): Suppose the implication fails, that is, that $z \leq_d x$ but $\exists A(x \in A \;\&\; \forall y(y' \in A \to y \in A) \;\&\; z \notin A)$. (So, $z \neq x$, and $z' \notin A$, nor is z'', etc.) Let F be the witness to $z \leq_d x$; that is, $F = \cap[G : \mathrm{Clos}_d(G, [z, x])]$.

Let $B = \{u : u \notin A \,\&\, u \in F\}$. B is finite, by WS-Sep, and $z \in B$ and $\forall y(y \in B \,\&\, y \neq x \rightarrow y' \in B)$; that is, $\mathrm{Clos}_d(B, [z, x])$. So, by definition of $z \leq_d x$, we have $x \in B$, that is, $x \notin A$, a contradiction. (Remark: Note the similarity to the Aczel proof, except that here we are "stepping forward" instead of "stepping back".)

$\leftarrow$ of (2): Now assume that we are in an N-structure, $\langle \mathbf{M}, 0, ' \rangle$, as in Theorem 1. We proceed by induction on z:

(i) For $z = 0$, the implication is trivial.

(ii) Let $z = y$. If $y = x$, the implication is trivial, as then $y' \nleq x$. Let $y \neq x$, and let H_y be the (minimal) witness to $y \leq_d x$, which we can suppose, by inductive hypothesis. We claim that $H_y - \{y\}$ is the minimal witness to $y' \leq_d x$. Since $y \in H_y$ and $y \neq x$, $y' \in H_y$ by $\mathrm{Clos}(H, [y, x])$, and so, $y' \in H_y - \{y\}$. If $u \in H_y - \{y\}$ and $u \neq x$, then, because $u \neq y$, $u \in H_y$, and so, $u' \in H_y$, whence $u' \in H_y - \{y\}$. (In the last step, we appeal to the minimality of H_y, which implies that $p(y) \notin H_y$, so that $u' \neq y$.) It remains to prove minimality of $H_y - \{y\}$ as witness to $y' \leq_d x$. Let G be such that $\mathrm{Clos}_d(G, [y', x])$, and suppose $\exists u$ such that $u \in H_y - \{y\}$ but $u \notin G$. Then, $u \in H_y$ but $u \neq y$, so $u \notin G \cup \{y\}$. But, $G \cup \{y\}$ contains y and meets the closure condition for H_y, viz. $\mathrm{Clos}_d(G \cup \{y\}, [y, x])$. Therefore, by minimality of H_y, $H_y \subseteq G \cup \{y\}$, contradicting the supposition of u. This proves the minimality of $H_y - \{y\}$ and completes the inductive step. ■

Thus, the Dummett-inspired construction of Theorem 1, as well as the Aczel construction, defines N-structures, provably in EFSC. And the orderings involved, $\leq_d$ and $\leq$, respectively, are extensionally equivalent in these structures.

Name Index

The editors would like to thank Kelly Becker for doing most of the work on this index.